IMMUNOLOGY

AF755303

IMMUNOLOGY

IMMUNOTOXICOLOGY, IMMUNOPATHOLOGY, AND IMMUNOTHERAPY

VOLUME 1

Edited by

M. A. Hayat

Distinguished Professor
Kean University,
Union, NJ, United States

ACADEMIC PRESS

An imprint of Elsevier

Academic Press is an imprint of Elsevier
125 London Wall, London EC2Y 5AS, United Kingdom
525 B Street, Suite 1800, San Diego, CA 92101-4495, United States
50 Hampshire Street, 5th Floor, Cambridge, MA 02139, United States
The Boulevard, Langford Lane, Kidlington, Oxford OX5 1GB, United Kingdom

Copyright © 2018 Elsevier Inc. All rights reserved.

No part of this publication may be reproduced or transmitted in any form or by any means, electronic or mechanical, including photocopying, recording, or any information storage and retrieval system, without permission in writing from the publisher. Details on how to seek permission, further information about the Publisher's permissions policies and our arrangements with organizations such as the Copyright Clearance Center and the Copyright Licensing Agency, can be found at our website: www.elsevier.com/permissions.

This book and the individual contributions contained in it are protected under copyright by the Publisher (other than as may be noted herein).

Notices
Knowledge and best practice in this field are constantly changing. As new research and experience broaden our understanding, changes in research methods, professional practices, or medical treatment may become necessary.

Practitioners and researchers must always rely on their own experience and knowledge in evaluating and using any information, methods, compounds, or experiments described herein. In using such information or methods they should be mindful of their own safety and the safety of others, including parties for whom they have a professional responsibility.

To the fullest extent of the law, neither the Publisher nor the authors, contributors, or editors, assume any liability for any injury and/or damage to persons or property as a matter of products liability, negligence or otherwise, or from any use or operation of any methods, products, instructions, or ideas contained in the material herein.

British Library Cataloguing-in-Publication Data
A catalogue record for this book is available from the British Library

Library of Congress Cataloging-in-Publication Data
A catalog record for this book is available from the Library of Congress

ISBN: 978-0-12-809819-6

For Information on all Academic Press publications
visit our website at https://www.elsevier.com/books-and-journals

ELSEVIER Book Aid International Working together to grow libraries in developing countries

www.elsevier.com • www.bookaid.org

Publisher: Mica Haley
Acquisition Editor: Linda Versteeg-Buschman
Editorial Project Manager: Kathy Padilla
Production Project Manager: Kiruthika Govindaraju
Cover Designer: Greg Harris

Typeset by MPS Limited, Chennai, India

Contents

6. Signaling by the Toll-Like Receptors Induces Autophagy Through Modification of Beclin 1: Molecular Mechanism

NEEL R. NABAR, CHONG-SHAN SHI AND JOHN H. KEHRL

7. Molecular Mechanisms Underlying the Action of Ipilimumab Against Metastatic Melanoma

LUCIA FESTINO, VITO VANELLA, MARTINA STRUDEL AND PAOLO A. ASCIERTO

8. To Help or to Harm: Dynamic Roles of $CD4^+$ T Helper Cells in Solid Tumor Microenvironments

SHANNON M. LIUDAHL AND LISA M. COUSSENS

9. Development and Function of Natural Killer Cells and Its Importance in Cancer Immunotherapy

SOURAV PAUL AND GIRDHARI LAL

II ROLE IN DISEASE AND TREATMENT

10. Varicella Zoster Vaccine Experience in HIV-Infected Children and Adolescents

WILLIAM BORKOWSKY AND ANNE A. GERSHON

11. Development of an Effective Alzheimer's Vaccine

DANTE J. MARCIANI

12. Obesity-Induced Defects in Dendritic Cell and T Cell Functions: Implications for Immunotherapeutic Efficacy in Cancer

RACHAEL ORLANDELLA AND LYSE NORIAN

13. Redox-Dependent Circuits Regulating B Lymphocyte Physiology

MILENA BERTOLOTTI AND ROBERTO SITIA

14. Immunomodulatory Attributes of Aged Garlic Extract and Its Components

YELDUR P. VENKATESH

15. MUC1 and Cancer Immunotherapy

CHUANWEI YANG, JAMES L. MURRAY AND NUHAD K. IBRAHIM

Preface

This is volume one of the multivolume series discussing many aspects of Immunology.

Cardiovascular function and cancer remain the major causes of morbidity and mortality for the elderly. Infectious diseases are also major causes of serious disease in older persons; infections caused by virus (HIV), bacteria (influenza, pneumonia), fungus (candida infection, ringworm) and other sources also take their toll on the lives of people, young and old. These conditions together are the sixth leading cause of death of the elderly in the United States. Medicare and Medicade for inpatient treatment of pneumonia was ~$14 billion in the United States.

Immunization is supposed to be most effective in preventing infection. However, the elderly respond poorly to currently available vaccinations. One of the reasons for this poor response might be that aging is associated with reduced thymic output of T cells. Both CD8 T cells and CD4 T cells are involved in defense.

The intracellular presence of pathogens is detected by the innate immune system that employs germline-encoded pattern recognition receptors (PRRs), thus initiating the immune response. The Toll-like receptors (TLRs) are the best studied PRRs, and they act by initiating an intracellular signaling cascade. The detailed molecular mechanisms through which TLR signaling primes macrophages for intracellular pathogen clearance by initiating autophagy are explained in this volume.

Another major protective function against infectious diseases is mediated by CD4 T cells through T helper-1 (Th1) type immunity. An example of this type of protective role is against bacterial tuberculosis. It is recognized that HIV-1 induced depletion of CD4 T cells predisposes to fatal tuberculosis. This study indicates that autophagy is a novel mechanism that increases BCG vaccine efficiency.

The contents of this volume are summarized below.

Autoimmunity affects approximately 10% of the global population. Examples include: systemic lupus, Graves' disease, rheumatoid arthritis, and multiple sclerosis. Current treatments of these diseases inhibit inflammation, but are nonspecific and tend to alter the immune system to respond to infectious and neoplastic insults. Multiple sclerosis is mediated by brain tissue-reactive lymphocytes. Both T cells and B cells play a role in the development of this disease. However, the latter have been clinically detected, but the former are expected to be also detected.

Aging invites autoimmune diseases which preferentially occur in the second half of one's adult life. As we age, immune efficiency is declined and thymic T cell generation begins to cease. These diseases develop when memory T and B cell response to a self-antigen is established. The ability of T cells to respond to infection decreases. For example, during the annual influenza epidemics, individuals older than 50 years show increased morbidity. In spite of the claims to the contrary, only a small percentage of vaccinated persons, especially older than 60 years, benefit from vaccination. However, consensus on the effect of vaccination is lacking, although there is unanimity that thymic activity begins to decline postpuberty.

Although autoimmunity is triggered by antibodies, B cells, T cells, and many cells of the innate immune system play a key role in the initiation of these diseases. Many autoimmune diseases occur in patients carrying specific major histocompatibility complex Class II molecules, and the human leukocyte antigen. The interaction between the $CD4^+$ T cells and the specific autoantigen is likely responsible for the development of this disease. There is a need to develop an antigen-specific therapeutic system that can effectively prevent the autoimmune attack, while leaving the rest of the immune system functioning normally.

Humoral immunity is of great importance in both health and disease, for it can be beneficial or deleterious. Molecular mechanisms underlying the impact of NKT cells on B cell activation are explained.

It is known that protective immunity against Alzheimer's disease decreases (weakens) with age. With this information in hand, we still have not been able to develop a successful vaccine using amyloid-β. An attempt is made in this volume to explain how a successful Alzheimer's vaccine can be produced. Presently available Alzheimer's vaccines have the wrong immunogen. It is suggested that a vaccine having immunogens of both Aβ and tau protein may be more successful. An effective vaccine should have an immunogen that has β and 7 cell epitopes.

Susceptibility and resistance to systemic autoimmunity are genetically regulated. One of the major goals of biomedical research is to determine the pathological mechanisms that lead to disease.

Two types of viruses, Human Cytomegalovirus (HCMV) and Varicella Zoster Virus (VZV), are discussed. The optimal vaccination dose for VZV-infected children and adolescents is recommended. Regarding HCMV, it can modulate the autophagy machinery for its own replication instead of being degraded by autophagy. A related subject discussed is the role of macroautophagy (autophagy) when it occurs inside T lymphocytes, resulting in autophagy-deficient T lymphocytes which, in turn, causes a number of abnormalities, including increased number of cell organelles, defects of memory T cell formation, and instability of the Treg cell lineage. Thus, autophagy plays multiple roles to regulate the physiological functions of T lymphocytes.

Obesity induces chronic low-grade inflammation which facilitates innate and adaptive dysfunctions. Obesity also increases the risk of several types of cancer. Current knowledge regarding the relationship between obesity, cancer, immunity, and immunotherapeutic efficacy is presented.

Garlic is known to be a biological response modifier as a stimulator of a variety of immune responder cells, including dendritic cells, lymphocytes, macrophages, and natural killer cells. The principle transformed organosulfur (OS) compound in garlic is allicin. Garlic is also an anti-tumor agent. The effects of *S*-allylcystein and *S*-allymethylcystein compounds in producing immunomodulation are detailed here.

This volume was authored by 35 colleagues representing six countries. I greatly appreciate the thoughtful help extended to me by Linda Versteeg-Buschman, the senior editor at Elsevier, before and during the publication of the series. I offer thanks to Phil Connelly and Dr. Dawood Farahi for recognizing the importance of medical research and publishing through an institution of higher education. I am thankful to Danielle Antonucci and my students for their contributions to the final preparation of this volume.

April 2017
M.A. Hayat

Abbreviations and Definitions

4-OHT	4-hydroxytamoxifen
7-AAD	7-aminoactinomycin D
ABC	an age-associate B cell subset in aged mice
AChR	acetylcholine receptor
ACPAs	anticitrullinated peptide antibodies
AD	Alzheimer's disease
ADAM	a disintegrin and metalloprotease; Zn^{2+}-dependent proteases that cleave membrane-bound proteins
ADCC	antibody dependent cellular cytotoxicity
a-dsDNA	antibodies to double-stranded DNA
adTRAIL	adenovirus-encoded TRAIL
AE	adverse event
AGE	aged garlic extract
AGS	Aicardi-Goutieres
AHSCT	allogenic hematopoietic stem cell transplantation
aHUS	atypical hemolytic uremic syndrome
AIH	autoimmune hepatitis
AIP	autoimmune pancreatitis
ALL	acute lymphocyte leukemia
ALP	alkaline phosphatase
ALT	alanine aminotransferase
AMA	antimitochondrial antibodies
AMAN	acute motor axonal neuropathy
AMD	age-related macular degeneration
AML	acute myelogenous leukemia
ANAs	antinuclear antibodies
ANCA	antineutrophil cytoplasmic Ab
anti-P	antiribosomal P proteins
AP	alternative pathway
APCs	antigen-presenting cells
APE1	apurinic/apyrimicinic endonuclease 1
APL	altered peptide ligand
APN	adiponectin
AR	androgen receptor
ARR	adjusted relative risk
ART	antiretroviral
AST	aspartate aminotransferase
AT	autoimmune thyroiditis
Atg/ATG	autophagy-related
ATL	adult T cell leukemia
ATM	ataxia telangiectasia muted
AUROC	area under the receiver operating curve

B cell (B lymphocyte)	a type of lymphocyte developed in bone marrow and circulates in blood and lymph. On encountering a foreign antigen, differentiates into a clone of memory cells that make the antibody on subsequent encounters. Any of the lymphocytes that develop into plasma cells in the presence of a specific antigen. Plasma cells produce antibodies that attack the antigen; this is called the humoral immune response
Bcl-2	B cell lymphoma 2
BCG	*Bacillus Calmette Guerin*
BER	base excision repair
BITE	bispecific T cell engager
BiTES	bispecific single chain monoclonal antibodies
BLyS	B-lymphocyte stimulator
BMI	body mass index
BMT	bone marrow transplantation
BSA	body surface area
BTLA	B- and T-lymphocyte attenuator
C4BP	C4b-binding protein
CAF	cancer-associated fibroblast
CAR	chimeric antigen receptor
CAR	coxsackie adenovirus receptor
CARD	caspase activation and recruitment domain
CASPR2	contactin-associated protein 2
CCL	CC chemokine ligand
CCR	CC chemokine receptor
CD30L	cluster of differentiation 30 ligand
CD4 T cell	white blood cell that plays a role in protecting the body from infection
CD44std	CD44 standard
CD44v6	CD44 variant 6
CD8 T cell	cytotoxic T cell; important for immune defense against intracellular pathogens and for tumor surveillance
CDNs	cyclic dinucleotide
CEA	carcinoembryonic antigen
CFA	Freund's complete adjuvant
CFP	cyan fluorescent protein
CHAPS	3-[(3-cholamidopropyl)dimethylammonio]-1-propanesulfonic acid
CHC	chronic hepatitis C
CI	confidence interval
CIA	collagen-induced arthritis
CIDP	chronic idiopathic demyelinating polyneuropathy
CIK	cytokine-induced killer
CK-7	cytokeratin-7
CLIP	class 11-associated invariant chain peptide
CLL	chronic lymphocytic leukemia
CM	central memory
cMET	hepatocyte growth factor receptor
CMPs	common myeloid progenitors
CMV	cytomegalovirus
CNS	central nervous system
CNV	copy number variation
CP	classical pathway
CPB	cardiopulmonary bypass
CR	complement receptor
CRP	C-reactive protein
CSF	cerebrospinal fluid
CSR	class switch recombination

CT	computed tomography
CTD	C-terminal domain
CTLA-4	cytotoxic T lymphocyte antigen 4
CTLs	cytotoxic T lymphocytes
DAT	dopamine transporter
DC	dendritic cell
DDR	DNA damage response
DETC	dendritic epidermal T cell
DHE	dihydroethidium
DHT	dihydrotestosterone
DIO	diet-induced obesity
DKO	double knockout
DLBCL	diffuse large B cell lymphoma
DN	double-negative
DP	differential pressure
DSS	dextran sodium sulfate
EAE	experimental autoimmune encephalomyelitis
EAM	experimental autoimmune myocarditis
EAMG	experimental autoimmune myasthenia gravis
EAU	experimental autoimmune uveoretinitis
ECDI	1-ethyl-3-(3-dimethylaminopropyl)-carbodiimide
ECM	extracellular matrix
eEF-2	elongation factor-2
EGF	epidermal growth factor
EIC	experimental induced colitis
ELISA	enzyme-linked immunosorbent assay
EM	effector memory
EM	electron microscopy
EpCAM	epidermal cell adhesion molecule
EQTL	expression quantitative locus
ER	endoplasmic reticulum
ER	estrogen receptor
ES	embryonic stem
ESR	erythrocyte sedimentation rate
FACS	fluorescence-activated cell sorting
FAIM3	Fas apoptotic inhibitory molecule
FAP	fibroblast activation protein
FasL	Fas ligand
FB	factor B
FCGR3B	Fc gamma receptor 3B
FCM	flow cytofluorometric
FD	factor D
FDA	Food and Drug Administration
FGF	fibroblast growth factor
FH	factor H
FI	factor I
FITC	fluorescin isothiocyanate
FLICA	fluorochrome inhibitor of caspases
FOXP3	forkhead box p3; master regulator of Treg development
FP	factor P (properdin)
FR	fludarabine refractory
GAD	glutamic acid decarboxylase
GBS	Guillain–Barre syndrome
GD	Graves' disease

GFP	green fluorescent protein
GGR	global genome repair
GGT	gamma-glutamyltransferase
GLUT4	glucose transporter type 4
GM-CSF	granulocyte–macrophage colony stimulating factor
GPI	glucose-6-phosphate isomerase
GrB	granzyme-B
GTP	guanosine triphosphate
GvT	graft versus leukemia effect
GWAS	genome-wide association studies
H&E	hematoxylin and eosin
HAART	highly acute antiretroviral therapy
HBV	hepatitis B virus
HCC	hepatocellular carcinoma
HCgp-39	human cartilage glycoprotein-39
HCMV	human cytomegalovirus
HCV	hepatitis C virus
HEK	human embryonic kidney cells
HER2	human epidermal growth factor receptor 2
HEVs	high endothelial venules
HGF	hepatocyte growth factor
HLA	human leukocyte antigen; present self- and nonself-peptides to the T cell receptor
HRM-PCR	high resolution melt-polymerase chain reaction
HSCs	hematopoietic stem cells
hsp60/65/90	heat shock protein of 60/65/90
HSV1	herpes simplex virus 1
HT	Hashimoto thyroiditis
HTLV-I	human T cell leukemia virus type I
IA2	insulinoma antigen 2
iACs	scaffold-free-induced-adipocyte cell-sheet
IAg	intracellular antigen
IAIHG	international autoimmune hepatitis group
IBD	inflammatory bowel disease
ICAM	intracellular adhesion molecule
ICE	intravenous immune globulin (10% caprylate-chromatography purified) for the treatment of chronic inflammatory demyelinating polyradiculoneuropathy
Id3	inhibitor of DNA-binding 3
IEC	intestinal epithelial cell
IEL	intraepithelial lymphocyte
IFN	interferon
IFN-I	type-1 interferon
IGF	insulin-like growth factor
IgG	immunoglobulin G
IgM	immunoglobulin M
IL	interleukin
IL-10	Interleukin 10; also known as human cytokine synthesis inhibitory factor (CSIF)
IL-2	Interleukin 2; part of the body's natural response to microbial infection
IL-4	Interleukin 4; induces differentiation of naïve helper T cells to Th2 cells
IL-7	Interleukin 7; stimulates differentiation of multipotent hematopoietic stem cells into lymphoid progenitor cells
IL-12	Interleukin 12; produced by monocytes, macrophages, and dendritic cells in response to bacterial products, such as lipopolysaccharide (LPS), intracellular pathogens, or upon interaction with activated T Cells
IL-15	Interleukin 15; similar to IL-2, it binds to and signals through a complex composed of IL-2/IL-15 receptor beta chain and the common gamma chain

Immune privileged organs	CNS, eyes, testes, and the fetomaternal interface
Immunoediting	tumor elimination, equilibrium, and escape
iNOS	inducible nitric oxide synthase
IPS-1	interferon-β promoter stimulator 1
IQR	interquartile range
IR	intrarenal(ly)
IRI	ischemia/ reperfusion injury
IRIS	immune reconstitution inflammatory syndrome
IRP	immune risk profile
ISGs	IFN-stimulated genes
isRNA	immunostimulatory RNA
ITAMS	tyrosine-based motifs
iTreg	induced regulatory T cell
IVIG	intravenous immunoglobulin
JIA	juvenile idiopathic arthritis
KGF	keratinocyte growth factor
KIR	killer immunoglobulin-like receptors
KLF4	Kruppel-like factor 4
KLRG1	killer cell lectin-like receptor G1
KO	knockout
LC3	microtubule-associated protein light chain 3
LCMV	lymphocytic choriomeningitis virus
LD	linkage disequilibrium; the nonrandom association of two or more loci resulting in segments of the genome being inherited as haplotype (blocks)
LepRb	leptin receptor b
LFA	leukocyte function-associated antigen
LGI1	leucine-rich glioma-inactivated 1
LKM-1	liver-kidney microsomal-1 antibodies
LN	lymph node
LP	lectin pathway
LPS	lipopolysaccharide
mAb	monoclonal antibody
MAC	membrane attack complex
MAL	MyD88-adaptor like
MAP2	microtubule-associated protein 2
MAPK	mitogen-activated protein kinase
MASP	mannose-binding lectin-associated serine protease
MAVS	mitochondrial antiviral signaling
MBL	mannose-binding lectin
MBP	myelin basic protein
MCMV	murine CMV
MDA5	melanoma differentiation-associated gene
mDCs	myeloid dendritic cells
MDD	major depressive disorder
MDSCs	myeloid-derived suppressor cells
MEF	mouse embryonic fibroblast
MFI	mean fluorescence intensity
MG	myasthenia gravis
MHC	major histocompatibility complex
MHC II	MHC class II
MI	memory inflation
MICA/B	MHC class I polypeptide-related sequence A/B
miRNA	micro RNA

MMN	multifocal motor neuropathy
MMP	matrix metalloproteinases
MOG	myelin oligodendrocyte glycoprotein
moi	multiplicity of infection
MRM	multiple reaction monitoring
MS	multiple sclerosis
MSC	mesenchymal stromal cell
MST	military sexual trauma exposure
mTNF	transmembrane TNF
mTOR	mammalian target of rapamycin
MUC-1	mucin MUC-1
MUM-1	multiple myeloma-1
MVP	mitral valve prolapse
MZ	marginal zone
NCR	natural cytotoxicity receptor
NeuN	neuronal nuclei
NFAT	nuclear factor of activated T cells
NF-κB	nuclear factor kappa B; also known as transcription factor p65
NHL	non-Hodgkin lymphoma
NK	natural killer cell
NKT	natural killer T cell
NLRs	NOD-like receptors
NMDAR	*N*-methyl-D-aspartate receptor
NNRTIS	nonnucleotide RT inhibitors
NOD	nonobese diabetic
NOG	NOD/Shi-*scid,IL-2Rg*^{null}
NOS	nitric oxide synthase
NPY	neuropeptide Y
NSCLC	nonsmall-cell lung cancer
OCB	oligoclonal band
OPN	osteopontin
OR	odds-ratio
OS compounds	organosulfur in garlic
P13K	phosphatidylinositol 3-kinase
P13K-AKT	P13K pathway that activates mTOR
PAMP	pathogen-associated molecular pattern
PANDAS	pediatric autoimmune neuropsychiatric disorders associated with streptococcal infection
PANVAC	vaccine containing human genes that cause production of CEA and MUC-1
PB	peripheral blood
PBMCs	peripheral blood mononuclear cells
pCR	pathologic complete response
PD-1	programmed death protein 1
PDCs	plasmacytoid dendritic cells
PDGF	platelet-derived growth factor
PD-K1	phosphoinositide-dependent protein kinase 1 gene
PD-L1	programmed death-ligand 1
PET	positron emission tomography
PeriNomS	peripheral neuropathy outcome measures standardization study
PFS	progression-free survival
PIDD	primary immunodeficiency diseases
PLEx	plasma exchange
PLO	peripheral lymphoid organ tissues
PLP	proteolipid protein
PMBCL	primary mediastinal B cell lymphoma

pMHC	peptide presented by the MHC
PNH	paroxysmal nocturnal hemoglobinuria
PR	progesterone receptor
PRD	protein rich domain
PRM	pattern-recognition molecule
PRNT	plaque reduction neutralization test
PRRs	pattern-recognition receptors
PSC	primary sclerosing cholangitis
Pso	psoriasis
PTSD	posttraumatic stress disorder
qPCR	quantitative polymerase chain reaction
RA	rheumatoid arthritis
RANKL	receptor activator for NF-κB ligand
RCA	regulator of complement activation
RCC	renal cell carcinoma
RegIII	regenerating islet-derived protein 3
RelA	gene that encodes transcription factor p65 (also known as NF-κB)
RIG-1	retinoic acid inducible gene-1
RNAPII	RNA polymerase II
R-ODS	Rasch-built Overall Disability Scale
RORγt	retinoid-related orphan receptor γt
ROS	reactive oxygen species
RP-HPLC	reversed-phase liquid chromatography
RRMS	relapsing-remitting multiple sclerosis
RSS	recombination signal sequence
RTL	recombinant T cell ligand
sAAIPI	second-line age-adjusted international prognostic index
SARM	sterile α-and armadillo motif-containing protein
SASP	senescence-associated secretory phenotype
SCIG	subcutaneous IgG
SCT	stem cell transplantation
SD	stable disease
SHM	somatic hypermutation
shRNA	short hairpin RNA
siglec-1	(sialoadhesin, CD169) is an adhesion molecule that plays a role in cell–cell interactions with T cells via CD43
SIRS	systemic inflammatory response syndrome
SLE	systemic lupus erythematosus
SL/LPA	soluble liver/liver–pancreas antibodies
SMA	smooth muscle antibodies
sMAgs	soluble MHC–antigen complexes
SMZL	splenic marginal zone lymphoma
SNP	single nucleotide polymorphism
SOCS-3	suppressor of cytokine signaling-3
spDC	splenic dendritic cell
SS	Sjögren's syndrome
ssRNA	single strand RNA
STAT	signal transducer and activator of transcription
STAT3	signal transducer and activator of transcription 3
sTILs	stromal tumor infiltrating lymphocytes
STING	stimulator of interferon genes
sTRAIL	soluble TRAIL
T cell	a lymphocyte produced or processed by the thymus gland; an active participant in the immune response

T1D	type 1 diabetes
T1DM	type 1 diabetes mellitus
TAAs	tumor-associated antigens
TAM	tumor-associated macrophage
TBK1	Tank-binding kinase 1
TCAAs	tumor-associated antigens
TCE	T cell clonal expansion
TCR	T cell receptor
TFs	master transcription factors
TGF	transforming growth factor
Th	T helper
TIDC	tumor-infiltrating $CD11c^{+}/CD11b^{+}/Gr\text{-}1^{neg}$ dendritic cell
TIL	tumor-infiltrating lymphocyte
TIR	toll/interleukin receptor
TLR	toll-like receptor
TLR-9	toll receptor 9
TMD	transmembrane domain
TNBC	triple negative breast cancer
TNF	tumor necrosis factor
TPO	thyroid peroxidase
TREC	T cell receptor excision circle
Treg	regulatory T cell
TRIF	TIR domain-containing adaptor protein inducing interferon β
TRPV	transient receptor potential vanilloid 1
TSDR	Treg-specific demethylated region
TSH	thyroid stimulating hormone
TSHR	thyroid stimulating hormone receptor
TUNEL	terminal deoxynucleotidyl transferase dUTP nick end labeling
UDCA	ursodeoxycholic acid
VEGF	vascular endothelial growth factor
VGKC	voltage-gated potassium channel
VIP	vasoactive intestinal peptide; a major immunoregulatory neuropeptide in the central and peripheral nervous system and control acute inflammation
VM	virtual memory
VNTRs	variable number tandem repeats
VSV	vesicular stomatitis virus
VZV	varicella zoster virus
WNV	West Nile Virus
WT	wild-type
YFP	yellow fluorescent protein

Contributors

Paolo A. Ascierto Melanoma, Cancer Immunotherapy and Innovative Therapy Unit, Istituto Nazionale Tumori Fondazione G. Pascale, Napoli, Italy

Milena Bertolotti Reth-Department of Molecular Immunology, BIOSS Centre for Biological Signaling Studies, University of Freiburg, Freiburg, Germany

William Borkowsky NYU School of Medicine, New York, NY, United States

Lisa M. Coussens Department of Cell, Developmental and Cancer Biology, Knight Cancer Institute, Oregon Health and Science University, Portland, OR, United States

Audrey Esclatine INSERM UMR984, Université Paris Sud, Faculté de Pharmacie, Châtenay-Malabry, France

Lucia Festino Melanoma, Cancer Immunotherapy and Innovative Therapy Unit, Istituto Nazionale Tumori Fondazione G. Pascale, Napoli, Italy

Anne A. Gershon Columbia University College of Physicians and Surgeons, New York, NY, United States

M.A. Hayat Kean University, Union, NJ, United States

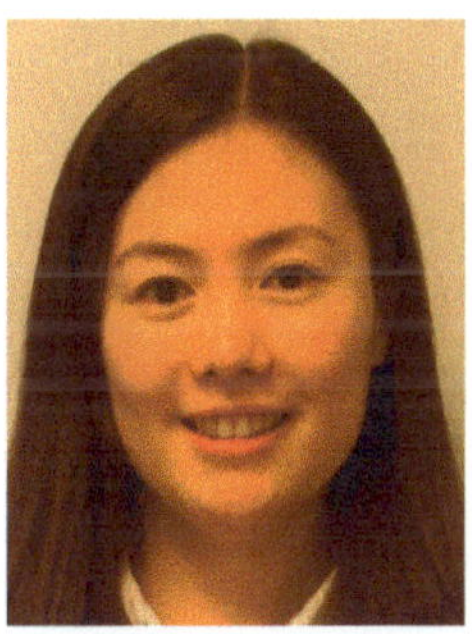

Ming-Xiao He Department of Immunology, Duke University Medical Center, Durham, NC, United States

You-Wen He Department of Immunology, Duke University Medical Center, Durham, NC, United States

Nuhad K. Ibrahim Department of Breast Medical Oncology, University of Texas MD Anderson Cancer Center, Houston, TX, United States

Chinnaswamy Jagannath Department of Pathology and Laboratory Medicine, University of Texas Health Services Center, Houston, TX, United States

Wei Jia Department of Immunology, Duke University Medical Center, Durham, NC, United States

Girdhari Lal National Centre for Cell Science, Savitribai Phule Pune University Campus, Pune, India

John H. Kehrl B Cell Molecular Immunology Section, Laboratory of Immunoregulation, National Institute of Allergy and Infectious Diseases, National Institutes of Health, Bethesda, MD, United States

Mark L. Lang Department of Microbiology and Immunology, University of Oklahoma Health Sciences Center, Oklahoma City, OK, United States

Arshad Khan Department of Pathology and Laboratory Medicine, University of Texas Health Services Center, Houston, TX, United States

Shannon M. Liudahl Department of Cell, Developmental and Cancer Biology, Knight Cancer Institute, Oregon Health and Science University, Portland, OR, United States

Marion Lussignol INSERM UMR984, Université Paris Sud, Faculté de Pharmacie, Châtenay-Malabry, France

Dante J. Marciani Qantu Therapeutics, Inc., Lewisville, TX, United States

Ian McLeod Department of Immunology, Duke University Medical Center, Durham, NC, United States

James L. Murray Department of Breast Medical Oncology, University of Texas MD Anderson Cancer Center, Houston, TX, United States

Neel R. Nabar B Cell Molecular Immunology Section, Laboratory of Immunoregulation, National Institute of Allergy and Infectious Diseases, National Institutes of Health, Bethesda, MD, United States

Lyse Norian Department of Nutrition Sciences, The University of Alabama at Birmingham, Birmingham, AL, United States

Rachael Orlandella Department of Nutrition Sciences, The University of Alabama at Birmingham, Birmingham, AL, United States

Sourav Paul National Centre for Cell Science, Savitribai Phule Pune University Campus, Pune, India

Pragya Rampuria Department of Microbiology and Immunology, University of Oklahoma Health Sciences Center, Oklahoma City, OK, United States

Chong-Shan Shi B Cell Molecular Immunology Section, Laboratory of Immunoregulation, National Institute of Allergy and Infectious Diseases, National Institutes of Health, Bethesda, MD, United States

Roberto Sitia Division of Genetics and Cell Biology, IRCCS San Raffaele Hospital, Università Vita Salute San Raffaele, Milan, Italy

Martina Strudel Melanoma, Cancer Immunotherapy and Innovative Therapy Unit, Istituto Nazionale Tumori Fondazione G. Pascale, Napoli, Italy

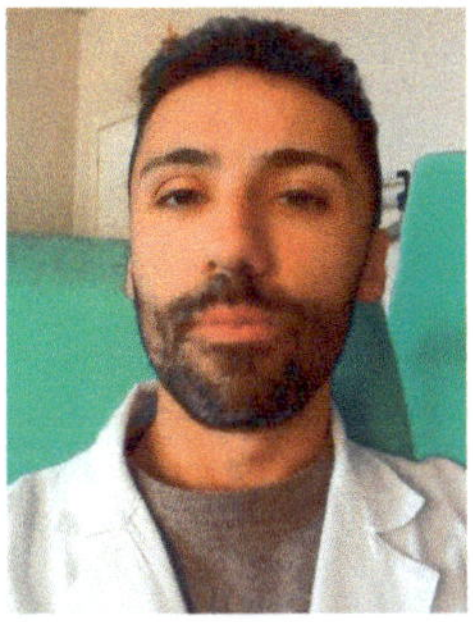

Vito Vanella Melanoma, Cancer Immunotherapy and Innovative Therapy Unit, Istituto Nazionale Tumori Fondazione G. Pascale, Napoli, Italy

Yeldur P. Venkatesh Department of Biochemistry and Nutrition, Central Food Technological Research Institute, Karnataka, India

Jin Wang Baylor College of Medicine, Houston, TX, United States

Chuanwei Yang Department of Breast Medical Oncology, University of Texas MD Anderson Cancer Center, Houston, TX, United States

PART I

MOLECULAR MECHANISMS

CHAPTER

1

Introduction to Immunology

M.A. Hayat

OUTLINE

Abstract

The success of standard treatments (chemotherapy, radiotherapy, surgery, and hormone therapy) in preventing/curing serious diseases is at best modest; moreover these treatments often carry unavoidable and devastating side effects. Therefore the immune system, the discipline of a healthy lifestyle, and the early detection of disease (using biomarkers) are our three friends in disease prevention and cure. With its aim of maintaining an exquisitely precise response to a vast array of potential native and foreign antigens, the immune system is a complex and powerful but imperfect system that needs to be given priority in the field of disease research.

In spite of the huge investment of time, effort, and finance, the progress in preventing and/or curing serious diseases by using standard treatments (chemotherapy, radiotherapy, surgery, and hormone therapy) has been, at best, only modest. The use of standard treatments is accompanied by unavoidable, devastating side effects. At the cost of being repetitious, it has to be emphasized that an improved understanding of the immune system, avoidance of unhealthy habits (e.g., smoking, intake of alcohol, perpetual stress, and lack of exercise), and early detection (using biomarkers) are the only three friends we have to at least delay the onset of serious diseases. Discipline is required to accomplish this goal. A better understanding of the immune system needs to be given priority in the field of disease research.

It is the immune system that informs the body of the difference between its own antigens and foreign antigens, with implications for every event related to us, including health, disease, transplantation, blood donation, and choice of mate. To adequately function, the immune system must maintain an exquisitely specific response to a vast repertoire of potential antigens. Such a diverse response is controlled by genes coding for the α and β chains

M.A. Hayat (ed): Immunology, Volume 1.
DOI: http://dx.doi.org/10.1016/B978-0-12-809819-6.00001-0

© 2018 Elsevier Inc. All rights reserved.

of the T cell receptors (TCRs), which undergo somatic recombination and nucleotide insertion, resulting in the potential generation of a large number of different receptors. However, the immune system is not perfect; occasionally it strays into exaggeration. As time is not in favor of the cell when it comes under the influence of dysregulation of the immune system, the sooner the correct autoregulation is reinstated, the better it is for the cell and the body—immunomodulation is the answer.

The body's defense system is an autoregulating process. It uses input from other systems to correct itself and create an effective defense against toxins and stresses of all types. This regulation occurs through the release of stimulating and inhibiting mediators. For each mediator there is an antagonist, indicating that the immune system is a subtle autoregulation process, which can be intervened using medications (immunomodulators).

The innate immune system is a powerful and evolutionarily well-preserved barrier to infectious pathogens. This system is also inherently capable of recognizing and destroying diseased—including cancer—cells. The immune system is a process of biological structures within an organism that protects it against diseases and tissue injury. In many species this system can be classified into subsystems, such as the innate immune system, adaptive immune system, and humoral immunity versus cell-mediated immunity. Under the innate immune system, we need to explain inflammation and natural killer cells, and under the adaptive immune system we should discuss killer T cells, helper T cells, B lymphocytes, and antibodies. Also, immune deficiencies, autoimmunity, hypersensitivity, and tumor immunology should be discussed. Cancer or tumors need to be discussed only in connection with immunology. Other subjects to be discussed include immune-suppression that leads to enhanced susceptibility to disease, and immunotoxicology that is an adverse or inappropriate change in the structure or function of the immune system after exposure to a foreign substance (xenobiotic).

Another subject to be discussed is inflammation. Both acute inflammation and chronic inflammation need to be explained. Inflammation is a signal-mediated response to cellular insult by infectious agents, toxins, and physical stress. While acute inflammation is important to the immune response, chronic inappropriate inflammation can cause tissue destruction (autoimmunity, neurodegenerative, and cardiovascular diseases).

Another relevant subject is the role of autophagy in various aspects of immunity. The immune system uses autophagy both as an effector mechanism to clear intracellular pathogens and as a mechanism to monitor its products for evidence of pathogen invasion and cellular transformation. Autophagy delivers intracellular material for lysosomal degradation, and its products are primarily loaded into major histocompatibility (MHC) class 1 and class 11 molecules for antigen presentation to $CD8^+$ T cells and $CD4^+$ T cells, respectively. The adaptive immune system depends on the orchestration by these cells as main effectors of cell-mediated immunity. These lymphocytes do not recognize antigens directly, but their T cell receptors detect antigen fragments on major MHC class 1 molecules for $CD8^+$ T cells, and MHC class 11 for $CD4^+$ T cells. Macroautophagy also influences lymphocyte differentiation and function. Based on this information, it is important to discuss the question of how autophagy assists in antigen presentation to the immune system.

Therapeutic use of immunotherapy in patients with neurodegenerative diseases (e.g., Alzheimer's disease, Parkinson's disease) should be discussed. Some information is available regarding the causes of these disorders. For example, the presence of misfolded and aggregated disease-specific protein alpha-synuelein into Lewy bodies and Lewy neurites is common in patients with Parkinson's disease, and accumulation of amyloid-beta and tau proteins is found in patients with Alzheimer's disease. Self-amplification, propagation, and transmission of these proteins are major contributing factors to disease onset and progression. Because these proteins seed to neighboring cells and brain regions, immunotherapy should provide a therapeutic approach for these patients. Considering the modest success of other therapies against serious diseases, especially cancer, it is becoming imperative to focus on the immune system. There is a growing need to harness a patient's immune system to fight diseases ranging from autoimmune to cancer. Immuno-based therapies are becoming increasingly critical in the fight against disease. There is an untapped opportunity in expanded effort to bring forward new, targeted, and effective immuno-based therapies, i.e., to program the immune system against disease.

One principle therapy is checkpoint inhibitors. These inhibitors are a class of monoclonal antibodies that inhibit the pathways responsible for blocking the response of T cells to antigens. Results from clinical trials of these therapeutics have been promising. On September 5, 2014, the treatments with these inhibitors were approved in the U.S. and Europe. One recent example is Merck's PD-1 inhibitor pembrolixumab (keytrude). Another principle therapy is the use of vaccines to prevent illness (e.g., smallpox, measles, and pertussis). Combination therapies are a route for these vaccines in the future. Another technology is adoptive cellular immunotherapy, which is used when syngeneic activated T cells are infused in patients to attack their cancer; examples are tumor infiltrating lymphocyte therapy, genetically engineered T cells bearing chimeric antigen receptors, and recombinant TCR technology. The improvement of these technologies is expected in the near future.

HUMAN IMMUNODEFICIENCY VIRUS

Although antiretroviral therapy allows HIV-infected persons a longer lifespan, it is not without negative side effects. No effective vaccine has been found, primarily because of the specific features of the viral envelope glycoprotein (Env). By virtue of its presence on the viral surface, it is the primary target of antibodies. HIV-1 Env promotes entry into target cells by recognizing cellular receptors and by fusion of virus and host cell membranes.

Env first engages cellular CD4, and this interaction leads to a conformational rearrangement in the Env, resulting in the presentation of the coreceptor binding site. This insertion with the coreceptor triggers the Env membrane-fusion domain to mediate virus entry. Env has evolved effective means of concealing functional centers from attacks by antibodies. In summary, viral glycoprotein molecule undergoes structural (conformational) changes induced by the cellular receptor, which allows the virus to fuse with the host cell membrane. This structural alteration drives the viral protein to enter and infect the cell.

AUTOPHAGY AND IMMUNITY

Roles of autophagy in infection, inflammation, immunity, and disease in general are well known. Different types of autophagy modulate innate immune response as well as adaptive immune response. Autophagy is central in the regulation of inflammation. Autophagy is involved in antigen presentation and lymphocyte differentiation as well as in the balance regulating survival/death and activation of lymphocytes. Autophagy protects, for example, against active tuberculosis by suppressing bacterial burden and inflammation. The role of autophagy in the etiology of autoimmune/autoinflammatory diseases should be explained. The link between pharmacological regulators of autophagy and inhibitors of immune diseases needs to be clarified. Future studies should focus on how each autophagic pathway regulates specific process, such as antigen presentation and cell activation.

It is well established that immunological memory is mediated by T- and B cells. However, mammalian innate immune cells (e.g., natural killer cells) also show immunological memory. The term *trained immunity* is used to describe such memory properties of innate immune cells. Trained immunity can be induced, for example, by anti-tuberculosis vaccine Bacillus Calmette-Guérin (BCG), through epigenetic reprogramming of monocytes. Overexpression of antigen Ag85B in BCG induces autophagy in antigen presenting cells and enhances immunogenicity, suggesting that vaccine efficacy can be increased by augmenting autophagy-mediated antigen presentation. BCG is the most widely used anti-tuberculosis vaccine in the world, and is also effective against some types of cancer (e.g., bladder cancer). An attempt should be made to explain the immunological mechanism involved in this process, autophagy being one of the participants.

Autophagy genes (pathways) are linked to diverse aspects of innate and adaptive immunity. In fact, autophagy is regulated by pathways critical for the function and differentiation of cells of the immune system, including Toll-like receptors, p47, kinases, and cytokines. A number of different immune signals either positively (e.g., GTPases, TLRs, tumor necrosis factor, PKR) or negatively (e.g., IL4, IL13) regulate autophagy. Indeed, autophagy genes, for example, are important for thymic selection, lymphocyte development and survival, antigen presentation, killing of intracellular pathogens, and tissue homeostasis. In addition to the known functions of autophagy genes in the immune system via classical autophagy mentioned above, information on the functions of these genes in the immune system based on some other possible mechanism is needed. For the convenience of the readers, the contents are classified into two subheadings: *Molecular Mechanisms* and *Disease and Treatment*. An extensive list of abbreviations and definitions is provided to assist the reader.

Autoimmunity affects approximately 10% of the global population. Examples include systemic lupus erythematosus, Graves' disease, rheumatoid arthritis, and multiple sclerosis. Current treatments for these diseases inhibit inflammation, but are non-specific and tend to alter the immune system such that it becomes less effective against infectious and neoplastic insults.

Although autoimmunity is triggered by antibodies, B cells, CD8 T cells, and many cells of the innate immune system (CD4 T cells) are mainly responsible for the initiation of these diseases. Many autoimmune diseases occur in patients carrying specific major histocompatibility complex (MHC) class II molecules, the human leukocyte antigen (HLA). The interaction between the CD4 T cells and the specific antigen is likely responsible for the development of this disease. There is a need to develop an antigen-specific therapeutic system that can effectively prevent autoimmune attack while leaving the rest of the immune system functioning normally.

The usual immune challenge faced by us is posed by pathogens, and vaccination usually protects us from infection. Sex-based disparity in immune response is well documented. It is based on the intrinsic genetic difference, XX status versus XY, and differential levels of sex hormones produced by males and females. Male hormones tend to suppress immune response. Clinical studies, for example, have indicated that androgen ablation boosts the immune response and increases the vaccination efficiency. Another example is testosterone. The immune-suppressive role of this hormone on both adaptive and innate immune responses is known. An increase in in vivo testosterone levels suppresses both innate and adaptive immune levels, and this effect is across species type. However, ironically, the majority of autoimmune affected individuals are females. Apparently, further studies are needed to elucidate the molecular mechanism(s) underlying hormone-induced immunosuppression.

In terms of aging, the world is changing exceedingly fast, especially in terms of the number of people in the age group of over 60 years in every country. The global population including ages 60 years and over is expected to reach 2 billion by the year 2050 (WHO), and people in the age group of over 60 years are increasing disproportionately more than people in other age groups in all countries. This difference in population is expected to continue, and we will have a problem unless we soon find means to prevent or cure old people's diseases discussed above or improve immune responses.

CHAPTER

2

Cytomegalovirus and Autophagy

Marion Lussignol and Audrey Esclatine

OUTLINE

Abstract

Autophagy is a well-described vacuolar degradative pathway which contributes to the homeostasis of the cell and to the response to environmental stresses. An increasing number of studies describe the modulation of autophagy by a large subset of viruses, either to avoid a host defense mechanism (potentially degrading viral particles before they can escape the host cell) or to hijack the autophagic machinery for their own benefit. Human cytomegalovirus (HCMV) has the capacity to fine tune the autophagic process during its own life cycle, and finally block the autophagic flux to promote its own replication and presumably avoid the degradation of cytoplasmic viral components. At least two anti-autophagic viral proteins encoded by HCMV have been described, but future studies may reveal new viral modulators. The purpose of this chapter is to discuss the significance of these findings in the light of the general knowledge on the autophagic pathway and the interplay between the viruses and this conserved homeostatic pathway. The importance of such information is supported by the recent publication of the 12-volume series (Elsevier) discussing autophagy.

M.A. *Hayat (ed): Immunology, Volume 1.*
DOI: http://dx.doi.org/10.1016/B978-0-12-809819-6.00002-2

© 2018 Elsevier Inc. All rights reserved.

INTRODUCTION

Autophagy is a cellular self-degradative process which ends into the lysosome, and acts in numerous processes such as cellular homeostasis, quality control, and cell survival. However, it also deals with aging, adaptive and innate immunity, or developmental processes. Dysfunction of autophagy has been implicated in numerous pathologies, among them neurodegenerative diseases; inflammatory bowel diseases such as Crohn's disease; diabetes; and cancer. Macroautophagy, the major form of autophagy (hereafter referred to as autophagy), relies on sophisticated machinery, which enables the building of a membrane cistern known as the phagophore; a double membrane vesicle called the autophagosome; and an autolysosome (Fig. 2.1). Numerous regulation systems allow autophagy to adapt easily to multiple environmental stresses (Lorin et al., 2013).

Autophagy-related proteins or ATG, were originally discovered in yeast and often have several counterparts in mammal cells. For example, in mammals, the Atg8 family comprises seven homologs, divided in two subfamilies: microtubules-associated protein 1 light chain 3 (MAP1LC3, hereafter referred to as LC3) and GABA receptor-associated protein (GABARAP). The LC3 subfamily includes LC3A, LC3B, LC3B2 and LC3C and the latter subfamily GABARAP, GABARAPL1 and GABARAPL2/GATE-16. LC3B is the most studied ATG8 homolog. To date, over 35 ATG proteins have been identified in yeast and among them, 18 are required for autophagosome formation. Two ubiquitin-like systems lead to the conjugation of LC3 with a lipid, and the conjugation of ATG5 with ATG12. The mammalian homolog of ATG6, Beclin 1 and the class III phosphatidylinositol 3-kinase (PtdIns3K) are organized into two complexes, which participate in the biogenesis and the maturation of autophagosomes, with other participants such as ATG14 or UVRAG (and Rubicon), respectively. The fusion between autophagosomes and lysosomes also requires ATG14 and a set of SNARE proteins (a family of proteins which mediates all intracellular membrane fusion events), including STX17, SNAP29 and VAMP8 (Diao et al., 2015).

Numerous DNA and RNA viruses, either enveloped or naked viruses, are able to modulate autophagy at different times of their life cycle (Faure and Lafont, 2013). First, some viruses can stimulate autophagy just after their fixation at the cell surface. Envelope glycoproteins of viruses belonging to numerous families (Paramyxoviridae, Orthomyxoviridae, Rhabdoviridae, Bunyaviridae, Herpesviridae) can induce autophagy by interacting with their receptors. Among these viruses, the vaccine strain of the measles virus (MeV) triggers autophagy by interaction with CD46 at the cell surface. Similarly, exposure of T cells to the human immunodeficiency virus 1 (HIV-1) gp120 can lead to autophagy activation, followed by cell death, after its interaction with CD4 and its co-receptor CXCR4 (Joubert et al., 2009). Transmitted from flies to mammals, Rift Valley fever virus (RVFV) virions interact with the pattern recognition receptors (PRRs) Toll-7 or mammalian toll-like receptors (TLR) at the cell surface to induce antiviral autophagy (Moy et al., 2014). Herpes simplex virus type 1 (HSV-1) activates autophagy by different signaling pathways, but recently it was reported that HSV-1 binding to THP-1 cells triggers a rapid and transient stimulation of autophagy by fixation to TLR2 (Siracusano et al., 2016). Below these receptors, activation of different cascade signaling pathways lead to stimulation of autophagy. For example, MyD88- or TRIF-dependent pathways trigger autophagy under TLRs activation. Similarly, autophagy is induced by GOPC after fixation of MeV to its receptor CD46.

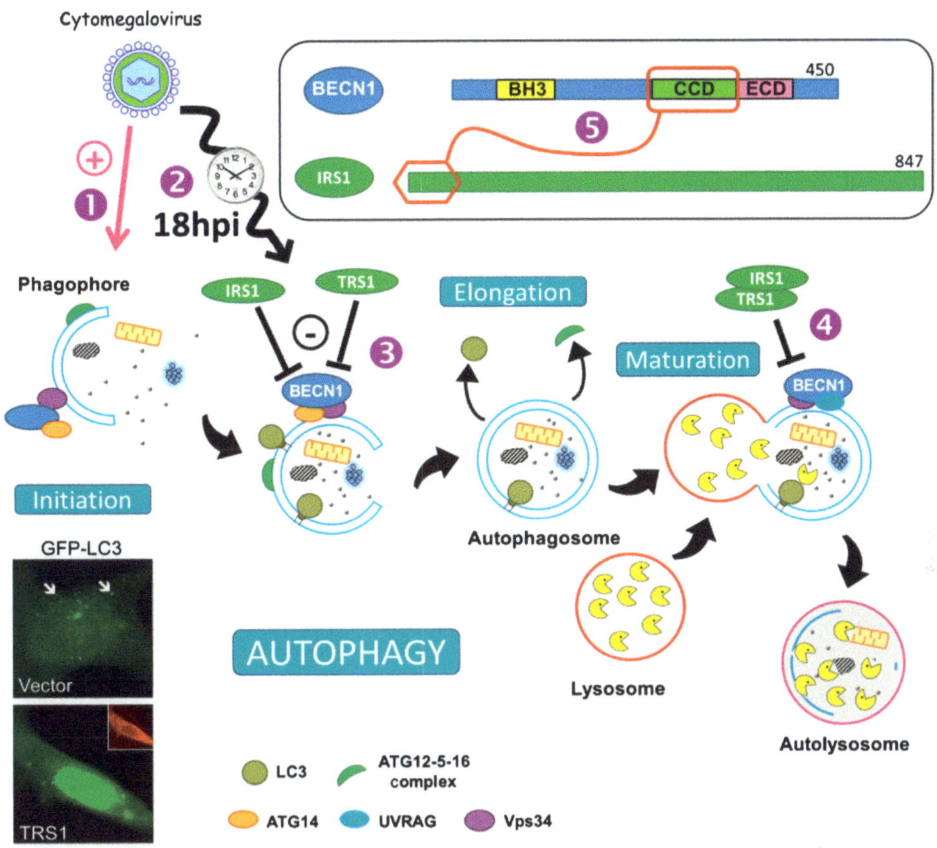

FIGURE 2.1 HCMV modulates autophagy in two opposite directions during infection of human fibroblasts. (1) Autophagy is induced by HCMV in fibroblasts independently of viral protein expression. (2) 18h post-infection, HCMV starts to block autophagy. (3) Two viral proteins, IRS1 and TRS1, are able to decrease autophagosome biogenesis by interacting with Beclin 1. (4) Co-expression of IRS1 and TRS1 blocks the autophagic flux. (5) The N terminal part of IRS1 interacts with the CCD of Beclin 1. Inset: Ectopic expression of TRS1 (red staining) decreases the number of autophagosomes (GFP-LC3) in HeLa cells. BECN1, Beclin 1; BH3, Bcl-2 homology domain 3; CCD, coiled coil domain; ECD, evolutionary conserved domain.

After viral entry into the cell, viruses may still be detected by the multiple intracellular recognition systems that activate antiviral defense mechanisms, including autophagy. In particular, DNA and RNA viral genomes or viral nucleic acids localized in the cytoplasm can be recognized. For example, Coxsackievirus B3 dsRNA induces autophagy through TLR3 activation (CVB3) (Gorbea et al., 2010). Viral nucleic acid, such as ssRNA of VSV, can be taken up into the endosome as a result of autophagy, and thus targeted to the TLRs localized in the endosomal compartment (Lee et al., 2007). Regarding DNA viruses, several studies reported that viral cytosolic DNA can be recognized by the cGAS-STING pathway (McFarlane et al., 2011; Rasmussen et al., 2011). Finally, HSV-1 produces dsRNA during infection, most likely as a result of hybridization of convergent overlapping mRNA. It has been previously reported that activation of dsRNA-activated kinase PKR also leads to autophagy induction by phosphorylation of eIF2α (Talloczy et al., 2002). In addition to viral genomes, numerous viral proteins are also able to trigger autophagy. Among them, we can cite MeV C protein; non-structural proteins 2 and 3 of poliovirus; hepatitis B virus X protein; and NS4B/NS5A/NS5B of hepatitis C virus (HCV) (Richetta and Faure, 2013). Rather than enabling recognition by the cell to impede viral infection, it seems that viral proteins modulate autophagy to the benefit of the virus. In fact, some viral proteins block the autophagosome maturation step and thus the degradative function of autophagy. We can cite P protein of parainfluenza virus 3, which interacts with SNAP29, or HIV Nef and influenza virus M2 which interact with Beclin 1 (Gannage et al., 2009; Kyei et al., 2009; Ding et al., 2014). Several viral proteins can also block the biogenesis of autophagosomes, notably in the Herpesviridae family (Lussignol and Esclatine, 2015).

Finally, stress resulting from viral infection can be indirectly responsible for autophagy activation, such as ER stress for Chikungunya virus (CHIKV) or HCV (Ke and Chen, 2011; Joubert et al., 2012), mitochondria depolarization or reactive oxygen species (ROS) release.

HCMV INDUCES AND BLOCKS AUTOPHAGY DURING ITS LIFE CYCLE

Human cytomegalovirus (HCMV) is a widespread virus, which belongs to the Herpesviridae family. HCMV commonly infects people of all ages, across the developed and developing world. Spread by close contact from person to person, through most bodily fluids such as saliva, blood, genital secretions, urine or breast milk, HCMV mostly causes asymptomatic infections. However, congenital infection by HCMV is the most prevalent viral cause of congenital disease and can result in severe disabilities or stillbirth. Moreover, HCMV is responsible for severe diseases in immunosuppressed patients, particularly solid organ and hematopoietic stem cell transplant recipients.

With a tropism restricted to human cells, HCMV is characterized by the largest genome of all those viruses which infect humans, and it has a relatively long replication cycle. The virus can be detected *in vivo* in many cell types including epithelial cells, connective tissue cells, hepatocytes, various leukocyte populations and vascular endothelial cells, indicating the involvement of a number of tissues as sites of productive infection.

Induction of Autophagy by HCMV in Fibroblasts Is Independent of Viral Protein Expression

A few hours after infection, the commonly used laboratory strain HCMV AD169, highly adapted to human fibroblasts, induces autophagy (Chaumorcel et al., 2012). Accumulation of autophagosomes can be easily detected in the cytoplasm of infected cells, using a GFP-LC3 probe, which is the most widely used marker for autophagosomes. Accumulation of LC3 form II, a form specifically associated with the internal and external membranes of the autophagosome, is also observed by immunoblot in the same conditions. Moreover, autophagy activation after HCMV contact is confirmed by an enhanced degradation of SQSTM1, an autophagy substrate (Fig. 2.1). What are the viral inducers responsible for this rapid and marked stimulation of autophagy? Induction of autophagy occurs very quickly, just a few hours after contact of the cell with the viral particle, and despite inactivation of viruses by ultraviolet treatment. Using a tandem fluorescent-tagged LC3 probe (mRFP-GFP-LC3) for monitoring autophagic flux, we clearly reported that HCMV and inactive HCMV both stimulate autophagy. This means that the autophagy inducer is very likely a component of the viral particle rather than a neo-synthesized viral protein. McFarlane et al. (2011) proposed that autophagy was triggered by recognition of cytosolic viral DNA by the cell. However, preliminary data from our laboratory suggest that TLR2 activation by envelope glycoproteins could also take part in this activation.

Inhibition of Autophagy and Autophagic Flux by HCMV

The life cycle of HCMV is extended and we previously reported in 2008 that HCMV infection is also able to inhibit autophagy in human fibroblasts (Chaumorcel et al., 2008). This inhibition occurs after 18–24h post-infection and depends on neo-synthesized viral protein expression, because it is never observed in cells in contact with UV-inactivated HCMV (Fig. 2.1). Not only is the initiation of autophagy impacted by HCMV, but the autophagic flux is also impaired in infected fibroblasts, as visualized with the tandem tagged LC3 probe. Indeed, inhibition of the autophagic flux means that the complete process is impaired, including the sequestration and degradation of the cargo. This blockade is confirmed by LC3 II accumulation throughout the HCMV infection (Chaumorcel et al., 2012; Mouna et al., 2016). A short study published in 2013 confirmed the inhibition of autophagy by HCMV in HEL cells (Zhao et al., 2013). Surprisingly, Belzile et al. (2015), after studying levels of autophagy during HCMV infection concluded that HCMV conversely stimulates autophagy in fibroblasts. However, they observed by immunoblot an accumulation of LC3 II during infection, compatible with a block of the autophagic flux and previously described by our group (Chaumorcel et al., 2012). Analysis of their data showed a decreased number of LC3 dots in cells expressing the immediate early viral proteins compared to the level in bystander cells. This was associated with a lack of red-stained autolysosomes with the tandem tagged mRFP-GFP-LC3 probe, which is again in agreement with our previous studies, and which clearly confirmed the inhibition of autophagy by HCMV. Nevertheless, it is less clear whether HCMV is able to modulate autophagy in human aortic endothelial cells and in developing neural cultures (Belzile et al., 2015).

In order to understand how HCMV impedes autophagy activation, the first idea that springs to mind is to look at autophagy regulation by cellular proteins. The Beclin 1–Bcl-2 complex is an important rheostat for autophagy, and it has been reported that interaction between Bcl-2 and Beclin 1 represses autophagy (Pattingre et al., 2005). During HCMV infection of human fibroblasts, Bcl-2 protein accumulates, which is clearly compatible with this hypothesis. However, this rheostat is finely tuned, in particular by phosphorylation mechanisms of Bcl-2 as much as Beclin 1 (Levine et al., 2015). In fact, the Bcl-2 form which accumulates after HCMV infection is phosphorylated by JNK kinase and does not interact with Beclin 1. Accumulation of Bcl-2 is therefore not involved in virus-induced inhibition of autophagy. Back to square one. Another possible track: Which are the suspected viral proteins? Not easy to guess when you know the number of viral proteins synthesized by HCMV. One clue lies in the fact that inhibition appears 18–24h post-infection, shifting our focus towards immediate early or early viral proteins. One additional clue is based on the fact that binding of PKR kinase to dsRNA leads to autophagy activation, and two HCMV proteins, TRS1 and IRS1, have been identified to directly counteract PKR activity, an important sensor of viral infections (Child et al., 2004).

TRS1 and IRS1 are Two Anti-Autophagic Viral Proteins

The HCMV genome is organized as two unique regions, flanked by large inverted repeats (TRL/IRL) and (IRS/TRS). Located in one set of inverted repeats, IRS1 and TRS1 ORFs encode two viral tegument proteins of, respectively, 847 and 795 amino acids, that localize to the nucleus and cytoplasm in infected cells. IRS1 and TRS1 proteins have identical N-terminal domains but divergent C-terminal regions (Romanowski and Shenk, 1997). They both bind to dsRNA and PKR, to block the PKR/eIF2α signaling pathway, and these interactions require their carboxy-termini (Hakki and Geballe, 2005; Hakki et al., 2006; Bierle et al., 2013). By inhibiting the phosphorylation of the translation initiation factor eIF2α, their essential role is to prevent the shutoff of protein synthesis that occurs upon infection (Braggin et al., 2016). These two proteins are complementary, since a deletion mutant virus lacking both of them is unable to grow, whereas the single deletion mutant viruses are able to grow efficiently.

TRS1 or IRS1 are both able to repress autophagy in epithelial cells and in murine fibroblasts (Mouna et al., 2016). Ectopic expression of each of them has the ability to block the initiation process. An automated quantification of GFP-LC3 dots using a high throughput microscope confirmed that there is no accumulation of autophagosomes observed in the cytoplasm of cells expressing either of the viral proteins. This was associated with a decrease of LC3-II and an accumulation of SQSTM1 by default of degradation following autophagy inhibition. TRS1 and IRS1 are also involved in the profound inhibition of the autophagic flux at later times of infection, and this blockade needs the complementary action of the two viral proteins. Indeed, only simultaneous expression of TRS1 and IRS1 leads to a block of the autophagic flux.

Confirming the involvement of these viral proteins in the context of viral infection, a mutant virus lacking TRS1 or a mutant virus lacking IRS1 are less efficient for blocking autophagy than the wild type, but still modulate the process. Nevertheless, a virus containing mutations on both TRS1 and IRS1 does not inhibit autophagy at all. However, it is possible that another anti-autophagic HCMV protein exists, such as for example KSHV, which expresses three different inhibitors, vBcl-2, vFLIP and K7 to block autophagy (Lussignol and Esclatine, 2015). Indeed, when all *de novo* HCMV protein expression is prevented by UV-inactivation, wild-type HCMV stimulates autophagy as assessed by accumulation of autophagosomes and of LC3 II, associated with activation of the autophagic flux. Similarly, infection with a mutant virus deleted for both IRS1 and TRS1 leads to an important induction of autophagy. As a matter of fact, this mutant virus, which cannot block PKR activity, leads to a general protein synthesis shutoff, affecting both cellular and viral proteins. Most viral proteins are therefore not expressed.

The mechanism by which the viral proteins modulate autophagy has been identified by our group. Although TRS1 and IRS1 are repressors of PKR activity, and the phosphorylation of eIF2α by PKR is an important inducer of autophagy, TRS1 and IRS1 inhibit autophagy independently of their activity on PKR. We have then turned to Beclin 1 for two reasons: (1) Several viral proteins that are able to modulate autophagy act by interaction with Beclin 1, such as M2 of influenza virus, vBcl-2 of KSHV or ICP34.5 of HSV-1; (2) Beclin 1 contributes to both biogenesis and maturation of autophagosomes, two steps impeded by the HCMV proteins. We unraveled the fact that TRS1 and IRS1 are able to bind to Beclin 1 in the context of HCMV infection. The coil-coiled domain of Beclin 1 interacts with the N terminal domain of IRS1 and this Beclin 1-binding domain is essential for autophagy inhibition (Fig. 2.1). These data suggest that individually, IRS1 and TRS1 may block autophagosome biogenesis through the PtdIns3K–Beclin 1–ATG14 complex, and coexpression of them may block the maturation process through the PtdIns3K–Beclin 1–UVRAG complex.

IMPACT OF THE MODULATION OF AUTOPHAGY ON VIRAL INFECTION

Whereas Autophagy Is Described as an Antiviral Mechanism, Several Viruses Hijack Autophagy for Their Own Benefit

Autophagy is now widely considered as an ancient antiviral mechanism, which can help cells to combat different viruses. Some studies have described how autophagy can restrict vesicular stomatitis virus (VSV), RVFV or chickungunya virus replication and mortality (Joubert et al., 2012; Moy et al., 2014). In these cases, stimulation of autophagy leads to a decrease of viral production and conversely, invalidated or even modestly impaired autophagy causes enhanced viral multiplication and pathogenicity. Autophagy also contributes to adaptive immunity, by allowing, for example, the presentation of EBNA-1 protein of EBV by major histocompatibility complex class II (MHC-II) (Paludan et al., 2005). The direct degradation of incoming microbial pathogens into autolysosomes is referred as xenophagy. Currently, xenophagy has been rarely detected for viruses, probably because viruses have developed strategies during evolution to combat this cell-autonomous defense. For example, viral particles of the mutant of HSV-1 lacking ICP34.5, its main anti-autophagic protein, are sequestered into autophagosomes and seem degraded by autophagy (Talloczy et al., 2006). This has not been observed with wild type HSV-1. However, neurons specifically rely on autophagy to limit HSV-1 replication both *in vivo* and *in vitro* (Yordy et al., 2012). Additionally, pharmacological induction of autophagy in murine fibroblasts seems to diminish HSV-1 infection (Yakoub and Shukla, 2015a) whereas, surprisingly, repression of HSV-2, closely related to HSV-1, is observed after autophagy suppression (Yakoub and Shukla, 2015b).

Other studies have demonstrated that a subset of viruses co-opts components of the autophagy pathway to promote their own replication. The autophagic process can be used, for example, to provide the energy necessary for viral replication. Dengue virus induces autophagic degradation of lipid droplets, a process called lipophagy, to release fatty acids with the aim of producing ATP (Heaton and Randall, 2010). MeV, by inducing autophagy, rather seeks to facilitate cell survival, until its replicative cycle is accomplished (Richetta et al., 2013). In the case of poliovirus, autophagy may also allow a non-lytic egress of the viral particle from the cell by exocytosis involving autophagic membranes, an unusual process for a naked virus (Bird et al., 2014). Some enteroviruses spread between cells via a process described as autophagosome-mediated exit without lysis (AWOL).

Regarding herpesviruses, autophagy can be hijacked by several of them, in different cell types (Nowag et al., 2014; Buckingham et al., 2016). It has been reported recently that two human herpesviruses, VZV and EBV, may use the autophagic protein LC3 to facilitate their secondary envelopment, which corresponds to the acquisition of their final envelope in the cytoplasm, from membranes which could be derived from autophagosomes or from amphisomes (formed by fusion of an autophagosome and an endosome). In both cases, autophagosomes accumulate in the cytoplasm of infected cells and extracellular purified viral particles contain the lipid-conjugated form of LC3. Another recent study reported a rapid but transient activation of autophagy in human monocytes infected with HSV-1, which facilitates viral multiplication, showing that cell type matters for HSV-1 (Siracusano et al., 2016). The precise mechanism of this proviral effect needs to be elucidated.

Proviral Activity of Autophagy in the Context of HCMV Infection

The inhibition of autophagy by HCMV during two steps of the autophagic process, repression of autophagosome biogenesis, and of autolysosome degradative and recycling functions, demonstrate that HCMV needs to control this cellular degradative pathway perfectly. The use of several mutant viruses, which modulate autophagy differently from the wild type virus, did not enable us to understand clearly the importance of this modulation. However, a pharmacological approach demonstrated that autophagy has a proviral activity on HCMV (Mouna et al., 2016). Indeed, induction of autophagy improved HCMV multiplication, whereas inhibition by Spautin 1, a potent and specific inhibitor of autophagy, repressed extra and intracellular HCMV production. shRNA-mediated stable knockdown of ATG16L1 also impedes the replication of HCMV. Another study, also using autophagy modulators, similarly suggested that autophagy facilitates HCMV replication (Zhao et al., 2013).

The life cycle of HCMV has been extensively studied in fibroblasts (Fig. 2.2) (Mocarski et al., 2013). HCMV virions bind to cell surface receptors and the virion envelope fuses with the plasma membrane, releasing the nucleocapsid and tegument proteins into the cytoplasm. Then the capsid is transported to the nuclear pore where the viral DNA is released into the nucleus. Transcription and replication of the viral genome with assembly of new capsids take place in the nucleus. Viral gene expression is coordinately regulated and can be roughly divided into three groups: immediate early, early and late gene products. DNA is encapsidated into the nucleus and the nucleocapsid buds through the inner nuclear membrane. Assembly of the viral enveloped particle occurs in several stages by a complex two-stage envelopment and egress process that starts in the nucleus and leads to virion release by exocytosis at the plasma membrane. This primary envelopment is followed by a de-envelopment step at the outer nuclear membrane that releases the nucleocapsid into the cytoplasm. Nucleocapsids are most likely transported on microtubules. Final envelopment occurs in a unique structure known as the assembly compartment, consisting of expanded

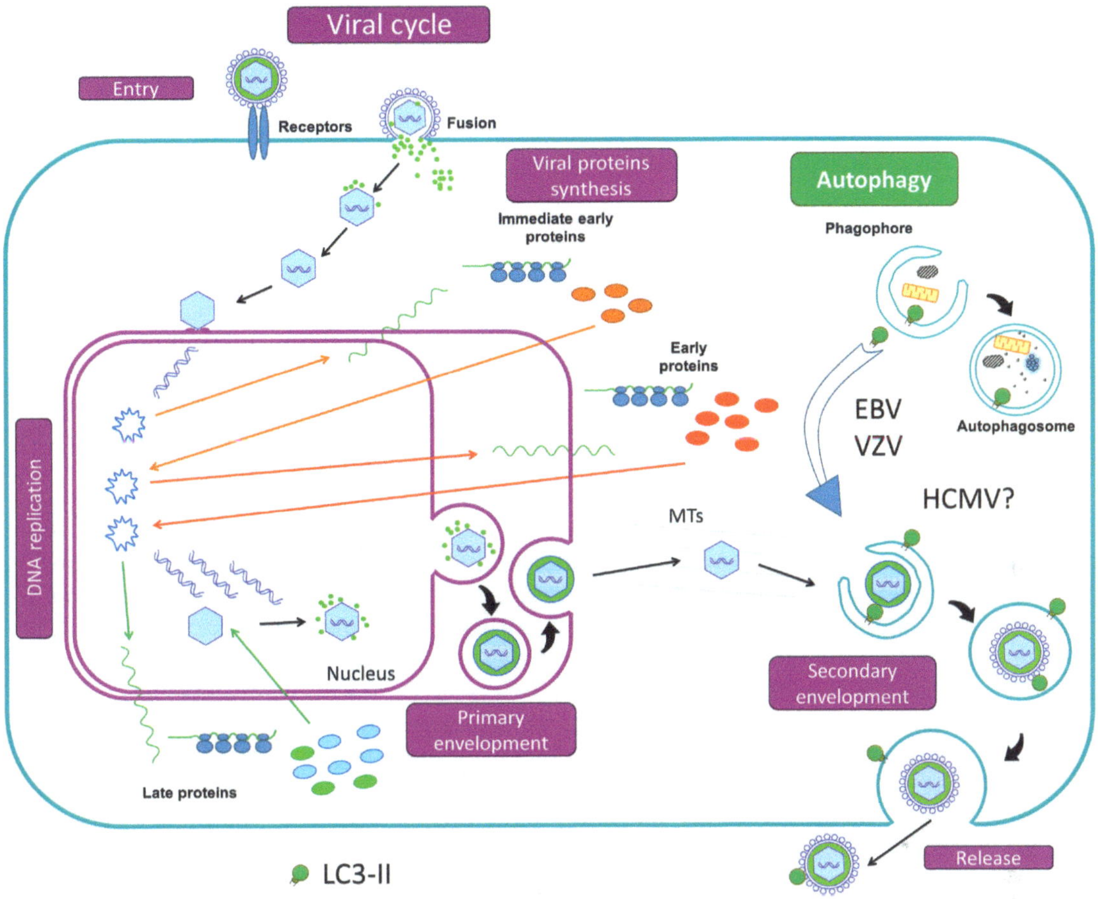

FIGURE 2.2 Viral replication cycle of herpesviruses and model for involvement of autophagy in secondary envelopment for EBV, VZV and possibly for HCMV. VZV and EBV may use lipidated-form of LC3 to facilitate their secondary envelopment, using membranes deriving from autophagosomes or from amphisomes. MTs, microtubules.

and reorganized ER and Golgi apparatus in which nucleocapsids and dense bodies accumulate (Mocarski et al., 2013). The entire process takes several days.

How can HCMV exploit autophagy machinery to its own advantage? It has been proposed that autophagy plays a role in viral release, for enveloped viruses such as HPIV3 or HBV, or for naked viruses such as poliovirus. In these cases, autophagy only influenced extracellular viral production (virions released into supernatants), but not intracellular viral production (cell-associated virions) or viral protein expression. It is interesting to note that, for HCMV, modulation of autophagy by inducers or inhibitors has the same impact on extracellular or intracellular viral production. Therefore, HCMV does not seem to exploit autophagy to exit the cells. The limited impact of autophagy on HCMV genome replication suggests that the machinery is hijacked by HCMV for some cytoplasmic steps. Based on recent data suggesting a role of LC3 in envelopment of EBV and VZV, this hypothesis needs to be explored for HCMV. Moreover, secondary envelopment of HCMV nucleocapsids shares clear similarities with the structure of a phagophore (Fig. 2.2).

Autophagy as a Host Antiviral Immunity Against HCMV

Few clues indicate that in different situations autophagy can restrict or control HCMV pathogenicity. First, autophagy has been reported to contribute to adaptive immunity by improving MHC presentation of viral proteins. In the case of HCMV, autophagy mediates the presentation of latency-associated UL138 protein epitopes by MHC class I to $CD8^+$ T cells (Tey and Khanna, 2012). This occurs independently of the ubiquitin–proteasome pathway in human fibroblasts and in human lymphoid cells deficient for TAP. In conventional MHC I antigen presentation, after degradation by the proteasome, endogenous proteins are transported to the ER and loaded onto MHC I molecules to interact with $CD8^+$ T cells. The contribution of the autophagolysosomal compartment to MHC I presentation of HSV-1 envelope glycoproteins has been previously reported in macrophages, although it is still dependent on proteasome and Golgi transport (English et al., 2009). However, a recent study reported that lack of ATGs results in elevated viral-specific $CD8^+$ T cell responses by default of the internalization of MHC I molecules, rather suggesting that autophagy inhibits MHC class I antigen presentation (Loi et al., 2016). Involvement of autophagy in HCMV antigen presentation therefore needs to be clarified.

Another study claimed that autophagy is a defense system against HCMV because treatment by a natural disaccharide called trehalose decreases viral replication in different cell types (namely fibroblasts, endothelial cells and cells derived from neural stem cells) (Belzile et al., 2015). Trehalose treatment for 24–72h clearly led to autophagosome accumulation and decreased both viral protein expression and viral production. The apparent discrepancy with the previous studies could be explained by the fact that the link between the upregulation of autophagy by trehalose and the antiviral activities has not been clearly established. Trehalose has been described as an inhibitor of the GLUT family of glucose transporters on the plasma membrane of hepatocytes, inhibiting glucose transport, and it is possible that the antiviral effect of trehalose is associated with this inhibitory activity rather than with autophagy (DeBosch et al., 2016). Preliminary data demonstrate an antiviral effect of trehalose in autophagy-deficient cells (Mouna and Esclatine, unpublished data). Interestingly, HCMV reprograms GLUT expression, inducing the overexpression of GLUT4 at the surface of infected cells, because the virus needs to increase its glucose uptake compared to mock-infected cells (Yu et al., 2011). Therefore, we hypothesize that the antiviral activity of trehalose can be related to its inhibitory function on GLUT.

CONCLUSION

As observed for several viruses, HCMV has the capacity to clearly modulate the autophagic process, at least in human fibroblasts. First, HCMV is able to stimulate autophagy by interaction of one component of its viral particle, which still needs to be identified. Second, the virus blocks autophagy and in particular the autophagic flux, using the combined action of at least two proteins, IRS1 and TRS1, which leads to degradation of the autophagosome content. Whereas Belzile et al. observed several common points with the results of Mouna et al., they concluded that HCMV stimulates autophagy at every time point. They also described an antiviral effect of trehalose, a treatment known to induce autophagy. However, whether trehalose repressed HCMV through autophagy needs to be clarified. Regulation of autophagy is probably cell-type dependent and extensive further study is needed using different permissive cells. Considering the potential of autophagy modulators for therapeutic purposes, it is essential that we understand clearly whether autophagy is pro or antiviral.

Acknowledgments

This work was supported by institutional funding from the Centre National de la Recherche Scientifique (CNRS), from Université Paris-Sud, and from Agence Nationale de la Recherche and Région Ile de France.

References

Belzile, J.P., Sabalza, M., Craig, M., et al., 2015. Trehalose, an mTOR-independent inducer of autophagy, inhibits human cytomegalovirus infection in multiple cell types. J. Virol. 90, 1259–1277.

Bierle, C.J., Semmens, K.M., and Geballe, A.P., 2013. Double-stranded RNA binding by the human cytomegalovirus PKR antagonist TRS1. Virology 442, 28–37.

Bird, S.W., Maynard, N.D., Covert, M.W., et al., 2014. Nonlytic viral spread enhanced by autophagy components. Proc. Natl. Acad. Sci. U.S.A. 111, 13081–13086.

Braggin, J.E., Child, S.J., and Geballe, A.P., 2016. Essential role of protein kinase R antagonism by TRS1 in human cytomegalovirus replication. Virology 489, 75–85.

Buckingham, E.M., Jarosinski, K.W., Jackson, W., et al., 2016. Exocytosis of varicella-zoster virions involves a convergence of endosomal and autophagy pathways. J. Virol. 90, 8673–8685.

Chaumorcel, M., Souquere, S., Pierron, G., et al., 2008. Human cytomegalovirus controls a new autophagy-dependent cellular antiviral defense mechanism. Autophagy 4, 46–53.

Chaumorcel, M., Lussignol, M., Mouna, L., et al., 2012. The human cytomegalovirus protein TRS1 inhibits autophagy via its interaction with Beclin 1. J. Virol. 86, 2571–2584.

Child, S.J., Hakki, M., De Niro, K.L., et al., 2004. Evasion of cellular antiviral responses by human cytomegalovirus TRS1 and IRS1. J. Virol. 78, 197–205.

DeBosch, B.J., Heitmeier, M.R., Mayer, A.L., et al., 2016. Trehalose inhibits solute carrier 2A (SLC2A) proteins to induce autophagy and prevent hepatic steatosis. Sci. Signal. 9, ra21.

Diao, J., Liu, R., Rong, Y., et al., 2015. ATG14 promotes membrane tethering and fusion of autophagosomes to endolysosomes. Nature 520, 563–566.

Ding, B., Zhang, G., Yang, X., et al., 2014. Phosphoprotein of human parainfluenza virus type 3 blocks autophagosome-lysosome fusion to increase virus production. Cell Host Microbe 15, 564–577.

English, L., Chemali, M., Duron, J., et al., 2009. Autophagy enhances the presentation of endogenous viral antigens on MHC class I molecules during HSV-1 infection. Nat. Immunol. 10, 480–487.

Faure, M., and Lafont, F., 2013. Pathogen-induced autophagy signaling in innate immunity. J. Innate Immun. 5, 456–470.

Gannage, M., Dormann, D., Albrecht, R., et al., 2009. Matrix protein 2 of influenza A virus blocks autophagosome fusion with lysosomes. Cell Host Microbe 6, 367–380.

Gorbea, C., Makar, K.A., Pauschinger, M., et al., 2010. A role for Toll-like receptor 3 variants in host susceptibility to enteroviral myocarditis and dilated cardiomyopathy. J. Biol. Chem. 285, 23208–23223.

Hakki, M., and Geballe, A.P., 2005. Double-stranded RNA binding by human cytomegalovirus pTRS1. J. Virol. 79, 7311–7318.

Hakki, M., Marshall, E.E., De Niro, K.L., et al., 2006. Binding and nuclear relocalization of protein kinase R by human cytomegalovirus TRS1. J. Virol. 80, 11817–11826.

Heaton, N.S., and Randall, G., 2010. Dengue virus-induced autophagy regulates lipid metabolism. Cell Host Microbe 8, 422–432.

Joubert, P.E., Meiffren, G., Gregoire, I.P., et al., 2009. Autophagy induction by the pathogen receptor CD46. Cell Host Microbe 6, 354–366.

Joubert, P.E., Werneke, S.W., de la Calle, C., et al., 2012. Chikungunya virus-induced autophagy delays caspase-dependent cell death. J. Exp. Med. 209, 1029–1047.

Ke, P.Y., and Chen, S.S., 2011. Activation of the unfolded protein response and autophagy after hepatitis C virus infection suppresses innate antiviral immunity in vitro. J. Clin. Invest. 121, 37–56.

Kyei, G.B., Dinkins, C., Davis, A.S., et al., 2009. Autophagy pathway intersects with HIV-1 biosynthesis and regulates viral yields in macrophages. J. Cell Biol. 186, 255–268.

Lee, H.K., Lund, J.M., Ramanathan, B., et al., 2007. Autophagy-dependent viral recognition by plasmacytoid dendritic cells. Science 315, 1398–1401.

Levine, B., Liu, R., Dong, X., et al., 2015. Beclin orthologs: integrative hubs of cell signaling, membrane trafficking, and physiology. Trends Cell Biol. 25, 533–544.

Loi, M., Muller, A., Steinbach, K., et al., 2016. Macroautophagy proteins control MHC class I levels on dendritic cells and shape anti-viral CD8(+) T cell responses. Cell Rep. 15, 1076–1087.

Lorin, S., Hamai, A., Mehrpour, M., et al., 2013. Autophagy regulation and its role in cancer. Semin. Cancer Biol. 23, 361–379.

Lussignol, M., and Esclatine, A., 2015. Chapter 8: Modulation of autophagy by herpesvirus proteins In: Hayat, M.A. (Ed.), Autophagy: Cancer, Other Pathologies, Inflammation, Immunity, Infection, and Aging, vol. 6 Elsevier, Academic Press, Amsterdam, pp. 145–158.

McFarlane, S., Aitken, J., Sutherland, J.S., et al., 2011. Early induction of autophagy in human fibroblasts after infection with human cytomegalovirus or herpes simplex virus 1. J. Virol. 85, 4212–4221.

Mocarski, E.S., Shenk, T., Griffiths, P., et al., 2013. Chapter 62: Cytomegaloviruses, sixth ed. In: Knipe, D.M.Howley, P.M. Fields Virology, vol. 2 Wolters Kluwer, Lippincott Williams & Wilkins, Philadelphia, PA.

Mouna, L., Hernandez, E., Bonte, D., et al., 2016. Analysis of the role of autophagy inhibition by two complementary human cytomegalovirus BECN1/Beclin 1-binding proteins. Autophagy 12, 327–342.

Moy, R.H., Gold, B., Molleston, J.M., et al., 2014. Antiviral autophagy restrictsRift Valley fever virus infection and is conserved from flies to mammals. Immunity 40, 51–65.

Nowag, H., Guhl, B., Thriene, K., et al., 2014. Macroautophagy proteins assist Epstein Barr virus production and get incorporated into the virus particles. EBioMedicine 1, 116–125.

Paludan, C., Schmid, D., Landthaler, M., et al., 2005. Endogenous MHC class II processing of a viral nuclear antigen after autophagy. Science 307, 593–596.

Pattingre, S., Tassa, A., Qu, X., et al., 2005. Bcl-2 antiapoptotic proteins inhibit Beclin 1-dependent autophagy. Cell 122, 927–939.

Rasmussen, S.B., Horan, K.A., Holm, C.K., et al., 2011. Activation of autophagy by alpha-herpesviruses in myeloid cells is mediated by cytoplasmic viral DNA through a mechanism dependent on stimulator of IFN genes. J. Immunol. 187, 5268–5276.

Richetta, C., and Faure, M., 2013. Autophagy in antiviral innate immunity. Cell Microbiol. 15, 368–376.

Richetta, C., Gregoire, I.P., Verlhac, P., et al., 2013. Sustained autophagy contributes to measles virus infectivity. PLoS Pathog. 9, e1003599.

Romanowski, M.J., and Shenk, T., 1997. Characterization of the human cytomegalovirus irs1 and trs1 genes: a second immediate-early transcription unit within irs1 whose product antagonizes transcriptional activation. J. Virol. 71, 1485–1496.

Siracusano, G., Venuti, A., Lombardo, D., et al., 2016. Early activation of MyD88-mediated autophagy sustains HSV-1 replication in human monocytic THP-1 cells. Sci. Rep. 6, 31302.

Talloczy, Z., Jiang, W., Virgin, H.W., et al., 2002. Regulation of starvation- and virus-induced autophagy by the eIF2alpha kinase signaling pathway. Proc. Natl. Acad. Sci. U.S.A. 99, 190–195.

Talloczy, Z., Virgin, H.W., and Levine, B., 2006. PKR-dependent autophagic degradation of herpes simplex virus type 1. Autophagy 2, 24–29.

Tey, S.K., and Khanna, R., 2012. Autophagy mediates transporter associated with antigen processing-independent presentation of viral epitopes through MHC class I pathway. Blood 120, 994–1004.

Yakoub, A.M., and Shukla, D., 2015a. Autophagy stimulation abrogates herpes simplex virus-1 infection. Sci. Rep. 5, 9730.

Yakoub, A.M., and Shukla, D., 2015b. Basal autophagy is required for herpes simplex virus-2 infection. Sci. Rep. 5, 12985.

Yordy, B., Iijima, N., Huttner, A., et al., 2012. A neuron-specific role for autophagy in antiviral defense against herpes simplex virus. Cell Host Microbe 12, 334–345.

Yu, Y., Maguire, T.G., and Alwine, J.C., 2011. Human cytomegalovirus activates glucose transporter 4 expression to increase glucose uptake during infection. J. Virol. 85, 1573–1580.

Zhao, J., Li, Z., Wang, M., et al., 2013. Manipulation of autophagy by HCMV infection is involved in mTOR and influences the replication of virus. Acta Biochim. Biophys. Sin. (Shanghai) 45, 979–981.

CHAPTER

3

The Role of Macroautophagy in T Cells

Wei Jia, Ian McLeod, Ming-Xiao He and You-Wen He

OUTLINE

Abstract

The highly conserved self-digestion pathway macroautophagy (hereafter referred to as autophagy) maintains the homeostasis of organelles, degrades long-lived proteins, and recycles amino acids under starvation conditions. Autophagy related molecules are expressed in T lymphocytes, and autophagy occurs inside T lymphocytes. Autophagy-deficient T lymphocytes, which are generated through the deletion of autophagy related molecules, show expansion of organelles content, increased cell death mediated by apoptosis, proliferation defects, instability of the Treg cell lineage, and defect of memory T cell formation. Several different mouse genetic deficiency models show very similar phenotypes. Thus, autophagy plays multiple roles in the regulation of the physiological functions of T lymphocytes. However, how TCR signaling contributes to the initiation of autophagy in T cells and how T cell functions will be affected if autophagy is modified still need to be further investigated.

M.A. Hayat (ed): *Immunology, Volume 1*.
DOI: http://dx.doi.org/10.1016/B978-0-12-809819-6.00003-4

© 2018 Elsevier Inc. All rights reserved.

INTRODUCTION

Autophagy is an evolutionarily conserved self-digesting pathway used for the degradation of long-lived proteins and other macromolecules, removal of aged or excess organelles, and recycling nutrients during starvation conditions (Mizushima et al., 2011). There are three types of autophagy, termed as macroautophagy, microautophagy, and chaperon-mediated autophagy (CMA) according to the differences in morphology, involvement in molecules, and mechanisms to enwrap cytosolic materials (Parzych and Klionsky, 2014). Macroautophagy (autophagy), which delivers cytoplasmic components to lysosomes for degradation in mammalian cells, is the most well studied major autophagy.

Starvation and other stresses trigger autophagy. During the induction phase of macroautophagy, two complexes are essential for autophagosome formation, which are the key steps in the autophagy processing pathway. The first complex is the serine/threonine protein kinase unc-51 like kinase (ULK, Atg1 in yeast) complex, including ULK1 or ULK2, autophagy related molecule 13 (Atg13), Atg101, and focal adhesion kinase family interacting protein of 200 kDa (FIP200, also known as retinoblastoma 1-inducible coiled-coil 1, RB1CC1). ULK1 is generally the dominant molecule in the ULK complex (Noda and Fujioka, 2015). Under nutrient-rich conditions, the ULK complex is inactivated by mammalian target of rapamycin complex 1 (mTORC1), which is a key protein kinase in nutrient-sensing pathway. ULK1/2 is dephosphorylated during starvation condition and then the ULK complex is released from mTORC1 complex. Following the release from mTORC1, ULK1/2 is activated, and then phosphorylates Atg13 and FIP200 to initiate autophagy. Both the ULK complex and Atg9a are essential in recruiting another complex—the autophagy-specific class III phosphatidylinositol 3-kinase (PI3K) complex (Yoshii and Mizushima, 2015). ULK1 phosphorylates serine 14 of Beclin-1 and promotes the kinase activity of Vps34 in the class III PI3K complex (Russell et al., 2013). An alternate class III PI3K complex includes Atg14 (also called Atg14L and Barkor), Beclin-1 (yeast Vps30/Atg6), PI3KC3/Vps34, and PI3KR4/p150 (Vps15 in yeast). Several other molecules are Beclin-1 associated proteins, which include UV irradiation resistance-associated gene (UVRAG), run domain protein as Beclin-1 interacting and cysteine-rich containing (Rubicon), Bcl-2, activating molecule in Beclin 1-regulated autophagy protein 1 (AMBRA1), SH3GLB1/Bif-1 (also known as Endophilin B1), and vacuole membrane protein 1 (VMP1). These Beclin-1 associated molecules positively or negatively regulate the functions of PI3K complexes (Parzych and Klionsky, 2014). PI3K complex can generate PI(3)P, which is required for autophagosome nucleation.

The activation of the above two complexes initiates autophagy, starting with a membrane structure termed the isolation membrane or autophore, which is derived from mitochondria (Hailey et al., 2010), endoplasmic reticulum (ER) (Hayashi-Nishino et al., 2009), Golgi (Takahashi et al., 2011), ER-Golgi intermediate compartment (ERGIC) (Ge et al., 2014), or plasma membrane (Ravikumar et al., 2010). The isolation membrane is then directed by autophagy related molecules through elongation steps to form a characteristic double membrane structure, called an autophagosome, to enwrap cytosol materials. The vacuole membrane protein 1 (VMP1) (Molejon et al., 2013), WD-repeat protein interacting with phosphoinositides (WIPI) (Dooley et al., 2014), and Atg2A/B (Velikkakath et al., 2012) proteins are essential for the maturation of autophagosome through this elongation step.

The elongation steps are directed through two ubiquitin-like conjugation systems, the Atg7–Atg12 conjugation system and the LC3 (Atg8 in yeast) conjugation system, to form autophagosomes. Atg7 is an E1 ubiquitin ligase-like molecule that participates in both conjugation systems. Atg10 is an E2-like molecule involved in Atg7–Atg12 conjugation system and Atg3 is an E2-like molecule which is involved only in the LC3 conjugation system. Finally, autophagosomes fuse with lysosomes to become autophagolysosomes, and engulfed material is degraded by lysosomal enzymes (Levine and Deretic, 2007). It has been reported that there is an Atg5- or Atg7-independent alternative autophagy pathway, in which typical autophagosomes are formed in etoposide-treated or starvation-induced Atg5- or Atg7-deficient mouse embryonic fibroblasts. The lipid form of LC3 (LC3-II) was not found during this Atg5/Atg7 independent autophagosome formation, but the ULK1 and Beclin-1 are required (Nishida et al., 2009).

Autophagy functions in both innate and adaptive immunity (Levine and Deretic, 2007; Munz, 2009). In innate immunity, autophagy is a mechanism to clear intracellular pathogens (Deretic et al., 2015), mediate the recognition ssRNA by TLR7 and IFN-α production in plasmcytoid DCs (Lee et al., 2007), as well as cross-present endogenous antigen to MHC-II molecules (Lee et al., 2010). Here we mainly focus on the functions of autophagy (macroautophagy) in adaptive immunity of T lymphocytes.

AUTOPHAGY OCCURS IN T LYMPHOCYTES

Both thymocytes and mature peripheral T cells express autophagy related molecules. The immature double negative (DN) thymocytes highly express the genes *Atg5*, *Beclin-1*, and *LC3* (*LC3a* and *LC3b*). Mature $CD4^+$ and $CD8^+$ T cells express these autophagy proteins as well and continue the expression after TCR stimulation (Pua et al., 2007). The expression levels of the above autophagy related molecules were analyzed in immature or mature T cells. T lymphocytes may express other autophagy related molecules. The expression of autophagy related proteins in T lymphocytes suggests that the highly conserved autophagy pathway plays important roles in T lymphocytes.

Autophagic activity is measured by several approaches, such as detection of the formation of the lipid form of LC3 (LC3-II) by Western blot, evaluation of LC3 puncta images (the expression of LC3-II on autophagosome), and measurement of the expression levels of substrates of autophagy, e.g., p62/Sequestosome 1(SQSTM-1) (Klionsky et al., 2016). Autophagosomes can be identified by the typical double membrane ultrastructure using transmission electron microscopy, which is one of the classical criteria to prove the presence of autophagy. Autophagy constitutively occurs in different stages of the development of thymocytes and mature T cells including both $CD4^+$ and $CD8^+$ T cells (Stephenson et al., 2009). Autophagosomes were found in activated primary $CD4^+$ Th2 polarized cells (Li et al., 2006). Interestingly, autophagy is promoted in T cells when stimulated by anti-CD3 antibodies in vitro, which mimics TCR stimulation (Pua et al., 2007). Autophagy can be induced in T cells under amino acid or growth factor withdrawal starvation conditions (McLeod and He, 2010).

Several conditional knockout mouse genetic deletion models were established to delete the *Atg* molecules in T lymphocytes lineages (Pua et al., 2007), specific T cell population or effector T cells, such as memory T or regulatory T cells (Treg) (Wei et al., 2016; Xu et al., 2014). Atg5 is involved only in Atg5 and Atg12 pathways, while Atg7 participates in both Atg5–Atg12 conjugation systems and LC3 pathways. Several research groups have reported that Atg5 or Atg7 is required for the autophagy in T lymphocytes (Hubbard et al., 2010; Pua et al., 2007, 2009; Stephenson et al., 2009). Atg3, an E2-like molecule, is involved only in the LC3 conjugation system. Autophagy is blocked when *Atg3* is deleted (Jia and He, 2011). Therefore, both the Atg5–Atg12 conjugation system and LC3 conjugation system are essential for the autophagy processing pathway in T lymphocytes.

Autophagy is regulated dynamically in antigen-specific T cells after viral infection as demonstrated in an LCMV infection model. Lower levels of flux were observed in the expansion phase of antigen-specific T cells at day 5 after LCMV infection, and autophagy flux was increased at day 8 after infection, which is the peak of the expansion phase. Autophagy flux was kept at higher levels during contraction phase (day 15 after LCMV infection) and memory T cell formation (day 30 and thereafter) (Xu et al., 2014). Different levels of autophagy were also observed in $CD4^+$ T cells. Compared to naïve $CD4^+$ T cells, Treg cells have higher levels of autophagy (Wei et al., 2016). Similar to conventional T cells, the activity of autophagy dynamically changes during the development of invariant natural killer T (iNKT) cells. The immature iNKT cells have relatively lower level of autophagy, while the mature iNKT cells have higher levels of autophagy (Salio et al., 2014). The dynamic regulation of autophagy in different phases of effector T cells suggests that autophagy is controlled intensively in T lymphocytes and autophagy plays essential roles in effector and memory T cells.

AUTOPHAGY REGULATES MULTIPLE FUNCTIONS IN T LYMPHOCYTES

Autophagy Developmentally Regulates the Homeostasis of Organelles in T Lymphocytes

One of the physiological functions of autophagy is to remove aged, damaged, or extra organelles. Corresponding to these functions, it is termed, e.g., as mitophagy and ER-phagy for selectively maintaining the homeostasis of mitochondria or ER, respectively. Selective autophagy is discussed in detail by Hayat in this volume. Autophagy is required to maintain organelles at certain levels during T cell development. In *Atg5-, Atg7-, Atg3-, or Vps34-*deficient T cells, the mitochondria and ER are expanded in mature T cells (Jia et al., 2011; Parekh et al., 2013; Pua et al., 2009; Willinger and Flavell, 2012). Autophagy regulates the homeostasis of mitochondria or ER differently. The content of mitochondria is relatively higher at DN, double positive (DP), and single positive (SP) stages than that in mature T cells during thymocyte development. The DN stage of thymocytes has highest content of ER, while DP, SP, or mature T cells have a relatively lower content of ER. Mature T cells must decrease the content of mitochondria before exiting from thymus to lower the level of ROS produced (Jia et al., 2011; Pua et al., 2009). As in conventional T cells, autophagy developmentally regulates the mitochondria homeostasis also in iNKT cells. The immature iNKT cells have higher mitochondrial contents and ROS production than those in mature iNKT cells, which is consistent with the lower level of autophagy activity in immature iNKT cells. The mitochondria contents and ROS production decrease in mature iNKT cells (Salio et al., 2014). The content of mitochondria and ROS are increased in Atg5-deficient iNKT cells (Pei et al., 2015). Autophagy maintains mitochondria or ER at a different level during T cell development. This may correspond to the difference of the functions of these organelles.

Role of Autophagy in the Development of T Lymphocytes

Mouse genetic models provide tools to explore the physiological function of autophagy in vivo. Using adoptively transferred d14 fetal liver cells, the *Atg5*-deficient thymocytes developed completely in congenic host mice according to the cell surface marker analysis of CD4, CD8, CD25, and CD44. These results indicated that autophagy may not regulate the development of T lymphocytes, and autophagy-deficient thymocytes can develop to mature T lymphocytes. However, the total thymocyte numbers were reduced by over 40% more than those of wild type fetal liver cell reconstituted thymocytes (Pua et al., 2007). Similar results were reported in studies using *Atg5*-deficient fetal liver cell transfer model or *Atg5$^{f/f}$ Lck-Cre* model (Hubbard et al., 2010). However, a *Beclin-1* deficient and *RAG1* knockout chimeric model demonstrated that autophagy is essential for maintaining thymic progenitor cells (Arsov et al., 2011). This may explain the reduced numbers of total thymocytes in *Atg5*-deficient fetal liver cell reconstituted mice. In other autophagy T cell lineage deficiency models, such as *Atg7* (Pua et al., 2009), *Atg3* (Jia and He, 2011), and *Beclin-1* (Kovacs et al., 2012), the peripheral mature CD4$^+$ and CD8$^+$ T cells are developed completely. However, the development of iNKT cells was blocked in early stages as shown in *Atg5$^{f/f}$*, *Atg7$^{f/f}$*, or *Vps34$^{f/f}$ CD4-Cre* models (Parekh et al., 2013; Pei et al., 2015; Salio et al., 2014). Therefore, different T cell lineage autophagy deficiency mouse models suggest that autophagy regulates the development of T lymphocytes.

Autophagy Promotes the Survival of T Lymphocytes and Memory T Cell Formation

According to the in vivo data from our lab and published data from other groups, autophagy is required for T cell survival and homeostasis. In *Atg5*-deficient fetal liver reconstituted mice, both CD4$^+$ and CD8$^+$ SP cells in thymocytes are statistically decreased. The cell numbers of both mature *Atg5*-deficient CD4$^+$ and CD8$^+$ T cells in the previously mentioned *Atg5*-deficient fetal liver transferred mice are dramatically reduced (Pua et al., 2007). The data from another genetic mouse model *Atg7$^{f/f}$ Lck-Cre*, in which the gene of autophagy related molecule *Atg7* is specifically deleted at genome level in T lineage cells, indicate that the peripheral mature *Atg7*-deficient CD4$^+$ and CD8$^+$ T cell numbers are dramatically reduced. In a BECN1 (*Beclin-1$^{f/f}$*) *CD4-Cre* model, both peripheral autophagy-deficient CD4$^+$ and CD8$^+$ numbers are reduced (Kovacs et al., 2012). Similar results were observed in an *Atg3$^{f/f}$ Lck-Cre* model (Jia and He, 2011) and *Vps34$^{f/f}$ CD4-Cre* model (Parekh et al., 2013; Willinger and Flavell, 2012). The regulation of the homeostasis of peripheral mature T cells by autophagy in a cell intrinsic manner has been demonstrated in mixed *Vps34$^{f/f}$ CD4-Cre* and wild type B6 BM chimera experiments (Willinger and Flavell, 2012). All these data from different autophagy T cell deficient models suggest that autophagy promotes the survival of T cells in vivo (McLeod et al., 2012).

Similar results were observed in that autophagy promotes the survival of effector T cells. Autophagy supports the survival and memory $CD8^+$ T cell formation in both *Atg7$^{f/f}$ Gzmb-Cre* and *Atg5$^{f/f}$ Gzmb-Cre* mouse models, in which the autophagy related molecules Atg7 or Atg5 were deleted in $CD8^+$-effector cells starting at the clonal expansion phase after LCMV infection (Xu et al., 2014). The survival of Atg7- or Atg5-deficient Treg cells is also defective. As shown in *Atg5$^{f/f}$ CD4-Cre* or *Atg7$^{f/f}$ CD4-Cre*, the survival of Atg5 or Atg7-deficient iNKT cell is defective (Pei et al., 2015; Salio et al., 2014). Therefore, autophagy promotes the survival and homeostasis of peripheral mature effector T lymphocytes.

The defective survival observed in autophagy-deficient T cells is related to enhanced apoptosis. Freshly isolated autophagy-deficient mature $CD8^+$ T cells showed increased apoptosis identified by Annexin V positive cells (Pua et al., 2007). Similarly, apoptosis is increased in both freshly isolated $CD4^+$ and $CD8^+$ mature *Atg7*-deficient T cells in an *Atg7$^{f/f}$ Lck-Cre* model (Pua et al., 2009). The naïve (identified as CD44low cells) T lymphocytes from *Atg3$^{f/f}$ Lck-Cre* mice showed increased Annexin V staining in both $CD4^+$ and $CD8^+$ compartments. Consistent with this finding, caspase-9 activity is increased in autophagy-deficient T cells. Increased apoptosis was also observed in *Beclin-1* deficient $CD4^+$ T cells upon TCR stimulation, which was measured by terminal transferase dUTP nick-end labeling (TUNEL) intensity when compared with that of wild type $CD4^+$ T cells (Kovacs et al., 2012). A similar phenotype was found in Vps34-deficient T cells (Parekh et al., 2013; Willinger and Flavell, 2012). Furthermore, autophagy-deficient Tregs showed increased caspase-3 activity. Consistent with the survival defect of autophagy-deficient Treg cells, the percentage of peripheral resident Treg cells was decreased with a cognate increase in infiltrating lymphocytes and myeloid cells in tissues (Wei et al., 2016). The apoptosis was increased in Atg5- or Atg7-deficient iNKT cells with higher levels of active activities of caspase-8 and caspase-3 (Pei et al., 2015; Salio et al., 2014).

Increased cell death in autophagy-deficient T cells is related to the abnormal degradation of proapoptotic molecules. Beclin-1 deficient $CD4^+$ T cells showed increased expression of proapoptotic molecules, such as procaspase-8, procaspase-3, and Bim at the protein level, but not at the mRNA level. Caspase-3 is colocalized with LC3 when autophagy is activated, which suggests it is being degraded via autophagy. Increased expression of Bak was found in *Atg7*-deficient T cells (Pua et al., 2009) and *Vps34*-deficient T cells (Parekh et al., 2013). These data suggest that the blockage of autophagy affects the degradation of the prodeath molecules. Knocking out *Bim* in *BECN1$^{f/f}$ CD4-Cre$^+$* mice rescues the cell numbers of both peripheral $CD4^+$ and $CD8^+$ T cells observed in *BECN1$^{f/f}$ CD4-Cre$^+$* mice. *BECN1/Bim* double-deficient $CD4^+$ T cells are more resistant to cell death, and the survival rates reached 60% of wild type $CD4^+$ T cells (Kovacs et al., 2012). The pan-caspase inhibitor zVAD, but not the Rip1 inhibitor necrostatin 1, can partially rescue the cell death in *BECN1$^{f/f}$, CD4-Cre$^+$*, and $CD4^+$ T cells. The pan-caspase inhibitor zVAD together with IL-7 could partially rescue the apoptotic *Atg3*-deficient T cells (Jia and He, 2011). More importantly, the increased cell death in autophagy-deficient T cells is mediated by apoptosis. Increased apoptosis is also attributed to the expansion of mitochondria and excessive ROS, which is toxic to T cells (Pua et al., 2009).

Several metabolic pathways, such as fatty acid oxidation or lipid synthesis pathways, are significantly impacted by the loss of autophagy in antigen-specific T cells. In an LCMV infection model, at day 8 after the infection, the metabolic pathway in antigen-specific autophagy-deficient $CD8^+$ T cells was altered from that of wild type $CD8^+$ T cells. Certain metabolic pathways, such as the carnitine shuttle and di-unsaturated fatty acid β–oxidation, are essential for T cell survival. Other metabolic pathways, such as the hexosamine pathway, are required to produce substrates for the glycosylation of cell surface receptors, including growth factor receptors (Xu et al., 2014). Autophagy may directly affect the maintenance of these growth factor receptors and finally contribute to the survival defect of autophagy-deficient T cells.

Treg survival in the absence of autophagy is also defective. Autophagy-deficient Tregs show decreased expression of Foxp3 and increased production of IFN-γ. This is mediated by increased mTORC1 activity and c-Myc expression. Autophagy suppresses the activation of mTORC1 and decreases the expression of c-Myc to maintain the stability of Foxp3 and Treg function (Wei et al., 2016).

There are conflicting theories about the function of autophagy in cell death and survival in T lymphocytes, with certain data indicating autophagy plays a proapoptotic role. HIV-infected human $CD4^+$ T cells express surface HIV envelope glycoproteins (ENV), and ENV binds bystander uninfected $CD4^+$ T cells through CXCR4 to induce autophagy and cause apoptotic cell death in these bystander cells. Autophagy may be the mechanism to mediate the HIV induced bystander T cell death (Espert et al., 2006). Additionally, in vitro data from primary T cells or T cell lines indicate that autophagy mediates the cell death during growth factor withdrawal (Li et al., 2006).

Autophagy Regulates T Cell Signaling

In T lymphocytes, TCR stimulation promotes autophagy induction and flux as detected by LC3 I/LC3 II conversion (Pua et al., 2007). At the same time, autophagy restrains TCR-mediated NF-κB activation in effector T cells, but not in naïve T cells. TCR activation initiates the K63 polyubiquitination of Bcl10, which is a regulatory adaptor molecule in NF-κB signaling, and induces selective binding of the autophagy adaptor p62/SQSTM-1 to Bcl10, leading to degradation of Bcl10 by autophagy. The degradation of Bcl10 is highly selective since its binding partner, Malt1, is not affected by autophagy. The degradation of Bcl10 is blocked in the absence of p62/SQSTM-1, or in Atg3-deficient effector T cells, which results in the production of more IL-2 (Paul et al., 2012). Higher IL-2 production was also observed in TCR-stimulated Atg7-deficient T cells (Jia et al., 2011).

Autophagy developmentally regulates the homeostasis of the ER in T lymphocytes. One of the major functions of the ER in T cells is to control calcium signaling during TCR activation. TCR triggered calcium signaling starts with the depletion of ER calcium stores and opening of Ca^{2+} release-activated Ca^{2+} (CRAC) channels on the surface of T cells, which is named store-operated Ca^{2+} entry (SOCE). The content of ER is expanded and higher calcium stores are found in *Atg7*-deficient T cells. The incomplete depletion of calcium stores in abnormally expanded ER causes the incomplete opening of CRAC and lower calcium influx in autophagy-deficient T cells. Although there is lower calcium influx in autophagy-deficient T cells, the production of IL-2 is higher, suggesting regulation at different levels, and the function of NF-κB may be more dominant in autophagy-deficient T cells (Jia et al., 2011, 2013).

CMA is essential to maintain T cells in an activated state after TCR stimulation. CMA degrades the ubiquitin ligase Itch and the calcineurin inhibitor RCAN1 to promote the activation of T cells triggered by TCR signaling. Defective CMA affects T cell responses to infection. The maintenance of T cell activation by CMA is impaired in aged mice (Valdor et al., 2014). This finding may explain the lower immunity in the elder population and enhancement of CMA may benefit their immune response.

Autophagy Is Essential for T Cell Efficient Proliferation

At the outset of the adaptive immune response, T lymphocytes are activated and undergo robust proliferation to expand the population of effector cells. Autophagy-deficient T cells can be activated as defined by the expression of the activation markers CD69 and CD25, when TCR is activated by anti-CD3 antibody stimulation. However, in the absence of autophagy, T cells fail to proliferate efficiently. Costimulatory signaling via CD28, or providing extra cytokines, such as IL-2, can only partially restore the defective proliferation in autophagy-deficient T cells, as shown in *Atg5*, *Atg7*, and *Atg3* knockout models (Jia et al., 2015; Jia and He, 2011; Pua et al., 2007). Interestingly, use of PMA plus ionomycin to stimulate autophagy-deficient T cells, a method bypassing the proximal TCR signaling, does not restore the proliferation compared to wild type T cells. Therefore, the proliferation defect of autophagy-deficient T cells rises from distal TCR signaling. This is further confirmed that the proximal TCR signaling remains intact in the absence of autophagy in T cells (Jia et al., 2011).

Consistent with this proliferation phenotype, fewer cells could enter into S-phase after TCR stimulation. The master regulator of cell cycle in T cells is $p27^{Kip1}$, which forms large polymers under physiological conditions and autophagy selectively degrades $p27^{Kip1}$ through p62/SQSTM-1 after TCR stimulation. Even the loss of one allele of $p27^{Kip1}$ could lower the $p27^{Kip1}$ expression level in *Atg3*-deficient T cells and restore the proliferation after TCR stimulation (Jia et al., 2015). One study reported that ATP production was inhibited when autophagy was blocked and the production of lactate and fatty acid metabolism were decreased (Hubbard et al., 2010). This may be another reason that autophagy-deficient T cells cannot proliferate efficiently.

DISCUSSION

Several different mouse genetic models lacking autophagy related-genes, such as *Atg5*, *Atg7*, *Atg3*, *Beclin-1*, or *Vps34*, showed very similar phenotypes. According to these phenotypes, autophagy developmentally regulates organelle homeostasis during thymocyte maturation, promotes mature T cell survival and memory cell formation, facilitates T cell proliferation, and maintains the stability of Tregs. All these data indicate that autophagy plays multiple physiological roles in T lymphocytes.

Presently, it seems that autophagy regulates the function of T lymphocytes through two main aspects. One function is that autophagy selectively degrades target molecules to modulate the functions of T cells. Autophagy degrades Bcl10 to restrain TCR-activated NF-κB signaling, and specifically degrades the cell cycle master regulator, $p27^{Kip1}$, to help T cells enter into S-phase. Autophagy degrades the proapoptotic molecules to support T cell survival. The other function of autophagy is to coordinate metabolic homeostasis in T cells to promote their survival. This metabolic homeostasis regulated by autophagy in T lymphocytes is also essential for the formation of memory T cells. In Tregs, autophagy inhibits the activity of mTORC1, decreases the expression of c-Myc, and inhibits glycolytic activity to support the stability of Treg cells.

There are several disagreements regarding the phenotypes from autophagy-deficient T cells derived in different groups. No proliferation defect was found in antigen-specific $CD8^+$ T cells from *Atg7^f/f^ Gzmb-Cre* mice or Treg from an *Atg7^f/f^ Foxp3-Cre* model. These differences may be explained by the differential functions of autophagy in specific T cell populations: or this may be due to differences of deletion models. Another potential concern is the accumulation effects after autophagy blockade as shown in our *Atg3^f/f^ ER-Cre* inducible model, in which the target molecule could be deleted after tamoxifen injection in vivo or 4-hydroxy tamoxifen treatment in vitro. In this *Atg3^f/f^ ER-Cre* mouse model, the Atg3 protein could be deleted at day 3 after tamoxifen treatment. After acute *Atg3* deletion in vivo or in vitro, T cells did not show increased cell death or organelle expansion as found in *Lck-Cre* model. However, in long-term tamoxifen injected mice (more than 3 weeks after tamoxifen injection in vivo) or long-term cultured 4-hydroxy tamoxifen treated T cells (cultured for 24 days in the presence of IL-7), T cells showed phenotypes consistent with that of *Lck-Cre* autophagy ablated T cells, such as increased apoptotic cell death and organelle expansion (Jia and He, 2011). Therefore, there is an accumulated effect after the deletion of autophagy genes in T lymphocytes.

Another disagreement is the production of IL-2 from autophagy-deficient T cells. IL-2 production was increased in T cells deficient in autophagy (Paul et al., 2012; Jia et al., 2011). However, another group reported that the production of IL-2 after TCR stimulation is defective in $CD4^+$ T cells through using *Atg7^f/f^ Lck-Cre* mice. Also using an *Atg7^f/f^ ER-Cre* mouse model, Th1 cells were first differentiated, followed by the deletion of Atg7, and in a similar manner to *Atg7^f/f^ Lck-Cre* T cells, the production of IL-2 and IFN-γ were reduced. The addition of a metabolic substrate, such as methyl pyruvate, can partially restore the production of IL-2 and ATP in autophagy-deficient T cells in *Atg7^f/f^ ER-Cre* mice under Th1 differentiation conditions (Hubbard et al., 2010).

Presently, all published data suggest that autophagy regulates the basic physiological functions of T lymphocytes. Other remaining unanswered questions include exactly how TCR signaling induces autophagy, how autophagy is regulated, and how more specific T cell functions will be affected if autophagy is enhanced or inhibited. The exploration of the function of autophagy may provide new clues for vaccinations and immune therapies in autoimmune diseases, chronic viral infections, or tumors.

References

Arsov, I., Adebayo, A., Kucerova-Levisohn, M., et al., 2011. A role for autophagic protein beclin 1 early in lymphocyte development. J. Immunol. 186, 2201–2209.

Deretic, V., Kimura, T., Timmins, G., et al., 2015. Immunologic manifestations of autophagy. J. Clin. Invest. 125, 75–84.

Dooley, H.C., Razi, M., Polson, H.E., et al., 2014. WIPI2 links LC3 conjugation with PI3P, autophagosome formation, and pathogen clearance by recruiting Atg12-5-16L1. Mol. Cell 55, 238–252.

Espert, L., Denizot, M., Grimaldi, M., et al., 2006. Autophagy is involved in T cell death after binding of HIV-1 envelope proteins to CXCR4. J. Clin. Invest. 116, 2161–2172.

Ge, L., Zhang, M., and Schekman, R., 2014. Phosphatidylinositol 3-kinase and COPII generate LC3 lipidation vesicles from the ER-Golgi intermediate compartment. Elife 3, e04135.

Hailey, D.W., Rambold, A.S., Satpute-Krishnan, P., et al., 2010. Mitochondria supply membranes for autophagosome biogenesis during starvation. Cell 141, 656–667.

Hayashi-Nishino, M., Fujita, N., Noda, T., et al., 2009. A subdomain of the endoplasmic reticulum forms a cradle for autophagosome formation. Nat. Cell Biol. 11, 1433–1437.

Hubbard, V.M., Valdor, R., Patel, B., et al., 2010. Macroautophagy regulates energy metabolism during effector T cell activation. J. Immunol. 185, 7349–7357.

Jia, W., and He, Y.W., 2011. Temporal regulation of intracellular organelle homeostasis in T lymphocytes by autophagy. J. Immunol. 186, 5313–5322.

Jia, W., Pua, H.H., Li, Q.J., et al., 2011. Autophagy regulates endoplasmic reticulum homeostasis and calcium mobilization in T lymphocytes. J. Immunol. 186, 1564–1574.

Jia, W., He, M.X., McLeod, I.X., et al., 2013. Autophagy, a novel pathway to regulate calcium mobilization in T lymphocytes. Front. Immunol. 4, 179.

Jia, W., He, M.X., McLeod, I.X., et al., 2015. Autophagy regulates T lymphocyte proliferation through selective degradation of the cell-cycle inhibitor CDKN1B/p27Kip1. Autophagy 11, 2335–2345.

Klionsky, D.J., Abdelmohsen, K., Abe, A., et al., 2016. Guidelines for the use and interpretation of assays for monitoring autophagy (3rd edition). Autophagy 12, 1–222.

Kovacs, J.R., Li, C., Yang, Q., et al., 2012. Autophagy promotes T-cell survival through degradation of proteins of the cell death machinery. Cell Death Differ. 19, 144–152.

Lee, H.K., Lund, J.M., Ramanathan, B., et al., 2007. Autophagy-dependent viral recognition by plasmacytoid dendritic cells. Science 315, 1398–1401.

Lee, H.K., Mattei, L.M., Steinberg, B.E., et al., 2010. In vivo requirement for Atg5 in antigen presentation by dendritic cells. Immunity 32, 227–239.

Levine, B., and Deretic, V., 2007. Unveiling the roles of autophagy in innate and adaptive immunity. Nat. Rev. Immunol. 7, 767–777.

Li, C., Capan, E., Zhao, Y., et al., 2006. Autophagy is induced in $CD4^+$ T cells and important for the growth factor-withdrawal cell death. J. Immunol. 177, 5163–5168.

McLeod, I.X., and He, Y., 2010. Roles of autophagy in lymphocytes: reflections and directions. Cell. Mol. Immunol. 7, 104–107.

McLeod, I.X., Jia, W., and He, Y.W., 2012. The contribution of autophagy to lymphocyte survival and homeostasis. Immunol. Rev. 249, 195–204.

Mizushima, N., Yoshimori, T., and Ohsumi, Y., 2011. The role of Atg proteins in autophagosome formation. Annu. Rev. Cell Dev. Biol. 27, 107–132.

Molejon, M.I., Ropolo, A., Re, A.L., et al., 2013. The VMP1-Beclin 1 interaction regulates autophagy induction. Sci. Rep. 3, 1055.

Munz, C., 2009. Enhancing immunity through autophagy. Annu. Rev. Immunol. 27, 423–449.

Nishida, Y., Arakawa, S., Fujitani, K., et al., 2009. Discovery of Atg5/Atg7-independent alternative macroautophagy. Nature 461, 654–658.

Noda, N.N., and Fujioka, Y., 2015. Atg1 family kinases in autophagy initiation. Cell. Mol. Life Sci. 72, 3083–3096.

Parekh, V.V., Wu, L., Boyd, K.L., et al., 2013. Impaired autophagy, defective T cell homeostasis, and a wasting syndrome in mice with a T cell-specific deletion of Vps34. J. Immunol. 190, 5086–5101.

Parzych, K.R., and Klionsky, D.J., 2014. An overview of autophagy: morphology, mechanism, and regulation. Antioxid. Redox Signal. 20, 460–473.

Paul, S., Kashyap, A.K., Jia, W., et al., 2012. Selective autophagy of the adaptor protein Bcl10 modulates T cell receptor activation of NF-kappaB. Immunity 36, 947–958.

Pei, B., Zhao, M., Miller, B.C., et al., 2015. Invariant NKT cells require autophagy to coordinate proliferation and survival signals during differentiation. J. Immunol. 194, 5872–5884.

Pua, H.H., Dzhagalov, I., Chuck, M., et al., 2007. A critical role for the autophagy gene Atg5 in T cell survival and proliferation. J. Exp. Med. 204, 25–31.

Pua, H.H., Guo, J., Komatsu, M., et al., 2009. Autophagy is essential for mitochondrial clearance in mature T lymphocytes. J. Immunol. 182, 4046–4055.

Ravikumar, B., Moreau, K., Jahreiss, L., et al., 2010. Plasma membrane contributes to the formation of pre-autophagosomal structures. Nat. Cell Biol. 12, 747–757.

Russell, R.C., Tian, Y., Yuan, H., et al., 2013. ULK1 induces autophagy by phosphorylating Beclin-1 and activating VPS34 lipid kinase. Nat. Cell Biol. 15, 741–750.

Salio, M., Puleston, D.J., Mathan, T.S., et al., 2014. Essential role for autophagy during invariant NKT cell development. Proc. Natl. Acad. Sci. U.S.A. 111, E5678–5687.

Stephenson, L.M., Miller, B.C., Ng, A., et al., 2009. Identification of Atg5-dependent transcriptional changes and increases in mitochondrial mass in Atg5-deficient T lymphocytes. Autophagy 5, 625–635.

Takahashi, Y., Meyerkord, C.L., Hori, T., et al., 2011. Bif-1 regulates Atg9 trafficking by mediating the fission of Golgi membranes during autophagy. Autophagy 7, 61–73.

Valdor, R., Mocholi, E., Botbol, Y., et al., 2014. Chaperone-mediated autophagy regulates T cell responses through targeted degradation of negative regulators of T cell activation. Nat. Immunol. 15, 1046–1054.

Velikkakath, A.K., Nishimura, T., Oita, E., et al., 2012. Mammalian Atg2 proteins are essential for autophagosome formation and important for regulation of size and distribution of lipid droplets. Mol. Biol. Cell 23, 896–909.

Wei, J., Long, L., Yang, K., et al., 2016. Autophagy enforces functional integrity of regulatory T cells by coupling environmental cues and metabolic homeostasis. Nat. Immunol. 17, 277–285.

Willinger, T., and Flavell, R.A., 2012. Canonical autophagy dependent on the class III phosphoinositide-3 kinase Vps34 is required for naive T-cell homeostasis. Proc. Natl. Acad. Sci. U.S.A. 109, 8670–8675.

Xu, X., Araki, K., Li, S., et al., 2014. Autophagy is essential for effector CD8($^+$) T cell survival and memory formation. Nat. Immunol. 15, 1152–1161.

Yoshii, S.R., and Mizushima, N., 2015. Autophagy machinery in the context of mammalian mitophagy. Biochim. Biophys. Acta 1853, 2797–2801.

CHAPTER

4

Increased Immunogenicity Through Autophagy

Chinnaswamy Jagannath, Arshad Khan and Jin Wang

OUTLINE

Abstract

For many infectious diseases, CD4 T cell mediated T helper-1 (Th1) type immunity is a major protective mechanism. CD4 T cells also play a protective role against tuberculosis since HIV-1 induced depletion of CD4 T cells predisposes to fatal tuberculosis. Indeed, tuberculosis due to *Mycobacterium tuberculosis* kills around 1.5 million people per year and infects a third of humankind. Both subunit vaccines and a live attenuated *Bacillus Calmette Guerin* (BCG) have been used inducing immunity against tuberculosis in humans. BCG vaccination is widely used during childhood in developing countries and is the safest vaccine known to humans. BCG induces CD4 and CD8 T cell dependent Th1 immunity and protects against childhood tuberculosis, but is poorly effective against adult tuberculosis. An intriguing property of intracellular BCG

M.A. Hayat (ed): *Immunology, Volume 1.*
DOI: http://dx.doi.org/10.1016/B978-0-12-809819-6.00004-6

© 2018 Elsevier Inc. All rights reserved.

is the "maturation arrest," where the vaccine is sequestered within an immature phagosome, which does not fuse with lysosomes. This affects the ability of antigen presenting cells (macrophages and dendritic cells; APCs) to produce peptide epitopes from BCG and reduces CD4 T cell responses. We describe here that rapamycin can be used to induce autophagy in APCs, enhancing the delivery of BCG vaccine phagosomes and their secreted Antigen 85B into autophagosomes and then to proteolytic lysosome compartments of APCs. This process leads to increased peptide presentation by APCs to CD4 T cells. Overexpression of secreted Ag85B in recombinant BCG leads to rapamycin independent activation of aggresomes in the cytosol, which also triggers autophagy, and boosts antigen presentation. Consistent with intracellular effects, rapamycin activated dendritic cells containing BCG are more effective as vaccines in mice against tuberculosis. Finally, coadministration of rapamycin with BCG given to mice enhances both CD4 and CD8 T cell responses, which protects better against a challenge with tuberculosis. Autophagy, therefore, is a novel mechanism to increase vaccine efficacy.

INTRODUCTION

Infectious diseases account for extensive morbidity and mortality in humans. Multiple vaccines have been developed for their prevention, and vaccination has made a great impact on the maintenance of human health. Vaccination during childhood, e.g., saved an estimated 2.5 million lives every year from diphtheria, tetanus, pertussis, and measles (World Health Organization). While neonatal vaccination has enormously helped to improve the survival and health of children, adult vaccinations like the seasonal flu vaccine is helping adults from developing serious complications. The ever increasing load of immunocompromised adults requires that newer vaccinations are optimized for their protection, especially against the challenges posed by aging (Black et al., 2015). HIV infected individuals, e.g., present a unique situation where many live vaccines cause unfavorable complications during HIV infection.

Many types of newer vaccines are being developed but they mainly include the subunit (protein or DNA), killed, and live attenuated vaccines. However, despite advances in vaccine technology, many vaccines need to be given more than once to induce a long lasting immune response. Some, like the small pox vaccine, can generate lifelong immunity, and the main goal of research in vaccines has been to induce long lasting and sustainable immunity. Often, the key to developing better vaccination strategies lies in understanding the molecular basis of protection induced by the components of the immune system.

In almost all infectious diseases, dendritic cells (DCs), macrophages, and either B cell or T cell subsets play a dominant role in mediating immunity. Given that individual infectious diseases require distinct mechanisms of protection, mechanistic research seems to be necessary at the cellular level to understand vaccine-induced protection. Tuberculosis seems to be an excellent example to study vaccine mediated protection since the disease caused by *Mycobacterium tuberculosis* kills over 1.5 million people each year, and is also a leading cause of death due to opportunistic infection in AIDS patients. Paradoxically, a live attenuated *Bacille Calmette Guerin* (BCG), derived from antigenically similar *Mycobacterium bovis*, has been widely used as a safe neonatal vaccine around the world. Regrettably, metaanalysis of vaccine efficacy trials indicate that the BCG vaccine generates variable levels of immunity against childhood tuberculosis, and offers poor protection against adult disease. Since BCG is a live attenuated vaccine that can persist in the human body for years, the molecular basis for the lack of uniform and sustainable long-term protection against tuberculosis remains unclear.

While both *M. tuberculosis* and BCG organisms readily infect macrophages and DCs of animal models and humans, they pose a unique challenge for the development of protective immune responses since they can replicate or persist in these APCs. Both animal and human studies indicate that T-helper (Th) immunity can protect against tuberculosis as evident by the hypersusceptibility of CD4 knockout mice to tuberculosis and the enhanced risk of CD4 T cell depleted HIV patients to tuberculosis. Although various other immune cells may play a role during protection, macrophages and DCs (antigen presenting cells; APCs) are pivotal to induce T-helper immunity since they are capable of generating peptide epitopes from *M. tuberculosis* and BCG vaccine, activating CD4 and CD8 T cells respectively through MHC-II and MHC-I pathways. It is generally accepted that upon APC mediated activation, CD4 and CD8 T cells secrete cytokines like IFN-γ and IL-2, which in turn amplify immune responses and activate the macrophages in a loop mechanism to kill *M. tuberculosis*. While other types of immune cells like neutrophils can kill *M. tuberculosis*, macrophages appear to be leading phagocytes killing *M. tuberculosis* using oxidants, defensin like peptides, and lysosomal degradation. The immune paradox here is that macrophages and perhaps DCs are the cells that have the ability to kill intracellular pathogens like *M. tuberculosis* and the antigenically similar BCG vaccine. However, they also provide a cellular niche for the persistence of *M. tuberculosis* and there is sufficient evidence that even BCG vaccine can persist within macrophages for prolonged periods.

The unique persistence of the pathogen *M. tuberculosis* in macrophages was found to be due to its ability to evade the phagosome–lysosome (PL) fusion (Armstrong and Hart, 1971). Subsequent research has shown that after the initial phagocytosis in macrophages, both *M. tuberculosis* and BCG vaccine are localized to phagosomes which then appear to use similar mechanisms to evade lysosomal fusion in macrophages. Thus, the *rab5*$^+$ *rab7*$^-$ BCG phagosomes in macrophages do not seem to fuse with lysosomes, most likely due to the bacterial *sapM* phosphatase that dephosphorylates host phophatidyl-inositol 3 phosphate, required for the docking of *rab*-effectors on phagosomes, facilitating maturation (Clemens and Horwitz, 1996; Deretic and Fratti, 1999; Russell, 2001). While many other mechanisms may contribute to the lack of PL fusion in macrophages containing *M. tuberculosis* and BCG, it is relevant to note here that lysosomal degradation is essential for the generation of peptide epitopes which are sorted into MHC-II loading compartments (MIIC). The peptide loaded MHC-II are then routed to the plasma membrane for the activation of CD4 T cells and generation of T-helper 1 (Th1) immunity (Harding, 1996b; Mellman, 2005; Vyas et al., 2008). Along these lines of thought, earlier studies proposed that the BCG vaccine containing APCs are sequestered from T cells, preventing the development of an efficient immune response (Pancholi et al., 1993a,b).

In this context, we hypothesized that the "maturation arrest" of BCG vaccine prevents efficient degradation of the vaccine within lysosomes. Since lysosomes are the pivotal sites in both macrophages and DCs that facilitate the production of peptide epitopes for the MHC-II pathway (Mellman, 2005), we proposed that inefficient production and presentation of peptides to CD4 T cells by BCG infected APCs is likely to be an underlying cause for its inability to induce stronger and longer lasting immunity. Indeed, our initial studies demonstrated that immature phagosomes of BCG vaccine did contain several proteases including Cathepsin-D, albeit an inactive form due to the lack of acidification (Singh et al., 2006). Such macrophages showed significantly reduced antigen presentation to CD4 T cells in vitro, and prior activation of macrophages increased antigen presentation, suggesting an increased lysosomal targeting for BCG vaccine (Singh et al., 2006; Katti et al., 2008). These studies set the trend towards enhancing lysosomal targeting for both BCG vaccine and *M. tuberculosis* through a combination of macrophages activation strategies and mycobacterial genetic modifications (Singh et al., 2006; Katti et al., 2008; Saikolappan et al., 2011). In this chapter, we describe our efforts to enhance the immunogenicity of the BCG vaccine through the induction of autophagy in macrophages and DCs.

Autophagy is a process undergone by most mammalian cells, during which a double membrane phagophore is initiated through either intracellular or external signaling. The phagophore engulfs the cytosol, misfolded proteins, and damaged organelles, forming an autophagosome, which subsequently fuses with the lysosome, resulting in an autophago-lysosome. Acidification of the autophago-lysosome facilitates the degradation of the cargo generating metabolic intermediates that can be recycled within the cell. The mechanisms of induction of autophagy are complex but we observed that when autophagy is induced in macrophages and DCs, BCG vaccine is localized to autophago-lysosomes. In this chapter, we demonstrate that autophagy induction in macrophages and DCs enhances lysosomal localization for BCG, in turn increasing the production of peptide epitope and their in vitro presentation to CD4 T cells. We also demonstrate that rapamycin coadministration with BCG vaccine has similar in vivo beneficial effects in increasing the immunogenicity of the vaccine against tuberculosis.

BCG VACCINE IS PROCESSED BY ANTIGEN PRESENTING CELLS TO ACTIVATE TH1 TYPE CD4 T CELLS

BCG vaccine induced protection against tuberculosis has been analyzed mostly in mice since mouse strains are susceptible to tuberculosis to varying degrees, and immunological reagents for evaluating protection are readily available. Several studies show that subcutaneous immunization with BCG vaccine or its subunit vaccine leads to an expansion of both CD4 and CD8 T cells in mice. Of these, CD4 T cells seem to play a major role since CD4 knockout mice are extremely susceptible to tuberculosis, and BCG fails to protect against tuberculosis challenge among these mice (D'Souza et al., 2000; Cowley and Elkins, 2003). However, in most mouse models, vaccination with BCG reduces the bacterial loads of *M. tuberculosis* at best by a few logs in the lungs after aerosol infection without achieving sterility (Orme, 2005). This suggested that BCG vaccine is perhaps less than effective in generating the immune responses needed to eliminate tuberculosis. BCG is a safe vaccine that is routinely used in many developing countries for neonatal vaccination, and is amenable to genetic manipulation; we hypothesized that efforts should be made to increase its immunogenicity by analyzing its fate in APCs.

An in vitro model to analyze antigen presentation. Both macrophages and DCs derived from the bone marrow of mice are able to present peptide antigens to CD4 T cells in vitro. A method to detect antigen presentation using the mycobacterial secreted antigen-85B (Ag85B) was described earlier (Ramachandra et al., 2001). Thus, mouse APCs infected in vitro with either BCG vaccine or *M. tuberculosis* present the Ag85B derived peptide-25 epitope to the Ag85B specific CD4 T cell hybridoma (BB7 clone; henceforth referred to as

CD4 T cells). Such activated CD4 cells secrete IL-2 which is then quantitated using either sandwich ELISA or bioassays. Ag85B is an immunodominant antigen expressed by both BCG vaccine and *M. tuberculosis*. Furthermore, when given as a subunit DNA or protein vaccine to mice, Ag85B protects mice against aerosol induced tuberculosis (Bennekov et al., 2006; Geluk et al., 2000; Mustafa Abu and Al-Attiyah, 2003). This model has therefore been used as an in vitro parameter for MHC-II dependent mycobacterial antigen presentation (Ramachandra et al., 1999; Noss et al., 2000; Singh et al., 2006; Soualhine et al., 2007).

BCG INDUCED MHC-II TYPE ANTIGEN PRESENTATION IS REDUCED IN APCs

During our preliminary studies, mouse macrophages infected with BCG vaccine were found to present Ag85B epitope to CD4 T cells in vitro, but prior activation of macrophages with IFN-γ and LPS increased antigen presentation (Fig. 4.1A). The assay was specific for Ag85B since a mutant of *M. tuberculosis* lacking Ag85B was unable to induce antigen presentation.

Increased antigen presentation after macrophage or DC activation indicated that perhaps BCG vaccine was inefficiently processed in APCs unless they are activated. It is well established that pathogens internalized into phagosomes via phagocytosis are usually targeted to lysosomes through a *rab* and SNARE (soluble *N*-ethylmaleimide-sensitive factor-attachment protein receptor) dependent sorting process (Luzio et al., 2009). This process is accelerated by macrophage activators like IFN-γ (Via et al., 1998). Interestingly, BCG vaccine resides in an "immature" phagosome compartment that seems to avoid lysosomal fusion (Sun et al., 2007; Via et al., 1997). Thus, they are enriched for *rab5*, which is an early endosome marker, but not *rab7*, which facilitates fusion with lysosomes. Since lysosome mediated degradation of mycobacterial antigens would lead to peptide production for MHC-II dependent pathways, it follows that BCG vaccine is "arrested" in an immunological sense. This concept is illustrated in Fig. 4.1B where the maturation arrest of vaccine phagosomes in APCs can be bypassed by prior activation with IFN-γ or LPS. Fig. 4.1C illustrates that *rfp*BCG vaccine phagosomes in macrophages readily stain for *rab5* marker while evading *rab7*. Furthermore, prior activation of macrophages with IFN-γ or LPS increases enrichment with *rab7* (Fig. 4.1D). This is consistent with the well-known observation that activation of macrophages with IFN-γ enriches more than 90 endosome trafficking proteins on the phagosomes, facilitates phagosome maturation, enhances microbial degradation, induces the macrophage immune response, and promotes antigen loading on to the major histocompatibility complex (MHC) molecules (Jutras et al., 2008). In summary, increased labeling of BCG vaccine phagosomes with *rab7*, which facilitated lysosomal targeting, correlated with better antigen presentation (Fig. 4.1A).

Despite maturation arrest, it has been observed that certain antigens of the BCG vaccine like the secreted Ag85B can be assembled in situ at the level of phagosomes, and exported to the plasma membrane for presentation to CD4 T cells via the MHC-II pathway (Ramachandra et al., 2001). Indeed, we demonstrated that this is likely due to the protease Cathepsin-D, which cleaves Ag85B within phagosomes (Singh et al., 2006). Antigen processing within phagosomes however appears to be less than efficient since the pH of BCG

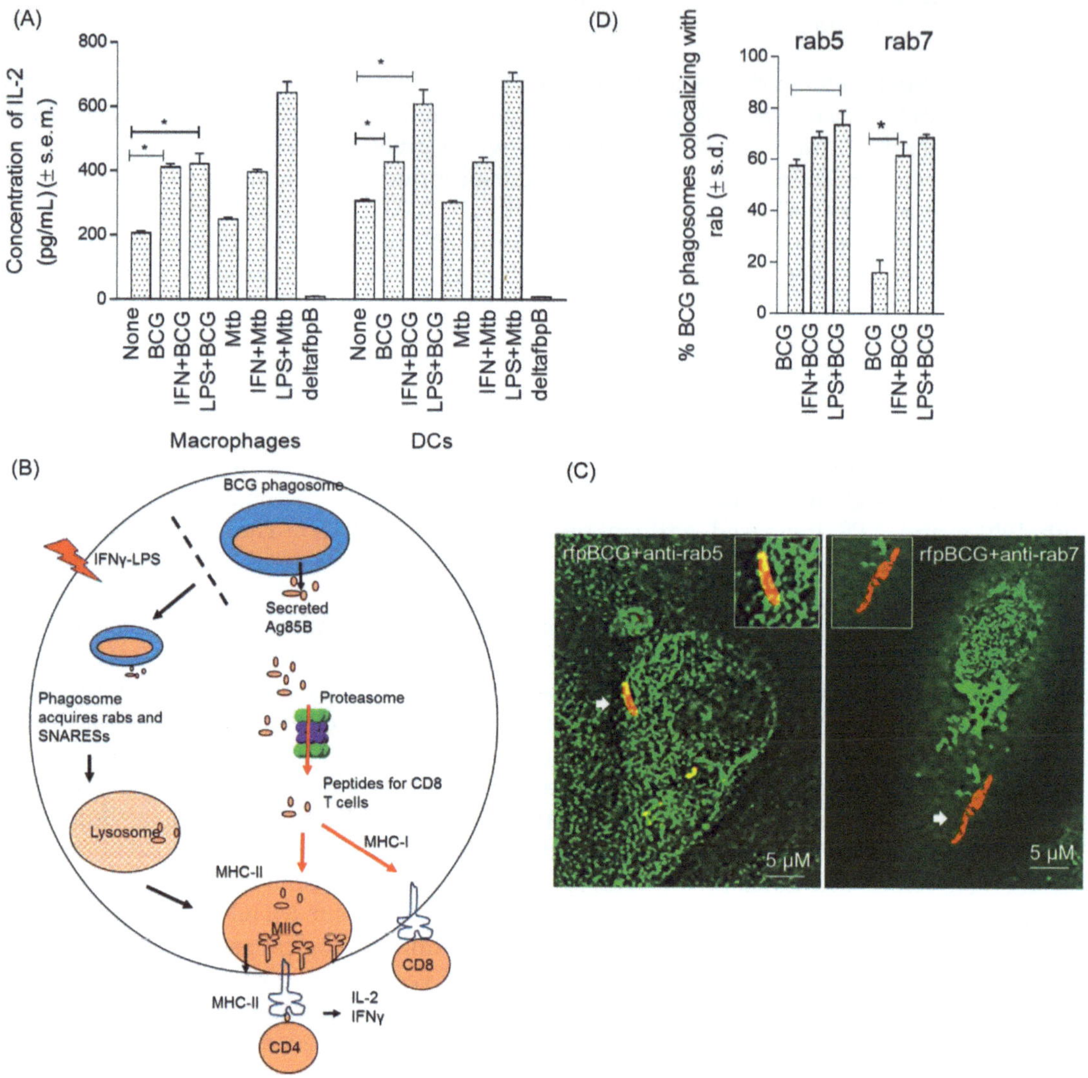

FIGURE 4.1 **MHC-II dependent antigen presentation is inefficient in mouse macrophages and dendritic cells (DCs; APCs) infected with live attenuated *Mycobacterium bovis* BCG vaccine.** (A) Antigen presenting cells (macrophages and DCs; APCs) were purified from the bone marrow of C57Bl/6 mice and were tested untreated or after 4-h treatment with 100 U/mL IFN-γ or 500 ng/mL lipopolysaccharide (LPS). APCs were washed and infected separately with BCG vaccine (MOI = 1) (Pasteur strain), *Mycobacterium tuberculosis* (H37Rv), or the ΔfbpB mutant of *M. tuberculosis* which lacks antigen 85B (Ag85B), followed by overlay with BB7 CD4 hybridoma T cells specific for Ag85B. Supernatants were tested for IL-2 using sandwich ELISA day 1 postinfection (2 experiments; *$p < .009$, t test). (B) Cartoon illustrates that BCG vaccine phagosomes are arrested at an "immature stage" and evade fusion with lysosomes and this event reduces the antigen presentation through MHC-II pathway. BCG vaccine also secretes Ag85B, which is perhaps processed through proteasomes, and such peptides are presented through MHC-I pathway. (C) Prior activation of BCG infected macrophages with either IFN-γ or LPS enhances the maturation process. Macrophages were infected with red fluorescent protein (*rfp*) labeled BCG and stained using antibodies to either *rab5* or *rab7* and anti-IgG FITC conjugates followed by deconvolution microscopy for enrichment of *rab* proteins on *rfp*BCG phagosomes. Panel illustrates the enrichment of *rab5* and lack of labeling with *rab7* on *rfp*BCG (inset). (D) The percent of BCG phagosomes acquiring *rab5* or *rab7* proteins, before and after activation were quantitated (50 macrophages counted each with 1–5 phagosomes, in triplicate for each experiment; 2 experiments; *$p < .01$, t test).

phagosome in macrophages is near neutral. In this direction, BCG secreted urease enzyme may alkalinize the phagosome lumen affecting antigen presentation (Sendide et al., 2004). Others have reported that infection of macrophages with a recombinant BCG vaccine secreting Cathepsin-S enhances Ag85B presentation to CD4 T cells (Soualhine et al., 2007).

It is pertinent to observe here that BCG vaccine secretes many other antigens (Ag85A, Ag85C, and TB10.4) along with the Ag85B, and most likely, they escape into the cytosol to be cleaved by the proteasome. This process produces peptide epitopes to be presented to CD8 T cells through the MHC-I pathway. Indeed, Mice vaccinated with BCG vaccine demonstrate CD8 T cells specific for TB10.4 antigen using tetramer stains. However, other studies suggest that the phagosome membrane for BCG vaccine may not be leaky enough to allow escape of protein antigens into the cytosol. Unlike *M. tuberculosis*, BCG lacks an entire RD1 operon which encodes for several secreted proteins including ESAT-6. The latter protein has been found to be a pore forming agent, and may help in permeabilizing the phagosome membrane for antigens to leak across to macrophage cytosol (Peng and Sun, 2016). Accordingly, investigators have cloned the pore forming toxin listeriolysin into BCG, and the vaccine construct has been found to generate better CD8 and Th17 T cell responses in mice (Desel et al., 2011; Hess et al., 1998). These observations together validate the notion that BCG vaccine is inefficiently processed by APCs, and alteration of the fusion capacity, leakiness, and pH of the vaccine phagosomes may increase antigen processing.

AUTOPHAGY IS A NOVEL APPROACH TO DELIVER BCG VACCINE TO LYSOSOMES

Autophagy is also a major recycling process in APCs, during which the double membrane phagophore can enclose either cytosol, damaged organelles, or microbes and their products, resulting in an autophagosome. Degradation of the contents ensues upon fusion with the lysosomes, forming an autophagolysosome. The lytic enzymes of the lysosomes, facilitated by the lumen acidification by the vacuolar proton ATPase (vATPase), degrade cytosol and organelles to generate metabolites (Fig. 4.2A). Interestingly, one consequence of degradation is the generation of peptide epitopes which are presented in the context of the MHC-II pathway. While the underlying mechanisms of "immune" autophagy are complex, it can be regulated by a variety of chemical inducers (e.g., rapamycin, tamoxifen, and Torin1), cytokines and various stress factors like oxidants. Multiple genes (*atgs*) regulate the phagophore formation that eventually results in autophagolysosomes (Fig. 4.2A).

The specificity of autophagy process can be ascertained by using APCs from mice that carry deletion in *atgs*. Thus, an *atg7* deficient mouse was found to develop exacerbation of infection when infected with BCG (Bonilla et al., 2013). Although several other *atg*-deficient mouse strains do not show enhanced susceptibility to tuberculosis, *atg5* seems to play an important role in regulating inflammation during tuberculosis of mice (Kimmey et al., 2015). Interestingly, *atg5* has also been found essential for MHC-II dependent listeria derived antigen presentation in DCs (Lee et al., 2010).

Since BCG vaccine is arrested at an immature stage, we investigated the hypothesis that targeting BCG vaccines to lysosomes through autophagy may enhance antigen degradation, and increased production of antigenic peptide epitopes (Jagannath et al., 2009). To

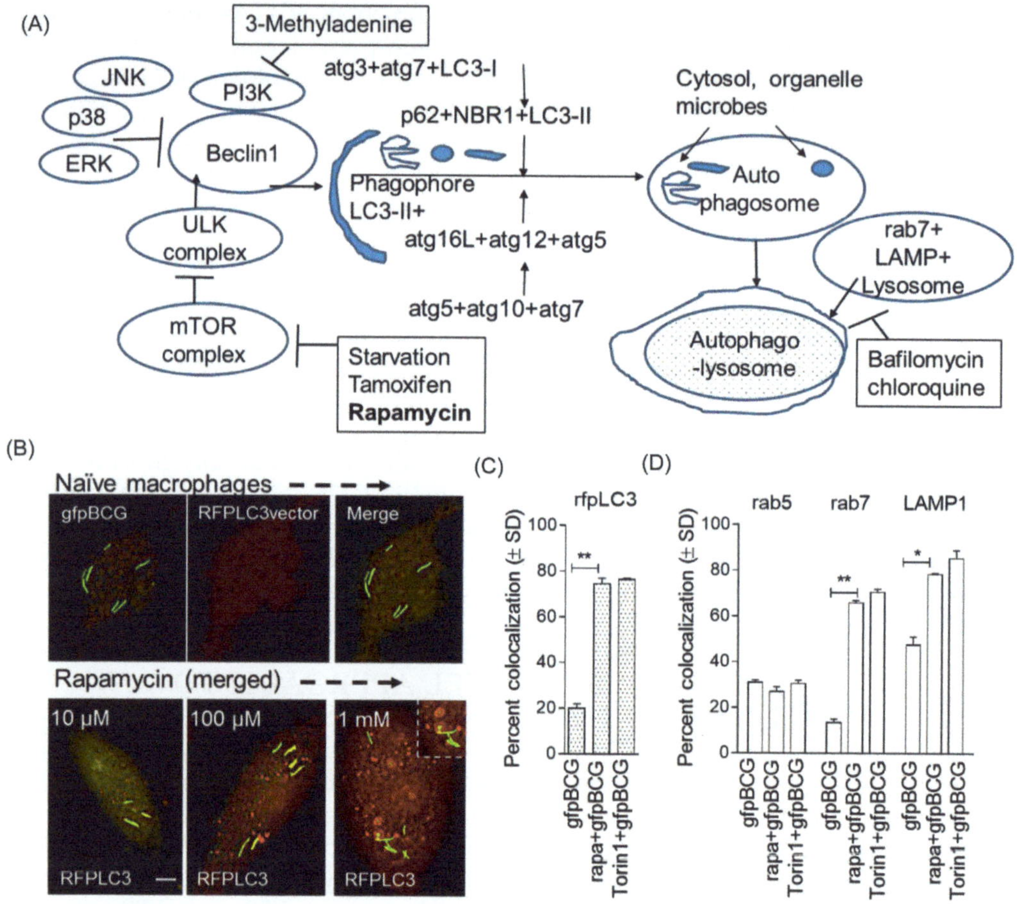

FIGURE 4.2 Rapamycin enhances the lysosomal localization for the BCG vaccine in mouse macrophages. (A) Cartoon illustrates that rapamycin and similar mTOR (mammalian target of rapamycin) repressors can trigger the formation of an autophagosome through the phagophore formation, dependent on several *atg*s. Autophagosomes fuse with lysosomes degrading cargo, and when vaccines are within the autophagosomes, peptide epitopes are produced in lysosomes to be sorted to MHC-II pathway. Key steps where inhibitors can affect autophagy are indicated. (B) Macrophages were purified from the bone marrow of C57Bl/6 mice and were tested untreated or transfected with *rfp*LC3 (*microtubule associated light chain-3*). Macrophages were then activated or not with a 4-h treatment with rapamycin followed by infection with *gfp*-expressing BCG vaccine. Fixed cells were imaged using deconvolution microscopy and percent macrophages (each containing 1–5 phagosomes) showing colocalization of *gfp*BCG phagosomes with *rfp*LC3 was determined as described elsewhere (Jagannath et al., 2009). (C) Quantitation of *gfp*BCG colocalization with *rfp*LC3 (3 experiments; ** $p < .007$ *t* test). (D) Untreated, rapamycin or Torin1 activated macrophages were infected with *gfp*BCG followed by fixation and staining with antibodies against early endosome (*rab*5) and late endosome lysosomes (*rab*7, LAMP1) followed by anti-IgG texas red conjugates and deconvolution imaging (50 macrophages counted each with 1–5 phagosomes, in triplicate for each experiment; 3 experiments; *,** $p < .009$ *t* test).

determine whether rapamycin enhanced targeting of the BCG vaccine to lysosomes, APCs were first transfected with *rfp*LC3 construct, then rested and activated with rapamycin, followed by infection with *gfp*BCG. Fixed macrophages were then examined using deconvolution fluorescence microscopy. As anticipated, rapamycin enhanced the colocalization of *gfp*BCG phagosomes with *rfp*LC3 (Fig. 4.2B and C). In additional experiments, naïve macrophages were treated with rapamycin, infected with *gfp*BCG and counterstained using specific antibodies against *rab*5, *rab*7, and lysosome associated membrane protein-1 (LAMP1), which are well-known early endosome or late endosome and lysosome markers. Rapamycin enhanced the staining of phagosomes with *rab*7, indicating targeting to lysosomes (Fig. 4.2D). The microcopy data were confirmed by western blot analysis where rapamycin induced strong lipidation of LC3 (Jagannath et al., 2009). These data support the observation that rapamycin can target BCG vaccine to lysosomes through autophagy (Jagannath et al., 2009; Saemann et al., 2009). This observation is consistent with the notion that repressors of mammalian target of rapamycin (mTOR) like rapamycin and Torin1 can affect multiple intracellular processes (Anderson and Zimring, 2006; Araki et al., 2009; Ferrer et al., 2010; Luo et al., 1994).

AUTOPHAGY ENHANCES PEPTIDE EPITOPE PRESENTATION THROUGH MHC-II PATHWAY

Previous studies have indicated that BCG vaccine localizes to *rab*5$^+$ immature phagosomes that do not fuse with *rab*7$^+$ lysosomes (Deretic et al., 1997; Via et al., 1997; Jagannath et al., 2009). On the other hand, rapamycin activation of macrophages led to the lysosomal delivery for BCG vaccine (Fig. 4.2). Since lysosomal degradation precedes the generation of peptide epitopes for the MHC-II pathway, it was hypothesized that rapamycin-induced autophagy would lead to enhanced antigen presentation from APCs (Harding, 1996b; Mellman, 2005; Vyas et al., 2008). The in vitro model of Ag85B presentation to CD4 T cells described above was used to measure the effects of autophagy on antigen presentation (Fig. 4.1A and B). Essentially, C57Bl/6 mouse derived cells were subjected to magnetic bead purification using CD11c beads, and the F40 positive adherent macrophages derived from DC depleted population were used as the primary source of DCs and macrophages, respectively. APCs were overlaid with BB7 CD4 T cells and antigen presentation was measured using IL-2 assays. Macrophages activated initially with either rapamycin, followed by infection with BCG vaccine and *M. tuberculosis*, induced a stronger antigen presentation to CD4 T cells than untreated macrophages (Fig 4.3A). DCs showed a similar pattern (Jagannath and Bakhru, 2012). Ag85B is shared by BCG and *M. tuberculosis*, and as anticipated, macrophages infected with the latter also induced stronger IL-2 secretion from CD4 T cells. To determine specificity of autophagy, macrophages were treated with siRNA versus beclin-1 and its missense control or 3-MA. Following these treatments, both BCG and *M. tuberculosis* infected macrophages showed significantly reduced antigen presentation. BCG vaccine contains Ag85B and antigen is also secreted. It was therefore anticipated that the antigen could escape into cytosol and be thereby accessible for rapamycin induced autophagic process. Thus, macrophages and DCs purified from the bone marrow of C57Bl/6 mice were endocytosed with purified, recombinant Ag85B followed by antigen presentation. Similar

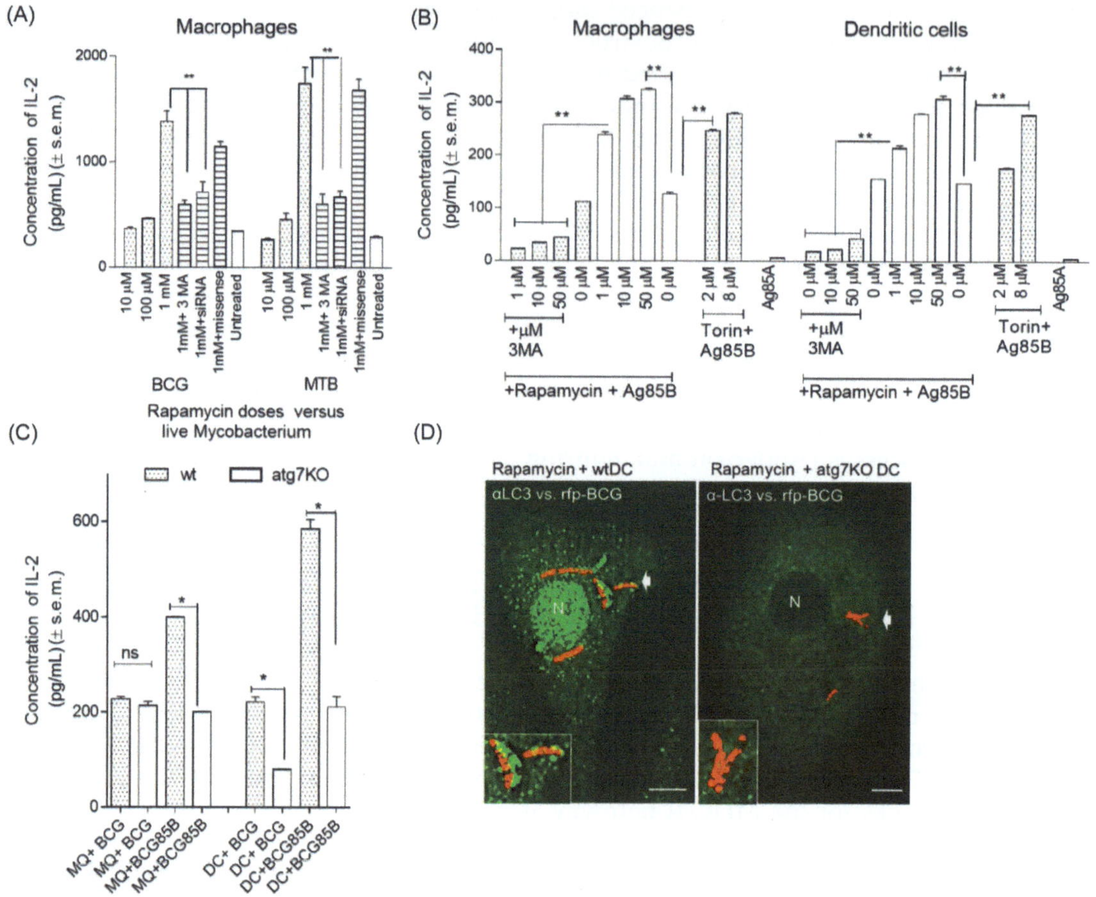

FIGURE 4.3 **Rapamycin induces autophagy and increases antigen presentation in BCG infected APCs.** (A) Macrophages were purified from the bone marrow of C57Bl/6 mice and were tested untreated, treated with siRNA or missence siRNA or 3-methyladenine (3-MA) and followed rapamycin or Torin1 at indicated doses. Macrophages infected with either BCG vaccine or *Mycobacterium tuberculosis* (MTB) (MOI = 1), were washed and overlaid with BB7 CD4 hybridoma T cells and supernatants collected 18 h later were tested for IL-2 using sandwich ELISA (2 experiments; ** $p < .009$, *t* test). (B) Macrophages and DCs were tested untreated, treated with 3-MA followed by either rapamycin or Torin1 before antigen presentation. APCs were pulsed with either recombinant Ag85B (positive control) or Ag85A (negative control) followed by T cell overlay for antigen presentation (2 experiments; ** $p < .009$, *t* test). (C) DCs purified from either wt-C57Bl/6 mice or mice with an *atg7* deletion in CD11C + DCs (*atg7*KODCs) were infected with BCG or BCG vaccine overexpressing Ag85B (BCG85B), followed by CD4 T cell overlay and IL-2 assay, Antigen presentation is decreased in *atg7*KODCs (2 experiments; *,** $p < .008$, *t* test). (D) wt-DCs and *atg7*KODCs were treated with 1 μM of rapamycin, infected with *rfp*BCG, fixed and stained for LC3+ autophagosomes using an antibody to LC3 followed by FITC-conjugates. Panel illustrates extensive colocalization of *rfp*BCG phagosomes (*arrow* and inset) with LC3 using deconvolution imaging (bar = 5 μM).

to the BCG vaccine, rapamycin enhanced the processing and presentation of Ag85B to CD4 T cells, and prior inhibition of autophagy using 3-MA reduced the antigen presentation (Fig. 4.3B). An irrelevant antigen, recombinant Ag85A, was not presented by APCs, confirming the specificity of the assay for Ag85B.

To further ascertain the specificity of link between autophagy and antigen presentation, both wild type C57Bl/6 mice and C57Bl/6 mice that carry a specific deletion in the *atg7* gene associated with the CD11c (specific for DCs) were used for evaluation of antigen presentation. These mice were generated using methods similar to those described for *atg7* deletion in B cells (Chen et al., 2014). Fig. 4.3C illustrates that DCs deficient in *atg7* showed a decreased antigen presentation when infected with BCG or BCG vaccine hyperexpressing Ag85B (aka BCG85B) (Jagannath et al., 2009). Furthermore, rapamycin activated wt-DCs induced *rfp*BCG phagosomes to colocalize with LC3, while *atg7*KO-CD11cDCs showed a lack of LC3 staining around *rfp*BCG phagosomes (Fig. 4.3D). These data validate the observation that autophagy is necessary in DCs for MHC-II dependent presentation of Ag85B. These observations are also consistent with the report that *atg5* and hence autophagy is essential for in vivo MHC-II dependent antigen presentation (Lee et al., 2010).

OVEREXPRESSION OF AG85B INDUCES AUTOPHAGY IN MOUSE APCs

Preceding studies demonstrate that prior activation of APCs with rapamycin enhances the lysosomal targeting of BCG vaccine, thereby increasing antigen presentation to CD4 T cells. Interestingly, when Ag85B was overexpressed in BCG vaccine, the vaccine construct (BCG85B) was able to trigger autophagy in both macrophages and DCs (Jagannath et al., 2009). Thus, BCG85B infected APCs showed an increased antigen presentation to CD4 T cells when compared to APCs infected with wt-BCG vaccine (Fig. 4.3C). To explain the mechanistic basis for this issue, secreted Ag85B was labeled with *rfp* and used for infection of macrophages, followed by staining with an antibody against LC3 and an anti-IgG-FITC conjugate. Fig. 4.4A illustrates that both the *rfp*85BBCG bacilli as well as secreted *rfp*Ag85B (visible as aggregates) colocalized with LC3+ autophagosomes. Several studies have shown that secreted antigens from pathogens like listeria and salmonella can form aggresome-like structures in the cytosol of macrophages or DCs (Canadien et al., 2005; Pierre, 2005; Wenger et al., 2012). Interestingly, autophagy is one mechanism through which aggresomes are targeted to lysosomes for degradation (Gamerdinger et al., 2011; Munz, 2006). Furthermore, aggresomes can be shuttled to lysosomes through a microtubule dependent pathway (Gamerdinger et al., 2011). To determine whether aggresomes of secreted *rfp*Ag85B are also shuttled to lysosomes through a microtubule dependent process, macrophages were treated with varying doses of the microtubule inhibitor nocodazole, followed by antigen presentation. Nocodazole showed a dose-dependent inhibition of antigen presentation supporting the hypothesis that *rfp*Ag85B aggresomes are targeted to lysosomes (Fig. 4.4B). Furthermore, a combination of nocodazole and lysosome acidification inhibitor bafilomycin almost completely abrogated antigen presentation, confirming the pivotal role of lysosome during peptide epitope production. To determine whether sorting of aggresomes to lysosomes was dependent on autophagy, antigen presentation was carried out using BCG85B infected

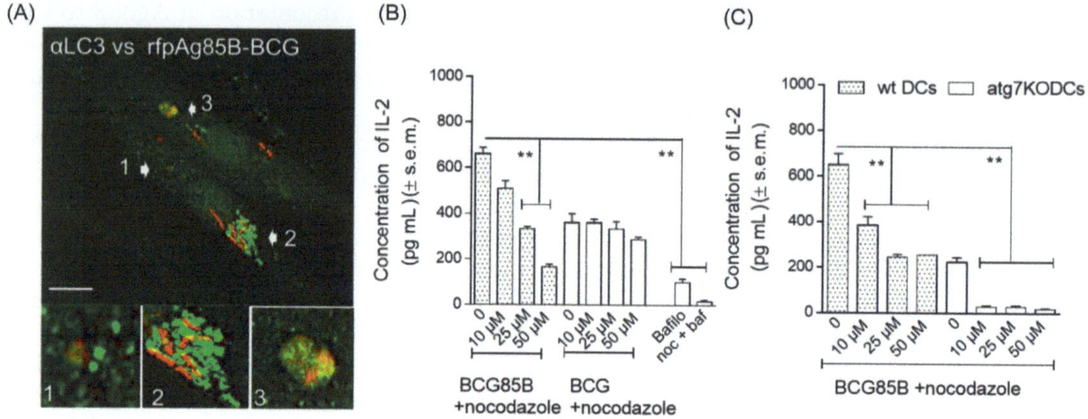

FIGURE 4.4 **Overexpression of Ag85B in BCG vaccine induces aggresome formation in DCs and increased antigen presentation through autophagy.** (A) C57Bl/6 bone marrow derived DCs were infected with *rfp* cloned into Ag85B and over expressed in BCG (*rfpAg85B*-BCG). DCs were fixed 18 h later and stained for LC3 using an antibody followed by FITC conjugates. Deconvolution image shows extensive colocalization of both *rfp*A85BBCG bacilli (*arrow* 2, inset) and *rfp*Ag85B (*arrows* 1 and 3, visible as aggregates, inset) with LC3+ vesicles. (B) wt-DCs were tested untreated or added with indicated doses of nocodazole or 1 μM of bafilomycin followed by infection with BCG85B and antigen presentation. Nocodazole inhibits antigen presentation by BCG85B that overexpresses Ag85B. Prolonged inhibition by nocodazole also inhibits antigen presentation in BCG infected DCs (not shown). Bafilomycin decreases antigen presentation. (C) wt-DCs are compared with atg7KODCs for antigen presentation. The latter show decreased antigen presentation even when infected with BCG85B that forms aggresomes. Nocodazole abrogates antigen presentation in atg7KODCs.

wt-DCs and *atg*7KODCs. Antigen presentation was decreased among the latter. Although the underlying autophagic mechanisms are yet to be unraveled, it is evident that secreted Ag85B finds its way into cytosol and forms aggresomes which then trigger autophagy and its routing to autophago-lysosomes.

RAPAMYCIN CONDITIONED DCs DELIVER BCG VACCINE WITH BETTER EFFICACY

It is well established that DCs generate and maintain T-helper immune responses by processing microbial antigens and presenting them via MHC-II, MHC-I, and CD1 dependent pathways to various subsets of T cells. Previous studies demonstrated that rapamycin augmented autophagy in DCs and enhanced peptide antigen presentation in vitro to the CD4 T cells. We therefore hypothesized that rapamycin activated DCs would serve as better vaccines than BCG vaccine alone. Thus, purified primary DCs were activated in vitro with rapamycin (rapaDC + BCG), infected with live BCG vaccine and washed DCs were administered as vaccines to mice along with untreated DCs containing BCG vaccine (DC + BCG) or untreated DCs alone (Fig. 4.5A). Mice were aerosol challenged 4 weeks after immunization with 100 CFU of *M. tuberculosis*, followed by sacrifice of mice 4 weeks later and determination of lung CFU counts as per standardized procedures (Jagannath and Bakhru, 2012;

Jagannath et al., 2009). Mice given the rapaDC + BCG vaccine showed a stronger immunity and a 1.5 $\log_{10}$ better protection against tuberculosis in lungs (Fig. 4.5B) compared to DC-BCG vaccine or mice given only naïve DCs. The lungs of these mice were also analyzed using flow cytometry for CD4 and CD8 T cells and the rapaDC + BCG vaccine induced a strong accumulation of both CD4 and CD8 T cells (Fig. 4.5C).

To determine whether vaccination with rapaDC + BCG correlated with an expansion of antigen specific CD4 T cells, mice were vaccinated as in Fig 4.5A, and lung T cells analyzed using flow cytometry, 3 weeks after aerosol challenge with *M. tuberculosis*. Once again the lungs of mice given rapaDC + BCG vaccine showed a significant increase in the numbers of Ag85B specific MHC-II tetramer positive CD4 T cells compared to mice given DC + BCG vaccine alone (Fig. 4.5D). These data confirm that rapamycin treatment of DCs decisively enhances the in vivo immunogenicity of BCG vaccine. These studies are consistent with those others who show that rapamycin activation of DCs increased costimulatory receptors and the expansion of CD8 T cells during autologous vaccination (Amiel et al., 2012). Interestingly, rapamycin has been found to uniquely alter the expansion of pathogen specific CD8 T cells in vivo depending upon whether the antigen is presented by the DCs or in the context of an allograft (Ferrer et al., 2010). In this context, it is pertinent to note that many vaccines are given subcutaneously, and we propose that rapamycin can perhaps be used to beneficially modulate the immunogenicity of APCs like Langerhans cells in the skin. Thus, additional research is needed on the use of rapamycin as an immunomodulating agent during local and systemic vaccination (Haidinger et al., 2010; Weichhart et al., 2015). In this direction, the in vivo effects of rapamycin on the efficacy of BCG vaccine against tuberculosis are described below.

RAPAMYCIN COMBINATION WITH BCG VACCINE ENHANCES PROTECTION AGAINST TUBERCULOSIS IN MICE

Rapamycin is a well-known drug with antineoplastic properties. By repressing the mTOR, it triggers a wide variety of biochemical signals within mammalian cells (Anderson and Zimring, 2006; Araki et al., 2009; Ferrer et al., 2010; Luo et al., 1994). Coinciding with our discovery that rapamycin enhances BCG vaccine efficacy through autophagy, the drug was found to prolong the longevity of CD8 T cell dependent memory in mice against viral infections (Jagannath et al., 2009; Araki et al., 2009). Subsequently, rapamycin has been shown to improve pathogen specific CD8 T cell response, and beneficially affect the immune responses of macrophages and DCs (Ferrer et al., 2010; Katholnig et al., 2013). More recently, rapamycin was found to regulate B cell memory and multiple aspects of innate immune signaling (Chen et al., 2014; Weichhart et al., 2015). These studies have led to the emerging concept that rapamycin is likely to improve the in vivo efficacy of vaccines when given in a dose and time dependent manner.

Fig. 4.6A illustrates this concept during the initiation and subsequent maintenance of immune response generated by a vaccine. Thus, rapamycin induces autophagy and increases antigen presentation through MHC-II pathway in both DCs and macrophages. Autophagy has also been associated with the release of the IL-1β cytokine which may amplify immune responses (Dupont et al., 2011). We have demonstrated that rapamycin

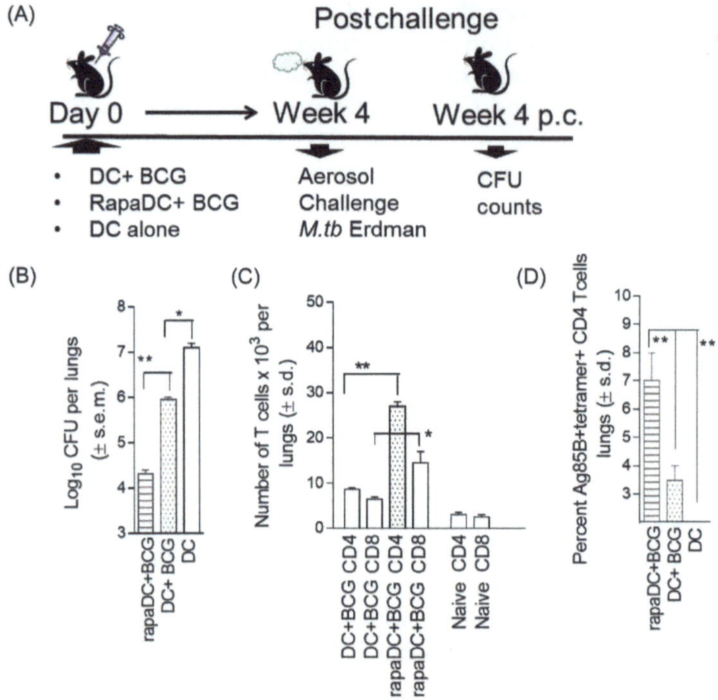

FIGURE 4.5 **Rapamycin activated BCG vaccine-infected dendritic cells induce better protection against tuberculosis in mice.** C57Bl/6 mice were immunized subcutaneously with rapamycin activated DCs (rapaDC + BCG) and untreated DCs preinfected with BCG (DC + BCG) or untreated DCs alone (5×10^6 DCs per mouse; 2 experiments). (A) 4 weeks later, mice were aerosol challenged with virulent *Mycobacterium tuberculosis* Erdman strain and colony forming bacterial unit (CFUs) counts of lungs determined by plating organ homogenates on 7H11 agar for additional 4 weeks. Untreated DC vaccinated mice permitted extensive growth of *M. tuberculosis* in the lungs (up to $7\log_{10}$). DC vaccines containing BCG vaccine offered significant protection. RapaDC + BCG induced a greater clearance of *M. tuberculosis* in the lungs of mice compared to DC + BCG vaccine (*,** $p < .009$ for groups compared; two-way ANOVA). (B) The lung cells were collected and analyzed using flow cytometry for T cell populations. RapaDC + BCG vaccine induced a better expansion of both CD4 and CD8 T cells in the lungs compared to DC + BCG vaccine (*,** $p < .01$ *t* test). (C) T cells of the lungs were also stained using a MHC-II tetramer specific for Ag85B (Emory University flow cytometry core). RapaDC + BCG vaccine induced a better expansion of Ag85B specific CD4 T cells than DC + BCG vaccine in mice.

increases the proportion of Ag-specific CD4 T cells in mice following vaccination (Fig. 4.5). Many studies show that rapamycin affects the longevity of CD8 memory T cells (Araki et al., 2009). Interestingly it also exerts a dual effect on the transcription factors T-bet and Eomes in mice thereby affecting the "effector-memory" and "central memory" T cell lineage, respectively (Rao et al., 2010). In summary therefore, rapamycin can potentially amplify multiple steps of a developing Th1 immune response.

With this concept in mind, we evaluated the effects of rapamycin during BCG vaccination against tuberculosis of mice, which have been described in detail elsewhere (Bakhru, 2012; Jagannath and Bakhru, 2012). BCG is the world's most frequently used live attenuated and safe vaccine. Although extensively used during childhood vaccination, it appears

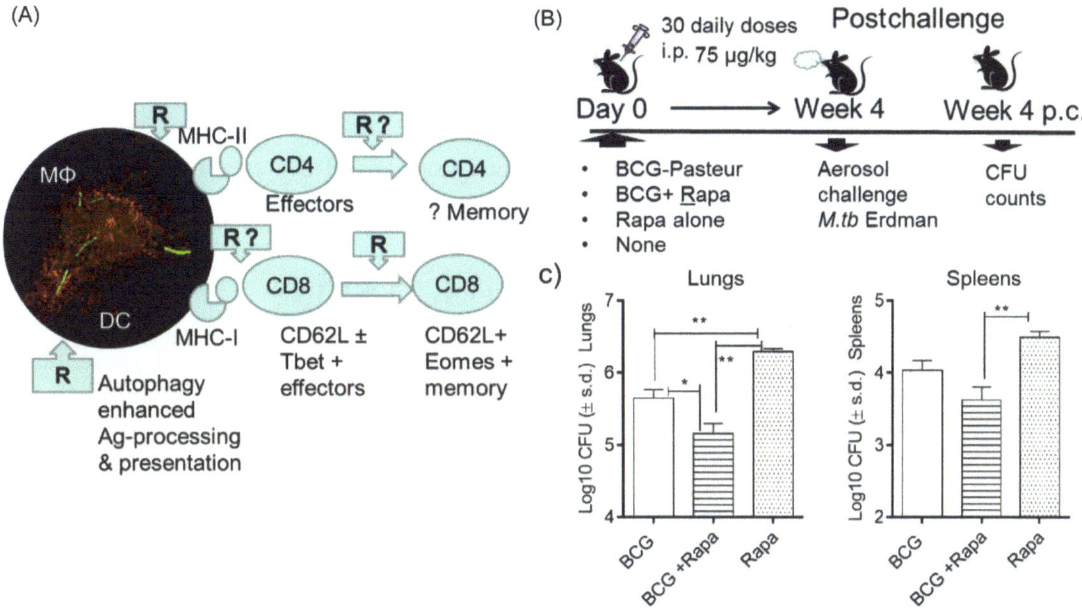

FIGURE 4.6 **Coadministration of rapamycin with BCG vaccine induces better protection against tuberculosis in mice correlating with increased CD4 and CD8 T cell responses.** (A) Cartoon illustrates that rapamycin is likely to affect multiple components of a developing Th1 immune responses. (B) C57Bl/6 mice were vaccinated with BCG (Pasteur strain) alone, BCG vaccine followed by 30 daily injections of rapamycin or given rapamycin alone (i.p. doses at 75 μg/kg). Four weeks later, they were aerosol challenged with 100 colony forming unit bacterial counts (CFU) per mouse of *Mycobacterium tuberculosis* to determine vaccine induced protection. Four weeks later, mice were sacrificed and organs (lungs and spleens) were homogenized and plated for CFU counts on 7H11 agar, which were expressed as $\log_{10}$ CFU per organ. (C) Rapamycin coadministration following BCG vaccine induced a better clearance of *M. tuberculosis* from both the organs (*,** p values determined using two-way ANOVA; $n = 5$ mice per group). Rapamycin treatment alone had no bactericidal effect on the organ counts of *M. tuberculosis* since these mice showed $>6\log_{10}$ growth of *M. tuberculosis* in the lungs and $>4.5\log_{10}$ in the spleens.

to protect children but not adults—suggesting that BCG induced protection may not last longer.

Since rapamycin exerted its beneficial effects on CD8 T cells at doses which are significantly below the toxic doses that can affect CD4 T cells, we examined the effect of low dose rapamycin on the immune responses generated in mice following BCG vaccination. We examined the hypothesis that at low doses, rapamycin is likely to have a dual effect on APCs and T cells, improving the overall quality of immune responses. We anticipated that activation of APCs may enhance CD4 T cell responses, while a direct effect of rapamycin on mTOR of CD8 T cells may expand its efficacy and longevity. A key experiment is shown Fig. 4.6B, where mice were given BCG vaccine followed by daily low dose rapamycin before aerosol challenge with *M. tuberculosis*. Mice were sacrificed for bacterial counts in the lungs 4 weeks later. Mice given rapamycin following BCG vaccination showed a better decrease in the numbers of *M. tuberculosis* in the lungs compared to mice given BCG alone, which indicated better protection. Flow cytometry analysis also confirmed that rapamycin induced a stronger expansion of CD4 and CD8 T cells in the lungs of the rapamycin treated and BCG vaccinated mice (Bakhru, 2012; Jagannath and Bakhru, 2012). Since low dose rapamycin (around 10 μg/mL) has no effect on the in vitro viability of *M. tuberculosis* (unpublished observations), and rapamycin alone did not affect bacterial counts of aerosol infected mice (Fig. 4.6C), these data suggest that rapamycin likely enhanced the vaccine efficacy at the level of both APCs and T cells (Fig. 4.6A).

DISCUSSION AND CONCLUSIONS

During vaccine induced immunity, an essential step is the processing and presentation of antigens from the vaccine to T cells. APCs use multiple pathways to accomplish this event. Thus, peptides are degraded in proteasomes and short chain peptides are routed to the MHC-I through a TAP-dependent pathway to CD8 T cells (Harding, 1996a). Likewise, peptides are degraded and epitopes are routed to the MIIC compartments for presentation in the context of MHC-II pathway to CD4 T cells (Moffat et al., 2013). Similar mechanisms appear to exist for the presentation for glycolipid antigens through CD1d in APCs.

We present the novel observations above that the most frequently used human vaccine BCG shows a maturation arrest which prevents its localization to lysosomes. This reduces its ability to present peptide antigens to CD4 T cells. While the relationship between reduced immunogenicity and lysosomal evasion by BCG vaccine has been characterized by us and other investigators, this issue could also be pertinent for other pathogens which evade lysosome fusion, like *Salmonella typhimurium*, *Legionella pneumophila*, and *Chalmydia trachomatis* (Carrol et al., 1979; Marra et al., 1992; Beatty and Stephens, 1994). In addition, many pathogens may localize to special vacuoles avoiding lysosomes, like *Brucella*, *Listeria monocytogenes*, and *Francisella tularensis* (Santic et al., 2005; Lam et al., 2012; Case et al., 2016). If live attenuated vaccines are derived from such pathogens, and they retain their ability to evade lysosomes, lack of degradation in lysosomes is likely to reduce the immunogenicity of macrophages and DCs infected with these vaccines and pathogens, similar to the behavior of BCG vaccine.

In this context, increasing the localization of vaccines to lysosomal compartment using autophagy appears to be a novel approach mainly because several inducers of autophagy are available which can be used as therapeutics in both animal models and humans, and the process of antigen presentation during autophagy is becoming better understood (Dengjel et al., 2005; Lee et al., 2010; Munz, 2006, 2009, 2012). It appears particularly intriguing to observe that rapamycin, which was initially observed to enhance antigen presentation in APCs (Jagannath et al., 2009), and to extend the longevity of CD8 memory T cells (Araki et al., 2009), has rapidly expanded its spectrum of action. Thus, rapamycin has been found to specifically expand pathogen specific CD8 T cells, regulate inflammatory responses of pathogen infected macrophages and DCs, induce an expansion of M1 phenotype of human macrophages, help to maintain B cell memory, and generally improve the functionality of innate and adaptive immune systems (Ferrer et al., 2010; Katholnig et al., 2013; Mercalli et al., 2013; Chen et al., 2014; Weichhart et al., 2015). Rapamycin has therefore made a rapid transition from an immunosuppressive into an immunotherapeutic, with a potential to improve vaccine responses in animal models and humans (Mannick et al., 2014).

Our detailed studies showed two key observations. The first lead was that DCs activated by rapamycin and containing BCG were better vaccines than BCG alone, supporting the notion that rapamycin activated intracellular antigen processing (Figs. 4.1–4.3). Since rapamycin has now been shown to beneficially augment the inflammatory responses of APCs, we now propose that vaccination strategies can incorporate rapamycin as a "local" activator of APCs, increasing their ability to process vaccines. We foresee liposomal formulations of rapamycin and vaccines in future. Increased immunogenicity, in turn, can perhaps minimize booster doses required for many subunit vaccines. In this direction, it is encouraging to note that in addition to regulating macrophage and DC immune responses, mTOR repressors may also affect Langerhans cells, a target cell during vaccination.

The second lead we observed was that coadministration of rapamycin with BCG vaccine enhanced its efficacy against tuberculosis of mice (Figs. 4.5 and 4.6) (Jagannath et al., 2009; Bakhru, 2012; Jagannath and Bakhru, 2012). Although monotherapy with rapamycin and BCG moderately increased the efficacy of the BCG vaccine, it should be noted here that various strains of mice are naturally susceptible to low dose aerosol tuberculosis, whereas nearly 90% of humans resist the development of tuberculosis even after aerosol inhalation. Furthermore, rapamycin has dose- and time-dependent effects and we propose that rapamycin-BCG vaccine combination regimen need to be further optimized, keeping in mind the immune profiles of humans. In this direction, it encouraging to note that mTOR repression and flu combination yielded better efficacy among the elderly, and that rapamycin therapy could expand M1 type proinflammatory macrophages among patients (Mannick et al., 2014; Mercalli et al., 2013).

Our study into the molecular mechanisms of autophagy also brought out newer strategies to induce autophagy in APCs. Although mTOR repressors like rapamycin show a promise to augment vaccines, it may not be feasible to administer rapamycin, even at low doses, to neonates and seriously immunosuppressed individuals. Unexpectedly, we observed that overexpression of certain protein antigens (Ag85B) in BCG led to induction of autophagy in APCs (Fig. 4.5). Interestingly, aggregates of proteins (aka. aggresomes) can form in the cytosol either due to metabolic disorders or through secretion of antigens from microbes. The "aggresomes" are usually removed through the ubiquitin dependent proteasomal degradation, but increasing evidence shows that autophagy plays a role in their removal through the lysosomes (Behl, 2011; Canadien et al., 2005). Overexpressed Ag85B indeed formed aggresomes, which triggered autophagy without the need for prior activation of APCs using rapamycin (Fig. 4.5) (Jagannath et al., 2009). Furthermore, recombinant BCG that overexpressed Ag85B showed an increased efficacy against tuberculosis of mice and its protection lasted longer (Jagannath et al., 2009).

In this direction, it is intriguing to note that autophagy can be triggered in situ by using a cell permeable Tat-beclin1 peptide (Shoji-Kawata et al., 2013). We therefore foresee that subunit vaccines or live attenuated vaccines can be genetically modified to induce the next generation "autophagy inducing" vaccines. There is therefore a strong need to further analyze the autophagy mechanisms to enhance delivery of vaccines to lysosomes. Indeed, many pathogens either secrete powerful toxins (e.g., listeriolysin) or proteins into the cytosol or localize to special "vacuoles" which may intersect the autophagic pathway (Santic et al., 2005; Lam et al., 2012; Case et al., 2016).

In conclusion, it seems clear that intracellular antigen processing is an essential step during vaccine-induced immunity. Sorting phagosome and cytosol derived antigens towards lysosomes is an important component of antigen processing, while many pathogens and vaccines derived from them have the ability to evade the lysosomes. As a consequence, processing and presentation of peptide epitopes through MHC-II pathway is reduced, as exemplified by the BCG vaccine. Inducing autophagy appears to be a novel strategy to deliver vaccines like BCG to lysosomes increasing their immunogenicity. Some mTOR repressors like rapamycin can exert beneficial effects on multiple components of the innate and adaptive immune system. Thus, newer immunotherapeutic autophagy based approaches to increase vaccine efficacy appear feasible.

Acknowledgments

This study was supported by AI49534 and AI78420. The authors acknowledge the gift of BB7 T cells from Dr. C.W. Harding, Case Western Reserve University, United States; and Dr. N. Tony Eissa, Baylor College of Medicine, United States for the generous gift of reagents during the autophagy studies.

References

Amiel, E., Everts, B., Freitas, T.C., et al., 2012. Inhibition of mechanistic target of rapamycin promotes dendritic cell activation and enhances therapeutic autologous vaccination in mice. J. Immunol. 189, 2151–2158.

Anderson, K.M., and Zimring, J.C., 2006. Rapamycin prolongs susceptibility of responding T cells to tolerance induction by CD8+ veto cells. Transplantation 81, 88–94.

Araki, K., Turner, A.P., Shaffer, V.O., et al., 2009. mTOR regulates memory CD8 T-cell differentiation. Nature 460, 108–112.

Armstrong, J.A., and Hart, P.D., 1971. Response of cultured macrophages to *Mycobacterium tuberculosis*, with observations on fusion of lysosomes with phagosomes. J. Exp. Med. 134, 713–740.

Bakhru, P., 2012. Induction of Stronger and Long Lasting Vaccine Immunity to Tuberculosis. University of Texas Graduate School of Biomedical Sciences, Houston, TX.

Beatty, P.R., and Stephens, R.S., 1994. CD8+ T lymphocyte-mediated lysis of Chlamydia-infected L cells using an endogenous antigen pathway. J. Immunol. 153, 4588–4595.

Behl, C., 2011. BAG3 and friends: co-chaperones in selective autophagy during aging and disease. Autophagy 7, 795–798.

Bennekov, T., Dietrich, J., Rosenkrands, I., et al., 2006. Alteration of epitope recognition pattern in Ag85B and ESAT-6 has a profound influence on vaccine-induced protection against *Mycobacterium tuberculosis*. Eur. J. Immunol. 36, 3346–3355.

Black, S., De Gregorio, E., and Rappuoli, R., 2015. Developing vaccines for an aging population. Sci. Transl. Med. 7 281ps288.

Bonilla, D.L., Bhattacharya, A., Sha, Y., et al., 2013. Autophagy regulates phagocytosis by modulating the expression of scavenger receptors. Immunity 39, 537–547.

Canadien, V., Tan, T., Zilber, R., et al., 2005. Cutting edge: microbial products elicit formation of dendritic cell aggresome-like induced structures in macrophages. J. Immunol. 174, 2471–2475.

Carrol, M.E., Jackett, P.S., Aber, V.R., et al., 1979. Phagolysosome formation, cyclic adenosine 3′:5′-monophosphate and the fate of *Salmonella typhimurium* within mouse peritoneal macrophages. J. Gen. Microbiol. 110, 421–429.

Case, E.D., Smith, J.A., Ficht, T.A., et al., 2016. Space: a final frontier for vacuolar pathogens. Traffic 17, 461–474.

Chen, M., Hong, M.J., Sun, H., et al., 2014. Essential role for autophagy in the maintenance of immunological memory against influenza infection. Nat. Med. 20, 503–510.

Clemens, D.L., and Horwitz, M.A., 1996. The *Mycobacterium tuberculosis* phagosome interacts with early endosomes and is accessible to exogenously administered transferrin. J. Exp. Med. 184, 1349–1355.

Cowley, S.C., and Elkins, K.L., 2003. CD4+ T cells mediate IFN-gamma-independent control of *Mycobacterium tuberculosis* infection both in vitro and in vivo. J. Immunol. 171, 4689–4699.

Dengjel, J., Schoor, O., Fischer, R., et al., 2005. Autophagy promotes MHC class II presentation of peptides from intracellular source proteins. Proc. Natl. Acad. Sci. U.S.A. 102, 7922–7927.

Deretic, V., and Fratti, R.A., 1999. *Mycobacterium tuberculosis* phagosome. Mol. Microbiol. 31, 1603–1609.

Deretic, V., Via, L.E., Fratti, R.A., et al., 1997. Mycobacterial phagosome maturation, rab proteins, and intracellular trafficking. Electrophoresis 18, 2542–2547.

Desel, C., Dorhoi, A., Bandermann, S., et al., 2011. Recombinant BCG DeltaureC hly+ induces superior protection over parental BCG by stimulating a balanced combination of type 1 and type 17 cytokine responses. J. Infect. Dis. 204, 1573–1584.

D'Souza, S., Denis, O., Scorza, T., et al., 2000. $CD4^+$ T cells contain *Mycobacterium tuberculosis* infection in the absence of $CD8^+$ T cells in mice vaccinated with DNA encoding Ag85A. Eur. J. Immunol. 30, 2455–2459.

Dupont, N., Jiang, S., Pilli, M., et al., 2011. Autophagy-based unconventional secretory pathway for extracellular delivery of IL-1beta. EMBO J. 30, 4701–4711.

Ferrer, I.R., Wagener, M.E., Robertson, J.M., et al., 2010. Cutting edge: rapamycin augments pathogen-specific but not graft-reactive $CD8^+$ T cell responses. J. Immunol. 185, 2004–2008.

Gamerdinger, M., Kaya, A.M., Wolfrum, U., et al., 2011. BAG3 mediates chaperone-based aggresome-targeting and selective autophagy of misfolded proteins. EMBO Rep. 12, 149–156.

Geluk, A., van Meijgaarden, K.E., Franken, K.L., et al., 2000. Identification of major epitopes of *Mycobacterium tuberculosis* AG85B that are recognized by HLA-A*0201-restricted $CD8^+$ T cells in HLA-transgenic mice and humans. J. Immunol. 165, 6463–6471.

Haidinger, M., Poglitsch, M., Geyeregger, R., et al., 2010. A versatile role of mammalian target of rapamycin in human dendritic cell function and differentiation. J. Immunol. 185, 3919–3931.

Harding, C.V., 1996a. Class I MHC presentation of exogenous antigens. J. Clin. Immunol. 16, 90–96.

Harding, C.V., 1996b. Class II antigen processing: analysis of compartments and functions. Crit. Rev. Immunol. 16, 13–29.

Hess, J., Miko, D., Catic, A., et al., 1998. *Mycobacterium bovis* Bacille Calmette-Guerin strains secreting listeriolysin of *Listeria monocytogenes*. Proc. Natl. Acad. Sci. U.S.A. 95, 5299–5304.

Jagannath, C., and Bakhru, P., 2012. Rapamycin-induced enhancement of vaccine efficacy in mice. Methods Mol. Biol. 821, 295–303.

Jagannath, C., Lindsey, D.R., Dhandayuthapani, S., et al., 2009. Autophagy enhances the efficacy of BCG vaccine by increasing peptide presentation in mouse dendritic cells. Nat. Med. 15, 267–276.

Jutras, I., Houde, M., Currier, N., et al., 2008. Modulation of the phagosome proteome by interferon-gamma. Mol. Cell. Proteomics 7, 697–715.

Katholnig, K., Linke, M., Pham, H., et al., 2013. Immune responses of macrophages and dendritic cells regulated by mTOR signalling. Biochem. Soc. Trans. 41, 927–933.

Katti, M.K., Dai, G., Armitige, L.Y., et al., 2008. The Delta fbpA mutant derived from *Mycobacterium tuberculosis* H37Rv has an enhanced susceptibility to intracellular antimicrobial oxidative mechanisms, undergoes limited phagosome maturation and activates macrophages and dendritic cells. Cell. Microbiol. 10, 1286–1303.

Kimmey, J.M., Huynh, J.P., Weiss, L.A., et al., 2015. Unique role for ATG5 in neutrophil-mediated immunopathology during *M. tuberculosis* infection. Nature 528, 565–569.

Lam, G.Y., Czuczman, M.A., Higgins, D.E., et al., 2012. Interactions of *Listeria monocytogenes* with the autophagy system of host cells. Adv. Immunol. 113, 7–18.

Lee, H.K., Mattei, L.M., Steinberg, B.E., et al., 2010. In vivo requirement for Atg5 in antigen presentation by dendritic cells. Immunity 32, 227–239.

Luo, H., Duguid, W., Chen, H., et al., 1994. The effect of rapamycin on T cell development in mice. Eur. J. Immunol. 24, 692–701.

Luzio, J.P., Parkinson, M.D., Gray, S.R., et al., 2009. The delivery of endocytosed cargo to lysosomes. Biochem. Soc. Trans. 37, 1019–1021.

Mannick, J.B., Del Giudice, G., Lattanzi, M., et al., 2014. mTOR inhibition improves immune function in the elderly. Sci. Transl. Med. 6 268ra179.

Marra, A., Blander, S.J., Horwitz, M.A., et al., 1992. Identification of a *Legionella pneumophila* locus required for intracellular multiplication in human macrophages. Proc. Natl. Acad. Sci. U.S.A. 89, 9607–9611.

Mellman, I., 2005. Antigen processing and presentation by dendritic cells: cell biological mechanisms. Adv. Exp. Med. Biol. 560, 63–67.

Mercalli, A., Calavita, I., Dugnani, E., et al., 2013. Rapamycin unbalances the polarization of human macrophages to M1. Immunology 140, 179–190.

Moffat, J.M., Mintern, J.D., and Villadangos, J.A., 2013. Control of MHC II antigen presentation by ubiquitination. Curr. Opin. Immunol. 25, 109–114.

Munz, C., 2006. Autophagy and antigen presentation. Cell. Microbiol. 8, 891–898.
Munz, C., 2009. Enhancing immunity through autophagy. Annu. Rev. Immunol. 27, 423–449.
Munz, C., 2012. Antigen processing for MHC Class II presentation via autophagy. Front. Immunol. 3, 9.
Mustafa Abu, S., and Al-Attiyah, R., 2003. Tuberculosis: looking beyond BCG vaccines. J. Postgrad. Med. 49, 134–140.
Noss, E.H., Harding, C.V., and Boom, W.H., 2000. *Mycobacterium tuberculosis* inhibits MHC class II antigen processing in murine bone marrow macrophages. Cell. Immunol. 201, 63–74.
Orme, I.M., 2005. Current progress in tuberculosis vaccine development. Vaccine 23, 2105–2108.
Pancholi, P., Mirza, A., Bhardwaj, N., et al., 1993a. Sequestration from immune $CD4^+$ T cells of mycobacteria growing in human macrophages. Science 260, 984–986.
Pancholi, P., Mirza, A., Schauf, V., et al., 1993b. Presentation of mycobacterial antigens by human dendritic cells: lack of transfer from infected macrophages. Infect. Immun. 61, 5326–5332.
Peng, X., and Sun, J., 2016. Mechanism of ESAT-6 membrane interaction and its roles in pathogenesis of *Mycobacterium tuberculosis*. Toxicon 116, 29–34.
Pierre, P., 2005. Dendritic cells, DRiPs, and DALIS in the control of antigen processing. Immunol. Rev. 207, 184–190.
Ramachandra, L., Chu, R.S., Askew, D., et al., 1999. Phagocytic antigen processing and effects of microbial products on antigen processing and T-cell responses. Immunol. Rev. 168, 217–239.
Ramachandra, L., Noss, E., Boom, W.H., et al., 2001. Processing of *Mycobacterium tuberculosis* antigen 85B involves intraphagosomal formation of peptide-major histocompatibility complex II complexes and is inhibited by live bacilli that decrease phagosome maturation. J. Exp. Med. 194, 1421–1432.
Rao, R.R., Li, Q., Odunsi, K., et al., 2010. The mTOR kinase determines effector versus memory $CD8^+$ T cell fate by regulating the expression of transcription factors T-bet and Eomesodermin. Immunity 32, 67–78.
Russell, D.G., 2001. *Mycobacterium tuberculosis*: here today, and here tomorrow. Nat. Rev. Mol. Cell Biol. 2, 569–577.
Saemann, M.D., Haidinger, M., Hecking, M., et al., 2009. The multifunctional role of mTOR in innate immunity: implications for transplant immunity. Am. J. Transplant. 9, 2655–2661.
Saikolappan, S., Estrella, J., Sasindran, S.J., et al., 2011. The fbpA/sapM double knock out strain of *Mycobacterium tuberculosis* is highly attenuated and immunogenic in macrophages. PLoS One 7, e36198.
Santic, M., Molmeret, M., Klose, K.E., et al., 2005. The *Francisella tularensis* pathogenicity island protein IglC and its regulator MglA are essential for modulating phagosome biogenesis and subsequent bacterial escape into the cytoplasm. Cell. Microbiol. 7, 969–979.
Sendide, K., Deghmane, A.E., Reyrat, J.M., et al., 2004. *Mycobacterium bovis* BCG urease attenuates major histocompatibility complex class II trafficking to the macrophage cell surface. Infect. Immun. 72, 4200–4209.
Shoji-Kawata, S., Sumpter, R., Leveno, M., et al., 2013. Identification of a candidate therapeutic autophagy-inducing peptide. Nature 494, 201–206.
Singh, C.R., Moulton, R.A., Armitige, L.Y., et al., 2006. Processing and presentation of a mycobacterial antigen 85B epitope by murine macrophages is dependent on the phagosomal acquisition of vacuolar proton ATPase and in situ activation of cathepsin D. J. Immunol. 177, 3250–3259.
Soualhine, H., Deghmane, A.E., Sun, J., et al., 2007. *Mycobacterium bovis* bacillus Calmette-Guerin secreting active cathepsin S stimulates expression of mature MHC class II molecules and antigen presentation in human macrophages. J. Immunol. 179, 5137–5145.
Sun, J., Deghmane, A.E., Soualhine, H., et al., 2007. *Mycobacterium bovis* BCG disrupts the interaction of Rab7 with RILP contributing to inhibition of phagosome maturation. J. Leukoc. Biol. 82, 1437–1445.
Via, L.E., Deretic, D., Ulmer, R.J., et al., 1997. Arrest of mycobacterial phagosome maturation is caused by a block in vesicle fusion between stages controlled by rab5 and rab7. J. Biol. Chem. 272, 13326–13331.
Via, L.E., Fratti, R.A., McFalone, M., et al., 1998. Effects of cytokines on mycobacterial phagosome maturation. J. Cell. Sci. 111, 897–905.
Vyas, J.M., Van der Veen, A.G., and Ploegh, H.L., 2008. The known unknowns of antigen processing and presentation. Nat. Rev. Immunol. 8, 607–618.
Weichhart, T., Hengstschlager, M., and Linke, M., 2015. Regulation of innate immune cell function by mTOR. Nat. Rev. Immunol. 15, 599–614.
Wenger, T., Terawaki, S., Camosseto, V., et al., 2012. Autophagy inhibition promotes defective neosynthesized proteins storage in ALIS, and induces redirection toward proteasome processing and MHCI-restricted presentation. Autophagy 8, 350–363.

CHAPTER

5

Regulation of Humoral Immunity by CD1d-Restricted Natural Killer T Cells

Pragya Rampuria and Mark L. Lang

OUTLINE

Abstract

Humoral immunity is of immense importance in health and disease, being both beneficial and deleterious. Vaccine- or infection-stimulated antibody-mediated protection against pathogens is critical in host defense, but pathogen-specific antibodies can also aid infectious processes or drive pathology. Loss of immunological tolerance is associated with production of self-reactive antibodies that can worsen the disease, and loss of growth regulation can lead to several types of B cell malignancies. For these reasons and others, it is incumbent on researchers to continue to understand how humoral immunity is regulated. In the past 10–15 years, it has become apparent that CD1d-restricted Natural Killer T (NKT) cells can enhance humoral immunity against T-dependent and T-independent antigens. Furthermore, NKT cells promote B cell memory and the

M.A. Hayat (ed): Immunology, Volume 1.
DOI: http://dx.doi.org/10.1016/B978-0-12-809819-6.00005-8

© 2018 Elsevier Inc. All rights reserved.

longevity of antibody-secreting plasma cells. NKT cells potentially have profound consequences for vaccine development and understanding the immune response to pathogens. This chapter introduces NKT cells and their known biological functions and discusses specific examples of NKT cells advancing humoral immunity. Available information on the mechanisms of action by which NKT cells impact B cell activation, and commitment to B cell memory or effector functions and plasma cell longevity are also reviewed.

INTRODUCTION

The humoral immune response to T-independent and T-dependent antigens may be described as "fairly well understood." However, important questions remain in regard to understanding how a desired B cell response may be programmed by vaccination. Efforts to understand B cell fates following vaccination are starting to bear fruit, with considerable advances recently in understanding transcriptional events in the activated B cell that cause selection into the B cell memory pool (reviewed in Igarashi et al., 2014). We now understand the particular types of classical T cell help that are required for programming B cell memory (reviewed in Crotty, 2011). Similar advances have taken place in understanding the factors that drive a long-lived plasma cell response (reviewed in Igarashi et al., 2014; Chu and Berek, 2013), another essential element of a good effective response to a vaccine or primary infection.

In parallel to these studies, several laboratories have advanced our understanding of how nonclassical T cells, including CD1d-restricted NKT cells influence primary and recall B cell responses as well as plasma cell longevity. Furthermore, there is a growing appreciation that NKT cells can influence T-dependent and T-independent humoral immune responses. In this chapter, we discuss the available knowledge on how NKT cells interface with the humoral immune response and what this may mean for vaccination against a wide range of pathogens.

THE HUMORAL IMMUNE RESPONSE

T-Independent Responses

During a humoral immune response the initial encounter of mature naive B cells with pathogen or vaccine antigen leads to their proliferation and clonal expansion. B cell clonal expansion is followed by the terminal differentiation of mature B cells into plasma cells that secrete IgM and then IgG antibodies (Ab or Immunoglobulins (Igs)) (Fig. 5.1). This antibody (Ab) response could be T cell-dependent or -independent and is determined by the antigen against which it is generated. T-independent antigens are of two types—Type I where the antigens include Toll like receptor (TLR) agonists such as lipopolysaccharide (LPS) and CpG that induce a polyclonal B cell activation, and Type II where the antigen is a polysaccharide that induces antigen-specific B cell responses. Type II T-independent antigens elicit robust long-lasting Ab responses.

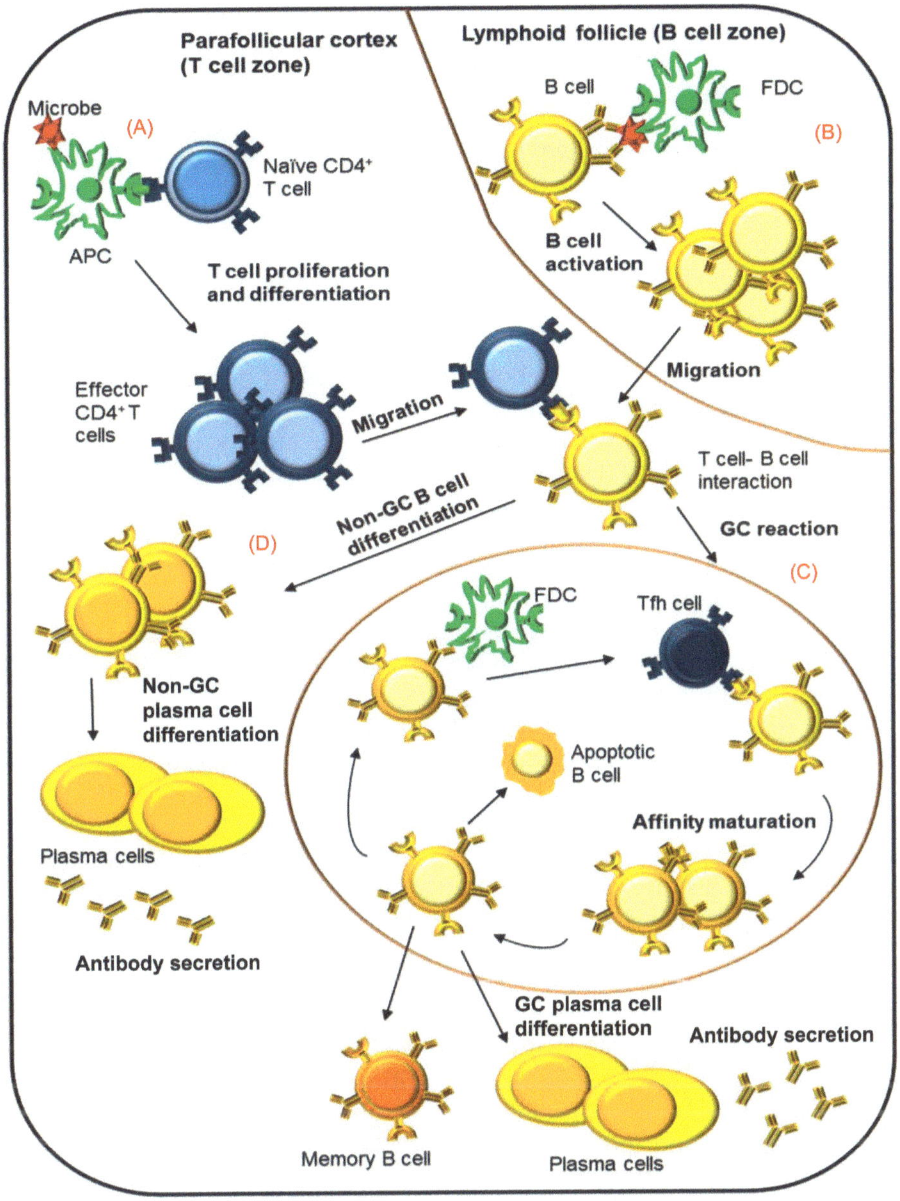

FIGURE 5.1 **Overview of the humoral immune response.** (A) An infection or vaccination induces maturation of antigen presenting cells (APC) and their migration to the T cell zones of secondary lymphoid organs. APCs that express peptide–MHC class II complexes interact with naive, antigen-specific $CD4^+$ T cells to induce their proliferation and differentiation into effector T helper (Th) cells. (B) Antigen-specific B cells become activated when follicular dendritic cells (FDC) or subcapsular sinus macrophages present antigen to naive B cells in the B cell zone. Activated B and T cells migrate to the T and B border where they engage each other to either enter the germinal center (GC) reaction or to produce short-lived antibody-secreting plasma cells. (C) Cognate contact between Tfh cells and antigen-primed B cells is required for multiple programming events. B cells undergo affinity maturation, somatic hypermutation and class switching in the GC reaction. Cells that lose the ability to bind to the presented antigen undergo apoptosis, while those that express a variant B cell receptor (BCR) with a higher affinity can compete for binding to antigen-specific GC Tfh cells. Some GC B cells exit the GC to differentiate into memory B cells and plasma cells that secrete high-affinity Abs. (D) In the extra-follicular regions B cells can undergo clonal expansion, antibody class switching and non-GC short-lived plasma cell development.

T-Dependent Responses

A special feature of the adaptive response is the generation of memory cells that "remember" the antigen they have encountered, so that in case of a reinfection the response is rapid and robust, clearing the infection promptly. For an excellent review of B cell memory please see (Good-Jacobson and Tarlinton, 2012). This recall or secondary response yields higher titers of high-affinity, class switched antibody. It is the recall immune response that forms the basis for vaccination, as it is the recall response that allows for rapid elimination of infection.

The memory B cells and long-lived bone marrow plasma cells are involved in the secondary response. These cells are usually generated in germinal centers (GCs) in a T cell-dependent manner. In the GC reaction B cells undergo affinity maturation and positive selection (Fig. 5.1). GC biology is reviewed in Allen et al. (2007). T cell help is usually required for class switching of antibody as well as GC formation, but T-independent class switching can occur if CD40 receptor on B cells is bound by C4BP (a complement protein), along with its classical ligand CD40L (CD154). The presence of different cytokines such as IL-4, IL-5, IL-21 or signaling through TLR activation with TLR ligands such as LPS also can result in T-independent class switching. Memory B cells may also form in the presence of T-independent antigen, but these might have low affinity for the antigen, as they do not undergo somatic hypermutation.

B and T cells primarily interact at the T–B border of the lymphoid follicle (Fig. 5.1). This interaction can lead to either the differentiation of B cells into extra-follicular plasmablasts or their movement into the follicle to form the GCs. GCs form 3–5 days after the activated B cells first move into the follicle. An important function of GCs is affinity maturation of B cells, wherein the B cells that have undergone somatic hypermutation and have an enhanced affinity for the antigen are selected to generate long-lived plasma cells and memory B cells. The extra-follicular plasmablasts ultimately form the short-lived plasma cells that secrete the early protective Abs against the antigen that activated both the B and T cells.

Chemokines and cytokines are important players in the humoral immune response. Chemokines play a significant role in directing cells to the light and dark zone of the GC. The light zone is an area rich in follicular dendritic cells and GC T follicular helper cells (Tfh) cells, and the dark zone is composed of proliferating B lymphocytes. Chemokine receptors CXCR4 and CXCR5, and their respective ligands CXCL12 and CXCL13, are important for maintaining the separation of the light and dark zones (Allen et al., 2007). Long-lived plasma cells, generated after GC reaction, exist in multiple lymphoid and nonlymphoid organs but their primary residence is the bone marrow. CXCL12 and CXCR4 are also important for the homing of plasma cells to the bone marrow. CXCL9, CXCL10, and CXCL11, which signal though CXCR3, have been implicated in helping homing of plasmablasts to the sites of inflammation. Plasma cell survival is maintained by cytokines like IL-6, B-cell activating factor (BAFF) and a proliferation-inducing ligand (APRIL). Different cytokine environments trigger different class switch recombination frequency, leading to the secretion of different isotypes of immunoglobulins, e.g., IFNγ is associated with IgG2b, IgG2c, and IgG3 class switch in mice and IgG3 class switch in humans, and IL-4 is associated with IgG1 and IgE in mice, and human IgG4 and IgE.

T Follicular Helper Cells

An important requirement for the generation and maintenance of GCs is a specialized subset of T helper (Th) cells known as T-follicular helper cells (Tfh). For an excellent review of this topic, please refer to Crotty (2011). Tfh cells were first described as a subset of CD4 T cells that expresses the chemokine receptor CXCR5 and provides help to B cells.

Important factors involved in the differentiation of Tfh cells in mice are the transcription factors B cell lymphoma-6 (Bcl-6), c-Maf, basic leucine zipper-activation transcription factor (BATF) and interferon regulatory factor (IRF-4) that regulate the gene expression of different cellular factors needed for differentiation into Tfh cells. The cytokines IL-6 and IL-21 promote differentiation of Tfh cells. Costimulatory factors CD28 and CD86 are also important for their differentiation and maintenance. ICOS–ICOSL interactions are required for induction of Bcl-6 and thus for the differentiation of Th cells to Tfh cells. Tfh cells are very important for the vaccine induced antibody response, but abnormal function of Tfh cells could trigger autoimmunity. Thus, negative regulators are in place to restrict Tfh activity. The primary transcription factor that negatively regulates Tfh development is B lymphocyte-induced maturation protein-1 (BLIMP-1). The cytokine IL-2 is also found to inhibit Tfh cells.

Tfh cells express a distinctive combination of effector molecules including high levels of the surface receptors inducible T cell costimulatory (ICOS), CD40 ligand (CD40L), OX40, Programmed cell death protein-1 (PD-1), B and T lymphocyte attenuator (BTLA) and CD84, the cytokine IL-21, and the cytoplasmic adaptor protein Signaling lymphocyte activation molecule (SLAM)-associated protein (SAP). These molecules are critical for stimulating the activation, differentiation, and survival of B cells and the CD4 T cells.

Tfh cells up-regulate CXCR5 expression while down regulating CCR7. This allows the Tfh cells to move from the T cell zone to the B cell follicles in response to CXCL13. IL-21 secreted by these cells enhances the differentiation of B cells, into plasma cells and thus in the secretion of all Ig isotypes. ICOS/ICOSL interactions between the Tfh cells and B cells induce the production of IL-10 and IL-21. SAP is necessary for Tfh cells to form stable conjugates with cognate B cells to promote their activation and differentiation.

Antibody Effector Functions

Abs are central mediators of immunity that perform several functions, and new perspectives on the range of these functions are reviewed in Casadevall and Pirofski (2012). They can directly neutralize pathogens and pathogen-derived products. They can also recruit both molecular and cellular immune effectors to eliminate infections and tumors. Abs have 5 main isotype classes: IgM, IgD, IgG, IgA, and IgE, classified based on their different constant regions in their heavy chains. The heavy chain isotype confers different functional properties on each antibody class, so that each isotype occupies specific sites in the body and has specific roles to play in defending the body against pathogens and their toxic products. IgM molecules are usually low affinity Abs secreted as pentamers or hexamers, so they have a high avidity for antigens with repetitive epitopes and mediate complement activation. However, pentameric IgM molecules cannot pass into the extravascular space, owing to their large size. By contrast, monomeric IgG, IgE, and monomeric or dimeric IgA molecules can be distributed systemically to tissues.

The different Ab isotypes have varied numbers of subclasses in different organisms. There are also different classes of Fc receptors (receptors for the crystallizable fragment of Igs), and the effector responses mediated via Ab interaction with these receptors include phagocytosis, endocytosis, antibody-dependent cell mediated cytotoxicity, the release of mediators of inflammation and the regulation of B-cell activation and antibody production (reviewed in Ravetch and Bolland, 2001).

NATURAL KILLER T (NKT) CELLS

NKT Cellular Subsets

NKT cells are lymphocytes of the T cell lineage that share properties of NK cells but cannot be distinctly classified as cells of the innate or adaptive immune system (reviewed in Van Kaer et al., 2011). Like adaptive immune cells, NKT cells express the antigen-specific T cell receptor (TCR) that is generated after VDJ recombination, but like innate immune cells, they do not develop immunological memory (Parekh et al., 2005), they rapidly elicit their effector functions, and are capable of auto-reactivity (Gapin, 2010).

NKT cells express T cell markers such TCR, CD40L, CD3, CD4, as well as some NK cell markers such as NK1.1, NKG2D, and CD16 (Godfrey et al., 2004). NKT cells constitutively express surface markers such as CD25, CD122, and are $CD69^{hi}$ $CD44^{hi}$ and $CD62L^{lo}$; all characteristics of effector or memory T cells (Ballas and Rasmussen, 1990). CD1d-dependent NKT cells are of at least two major types; Type I and Type II NKT cells. Type I NKT cells express a semiinvariant TCR with a canonical $V\alpha14/J\alpha18$ TCR α-chain and either $V\beta2$, 7, 8.1 or 8.2 β-chains in mice. Human Type I NKT cells express a $V\alpha24/J\alpha18$ and a $V\beta11$ β-chain (Porcelli et al., 1993). In contrast to Type I NKT cells, Type II NKT cells exhibit a diverse TCR and differ in their specificity for CD1d-associated lipids and glycolipids (Arrenberg et al., 2009).

NKT cells are present in different organs of mice in varying proportions: 30–50% of mature hepatic T cells, 0.2–0.5% of bone marrow cells, 0.3–0.5% of total thymocytes, 1–3% of total splenocytes, 0.1–0.3% of lymph node cells, 4% of circulating T cells, and 7% of lung T cells (Eberl et al., 1999).

CD1d

CD1 molecules are typically nonpolymorphic major histocompatibility complex (MHC) class I-like molecules. There are 3 groups of CD1 molecules in humans. Group 1 includes: CD1a, b, and c; Group 2 contains CD1d, and Group 3 includes CD1e (Porcelli and Modlin, 1999). Mice do not express group 1 or 3 molecules (Porcelli and Modlin, 1999) as they only express CD1d. While T cells recognize the peptide–MHC complex, NKT cells recognize lipid-containing antigens presented by the CD1d molecule. CD1d molecules can present both endogenous and exogenous ligands. CD1d is highly conserved across mammalian species. Mice and human CD1d have >90% sequence homology (Blumberg et al., 1995). It has been shown that human CD1d-reactive NKT cells recognize mouse CD1d/glycolipid complexes, and vice versa (Brossay et al., 1998).

NKT cells develop in the thymus from the same $CD4^+$ $CD8^+$ double-positive precursors as conventional T cells (Gapin et al., 2001). CD1d is expressed on cortical thymocytes, all antigen presenting cells (APCs) including B cells, monocytes, macrophages and dendritic cells, and also on activated T cells (Exley et al., 2000). Unlike Group 1 CD1 molecules, CD1d is also expressed on various nonhematopoietic cells in every organ; e.g., hepatocytes in the liver and intestinal epithelial cells express CD1d (Trobonjaca et al., 2001).

The crystal structure of the mouse CD1d molecule revealed that its Ag binding groove is deeper, narrower, and much more hydrophobic compared to related MHC class I and II molecules (Zeng et al., 1997). The hydrophobic nature of the antigen binding groove makes CD1d molecules ideal for the presentation of hydrophobic Ags, such as glycolipids (Zeng et al., 1997). De novo CD1d is loaded with self-lipid antigens in the endoplasmic reticulum (Dougan et al., 2005) before transport to the cell surface. CD1d molecules undergo rapid recycling from the cell surface. They are internalized via adaptor protein-2 (AP-2) and are localized to the late endosome or lysosome through AP-3 (Cernadas et al., 2003). CD1d encounters exogenous glycolipid antigens of engulfed pathogens in endocytic vesicles such as late endosome and lysosome. Here the self-lipids are replaced by exogenous glycolipid antigens through the involvement of saposin (Brigl and Brenner, 2004). The antigen-loaded CD1d molecules are then transported to the surface to present antigens to NKT cells.

α-Galactosylceramide

α-Galactosylceramide (α-GC or α-Galcer) is an adjuvant that has been used in experimental vaccines against several pathogens (Galli et al., 2007). α-GC was the first NKT ligand discovered (Kawano et al., 1997; Morita et al., 1995). It was found during a search for anticancer treatments at Kirin Pharmaceuticals. The study found that extracts from *Agelas mauritianus*, a marine sponge collected in the Okinawan sea, extended survival of mice bearing B16 melanoma (Morita et al., 1995). The active ingredient of the extract was identified as an α-branched galactosylceramide (Fig. 5.2). It is likely that α-GC is derived from the *Novosphingobium* bacteria that colonize the sponges, as multiple NKT cell-activating glycolipids with structural similarity to α-GC have been identified in the cell wall of the *Novosphingobium* species (Kinjo et al., 2005; Mattner et al., 2005). This was slightly modified to produce a compound termed KRN7000, also referred to as α-GalCer or α-GC (Kobayashi et al., 1995). Investigation of the role of α-GC against cancer led to the identification of Vα14 NKT cells (Kawano et al., 1997; Kobayashi et al., 1995).

α-GC is a glycolipid compound, that is presented to NKT cells by CD1d, independent of MHC class I or class II (Bendelac et al., 2007). α-GC has been used to generate the first CD1d tetramers specific for mouse and human NKT cells (Benlagha et al., 2000; Matsuda et al., 2000). The affinity of interaction, with a Kd value of ~100 nM, between CD1d-αGC and mouse TCRs, is one of the highest ever recorded for natural TCR/ligand pairs (Gadola et al., 2006). A unique quality of α-GC is that it triggers enhanced antibody production, even in mice lacking MHC II, suggesting that NKT cells can, to a certain extent, substitute for conventional $CD4^+$ T cells and provide assistance for activation of B cells (Lang et al., 2006).

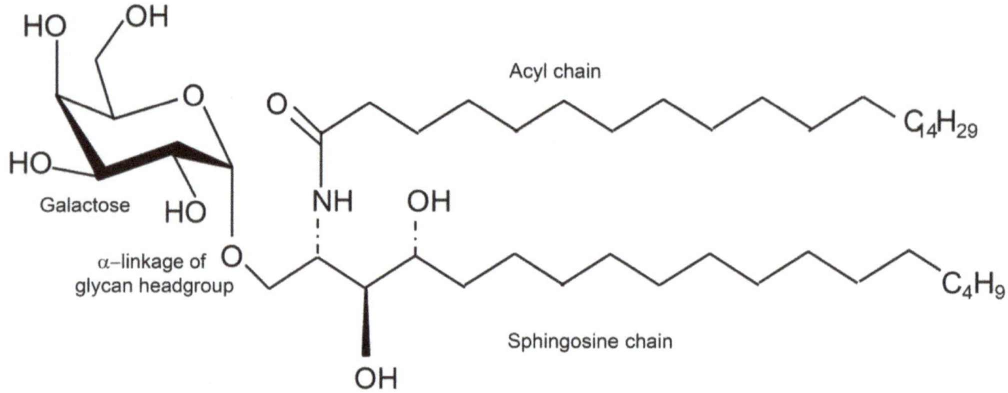

FIGURE 5.2 **Structure of α-galactosylceramide (α-GC).** Chemical name of α-GC: (2S,3S,4R)-1-*O*-(α-D-galactosyl)-*N*-hexacosanoyl-2-amino-1,3,4-octadecanetriol. α-GC is composed of a galactose head group that is linked through the α-hydroxyl to the sphingosine chain (18 carbons). The sphingosine chain is further linked to the fatty acyl chain (26 carbons). The head group of α-GC projects from the binding groove of CD1d and is recognized by the α-chain of the TCR of NKT cells. The sphingosine chain of α-GC binds to the F′ pocket of CD1d, whereas the acyl chain binds to the A′ pocket.

NKT Cell Activation and Function

NKT cells are activated directly by engagement of the invariant TCR with exogenous or endogenous glycolipid antigens complexed to CD1d, or indirectly via activated APCs and cytokines (Fig. 5.3). Exogenous antigens are derived from several pathogens. *Borrelia burgdorferi*—a bacterium that causes Lyme disease—contains diacylglycerol that is an NKT cell activating ligand. Apart from glycolipids, phospholipids (Jeannin et al., 2000), lysophospholipids (Fox et al., 2009), a tetra-mannosylated form of phosphatidylinositol termed PIM4 derived from *Mycobacterium bovis* (Fischer et al., 2004), a lipophosphoglycan from *Leishmania donovani* (Amprey et al., 2004), and a lipopeptidophosphoglycan from the membrane of *Entamoeba histolytica* (Lotter et al., 2009) are also pathogen-derived ligands that activate NKT cells.

Microorganisms that lack glycolipids or any CD1d-binding lipid ligand can also activate NKT cells indirectly (Fig. 5.3) (Matsuda et al., 2008). This mechanism of NKT cell activation was initially studied with *Salmonella typhimurium* that contain LPS in their cell wall (Brigl et al., 2003). LPS from *Salmonella* activates DCs via TLR4, to produce IL-12. IL-12 synergizes with self-antigens presented via CD1d and leads to activation of NKT cells. There are several variations of this mechanism that have been shown to activate NKT cells (Gapin, 2010; Matsuda et al., 2008). In case of *Escherichia coli*, both cytokines IL-12 and IL-18 seem to be important but CD1d expression was not required (Nagarajan and Kronenberg, 2007). Microbes like *Listeria monocytogenes* induce CD1d expression on macrophages and DCs in the presence of IFN-β, which leads to NKT cell activation.

NKT cells produce cytokines including IL-2, IL-4, IL-5, IL-6, IL-10, IL-13, IL-17, IL-21, TNF-α, IFN-γ, TGF-β, and GM-CSF (Van Kaer et al., 2011). NKT cells can drive Th1 or Th2 immunity very rapidly on activation, as they have high levels of IL-4 and IFN-γ mRNA stored (Stetson et al., 2003). One of the factors influencing the cytokine profile of NKT cells is

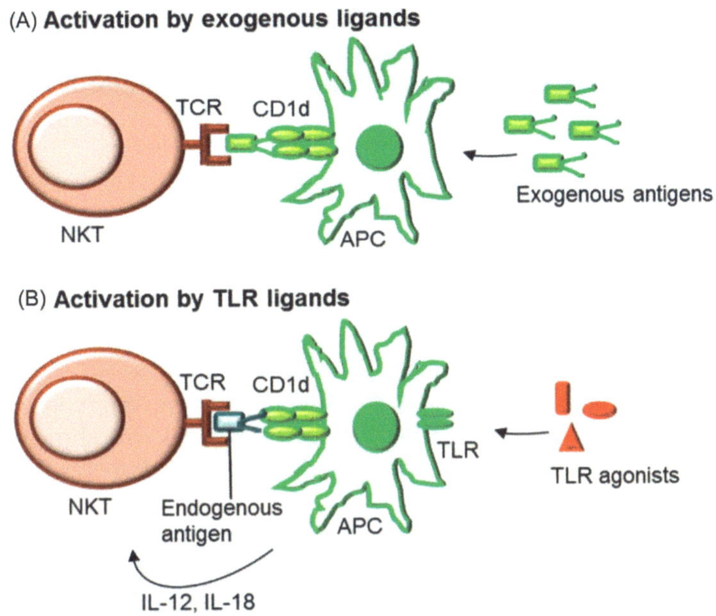

FIGURE 5.3 **Modes of NKT activation.** (A) In the direct method of activation, APCs present exogenous glycolipid antigen via CD1d to the semiinvariant TCR on NKT cells. (B) In the indirect method, APCs activated by TLR (Toll like receptor) ligands activate NKT cells by presenting endogenous ligands on CD1d in the presence of cytokines like IL-12 and IL-18.

the quality of the TCR signal, which is dependent on the ligand presented on CD1d. While a glycolipid analogue of α-GC like C-glycoside stimulates a Th-1 cytokine bias leading to IFN-γ, another analogue OCH stimulates a Th-2 bias with more IL-4 secretion (Schmieg et al., 2003). NKT cells, as evident from the list of cytokines they can secrete, can be both pro- and antiinflammatory. This depends on various factors, including the mode, duration, and strength of activation and the location and cytokine environment (Godfrey and Kronenberg, 2004). Cytokines secreted by NKT cells in turn help in the transactivation of other immune cells including NK cells, B cells, and T cells.

NKT cells can regulate the colonization of the intestine of mice by commensal microorganisms. $CD1d^{-/-}$ mice have increased colonization of bacteria when compared to B6 wild type mice. $CD1d^{-/-}$ mice also exhibit a defect in the structure and function of Paneth cells, which are vital for the maintenance of the epithelium in the small intestine (Nieuwenhuis et al., 2009). Conversely, commensal microbiota can also control mucosal NKT cell development, their homing and maturation (An et al., 2014; Olszak et al., 2012). Thus, there is bidirectional crosstalk between the microbiota and CD1d-restricted NKT cells and disruption of these processes can contribute to the pathogenesis of disorders such as asthma and inflammatory bowel disease (Zeissig and Blumberg, 2014).

NKT cells not only express markers of NK cells but also share some functions with NK cells. NKT cells can secrete perforin and granzyme B, which allows them to function as cytotoxic effector cells (Arase et al., 1994). The secretion of IFN-γ and TNF-α by NKT cells can transactivate the cytotoxic effects of NK cells and the CD8 T cells (Smyth et al., 2005). NKT cells enhance humoral immune responses against various microbial agents, which include—but are not limited to—H3N2 from human influenza virus, protective antigen (PA) from *Bacillus anthracis*, diphtheria, and tetanus toxins (Belperron et al., 2005; Devera et al., 2010; Galli et al., 2007; Kumar et al., 2000; Lang et al., 2006; Leadbetter et al., 2008).

NKT Cells and Disease

NKT cells exert both pro- and antiinflammatory effects and therefore have a beneficial and a harmful impact during infectious disease, inflammatory diseases, and autoimmunity.

Noninfectious Diseases

Qualitative and quantitative defects in NKT cells have been observed in various autoimmune disorders. The suppressive role of NKT cells in diabetes was shown in nonobese diabetic (NOD) mice where repeated injections of α-GC inhibited diabetes development (Hong et al., 2001; Sharif et al., 2001). This suppressive role was attributed to a NKT-induced Th2 profile (IL-4 and/or IL-10) (Hammond et al., 1998). α-GC treatment can either protect or aggravate experimental autoimmune encephalomyelitis, depending upon the time of administration of α-GC and the cytokine secreted (Jahng et al., 2001). In relapsing-remitting multiple sclerosis patients reduced numbers of NKT cells have been observed (Gausling et al., 2001). Systemic Lupus Erythematosus prone mice show reduced numbers of NKT cells before the disease develops (Takeda and Dennert, 1993). However it is still controversial whether treatment with α-GC can promote autoimmune like inflammation (Morshed et al., 2002).

NKT cells have been associated with several different graft survival models. α-GC treatment has been shown to prolong islet graft survival in an IL-10-dependent manner in NOD mice (Zeng et al., 1999). NKT cells have been implicated in tolerance induction in hepatic, cardiac, and xenogeneic pancreatic islet grafts, which may be dependent on IFN-γ and IL-4 production (Konieczny et al., 1998).

CD1d expression by tumors does not appear to be necessary for the antitumor activity of α-GC (Wu and Van Kaer, 2011). NKT-derived IFN-γ that led to further secretion of IFN-γ from NK and CD8 T cells was found to be protective against tumors. A trial with patients with nonsmall lung cancer using repeated treatment with α-GC-stimulated peripheral blood mononuclear cells has been shown to have promising clinical benefit to patients (Motohashi et al., 2009).

NKT cells have also been shown to have some adverse effects on health. NKT cells can contribute to the development of allergy-induced airway hyperactivity (AHR) (Lisbonne et al., 2003). NKT-derived IL-4 and IL-13 recruit Th2 cells to the pulmonary compartment, which induces AHR (Akbari et al., 2003).

Infectious Diseases

As mentioned earlier several microbial lipids can activate NKT cells. Mycobacterial phosphatidylinositolmannosides (PIMs) present in the bacterial cell wall stimulate NKT cells and lead to the formation of granulomas. This suggests a protective role of NKT cells, as granulomas prevent dissemination of the pathogen (Apostolou et al., 1999). However, whether NKT cells are protective against *Mycobacterium tuberculosis* still remains controversial.

$CD1d^{-/-}$ mice do not have any difference in infection rates when compared to wild type controls, but administration of α-GC decreases mycobacterial burden, reduces pulmonary injury, and enhances survival after infection (Chackerian et al., 2002). $CD1d^{-/-}$ mice were more susceptible to infection with *B. burgdorferi* (Kumar et al., 2000). Activation of NKT cells with α-GC induced IFN-γ secretion which enhanced phagocytosis and TNF-α secretion by macrophages that in turn led to rapid clearance of *Pseudomonas aeruginosa* (Nieuwenhuis et al., 2002). Moreover, immunization with α-GC and PA from *B. anthracis* leads to enhanced antibody responses against PA and to enhanced survival when mice were challenged with anthrax lethal toxin from *B. anthracis* (Devera et al., 2011).

Viral infections are also impacted by NKT cells. For example, NKT cells activated by α-GC inhibited replication of hepatitis B in HBV transgenic mice by recruiting NK cells via NKT-derived IFN-γ and IFN-α/β (Kakimi et al., 2000). α-GC-activated NKT cells recruit $CD45^+$ and $NK1.1^+$ cells to the vagina and cause elevated levels of IFN-γ in the vagina and decrease susceptibility to herpes simplex virus-2 (HSV-2) infection (Iversen et al., 2015). $CD1d^{-/-}$ mice that lack NKT cells have been found to be more susceptible to the picornavirus, diabetogenic encephalomyocarditis virus (EMCV-D), respiratory syncytial virus (RSV), and herpes simplex virus-1 (HSV-1) (Exley et al., 2003; Grubor-Bauk et al., 2008; van Dommelen et al., 2003). This protection may be attributed to the rapid cytokine (IFNs and TNF-α) production by activated NKT cells which induces bystander activation of NK cells and T lymphocytes, which are important for viral clearance (Carnaud et al., 1999). It has not yet been well established whether the depletion of NKT cells seen in several viral infections is due to a microbial survival strategy or due to activation induced cell death (van der Vliet et al., 2004). In case of Human Immunodeficiency Virus (HIV-1), the protein Nef produced by HIV-1 has been shown to downregulate CD1d expression by accelerating its internalization which in turn reduces NKT cell activation (Chen et al., 2006).

Regarding parasitic infections and eukaryotic pathogens, activated NKT cells have been shown to inhibit the intrahepatic but not the blood stage of *Plasmodium yoelii* and *Plasmodium berghei* (Gonzalez-Aseguinolaza et al., 2000). NKT cell-derived IFN-γ can lead to recruitment of CD4 and CD8 T cells which aid in eliminating parasites from the liver when mice were immunized with α-GC before live-sporozoite challenge (Gonzalez-Aseguinolaza et al., 2000). NKT cells can also limit *Trypanosoma cruzi* parasitemia during acute infection via NKT cell-derived IFN-γ or cytolytic activity and enhanced antibody responses (Duthie et al., 2002). NKT cells induced Th1 response also plays an important role in developing resistance against the fungus *Cryptococcus neoformans* (Kawakami et al., 2001).

The ability of NKT cells to help enhance immune responses has also been employed for developing vaccines against several pathogens (Kim et al., 2008). This protective role of NKT cells against several different pathogens is partly because of their ability to enhance the humoral immune response.

NKT CELLS AND HUMORAL IMMUNITY

In general, B cells form cognate interaction with Tfh cells in the GC of lymphoid follicles to generate long-lived PCs and memory B cells that produce protective antibodies. It is now well established that B cells can also receive additional support from CD1d-dependent

NKT cells. NKT cells aid not only in enhancing the primary antibody response but also in improving the recall response by influencing the generation of long-lived PCs and memory B cells (Devera et al., 2010; Galli et al., 2007; Lang et al., 2006; Lang et al., 2011). In addition to enhancing and complementing the B cell helper activity of canonical T cells, NKT cells can also help initiate rapid innate-like antibody responses against T cell-independent antigens at the mucosa and in the marginal zone of the spleen (Dellabona et al., 2014). The exact mechanism by which these NKT cells help the B cell response is not well understood.

There are two different ways NKT cells may communicate with B cells—via cognate (direct) interactions and/or by noncognate (indirect) interactions. In case of T-independent antigens like carbohydrate and glycolipid Ags the interactions appear to be direct (Lang et al., 2008). However, the interactions between NKT and B cells are more complex for T-dependent Ags as they rely on both cognate and noncognate interactions (Tonti et al., 2009, 2012).

T-Independent Antigens

In case of glycolipid Ags, Lang et al. (2005) showed that BCR targeting of a biotin-modified version α-GC leads to enhanced murine CD1d-mediated presentation to NKT cells by 1000 fold when compared with presentation of nontargeted biotin-α-GC. This suggested cognate interaction between NKT and B cells. Furthermore, Leadbetter and colleagues used a model hapten-conjugated lipid molecule, 4-hydroxy-3-nitrophenyl-α-GalactosylCeramide (NP-α-GalCer), to demonstrate that NKT cells provide cognate help to lipid antigen-presenting B cells (King et al., 2012; Leadbetter et al., 2008). They also showed that this help was dependent on CD1d and B7.1 expression by B cells.

A subset of NKT cells called NKT follicular helper cells (NKTfh) was discovered in 2012 by Chang and colleagues. NKTfh cells, like the conventional Tfh cells, express the chemokine receptor for CXCL13-CXCR5, PD1—an immunoreceptor which favors strong cognate interactions with B cells in the GC and costimulatory molecules like ICOS, CD40L (Chang et al., 2012; Legler et al., 1998). NKTfh formation is dependent on the costimulatory molecule-CD28 and the transcription factor Bcl-6 (Chang et al., 2012). Also like the Tfh cells, NKTfh cells secrete IL-21 (Chang et al., 2012). CXCR5 on Tfh cells help direct them towards B cell follicles and GCs. Over 80% of the total NKTfh cells express CD4, suggesting a skewing towards a Tfh-like phenotype. NKTfh cells also have reduced expression of CD127 (IL-7 receptor-α) and up-regulated expression of the inhibitory receptor CD272 (BTLA) (Chang et al., 2012). NKTfh cells have been implicated in regulating the antibody response. NKTfh cells lead to rapid formation of GCs and generation of higher affinity antibodies against lipid antigens (Chang et al., 2012; King et al., 2012).

Carbohydrate Ags usually stimulate a T-independent Ab response wherein IgM is the dominant Ab class generated and there is little or no isotype switching and/or differentiation of memory B cells. When carbohydrate Ag like NP-Ficoll is coadministered with α-GC, there is a CD1d/NKT dependent immunoglobulin class switch to IgG1 and IgG3 and production of memory B cells similar to classical T-dependent humoral immunity (Lang et al., 2006). Bendelac and coauthors used nanoparticles coexpressing pneumococcal capsular polysaccharides and a cell wall lipid antigen analog to demonstrate that CD1d-restricted cognate NKT and B cell interactions led to class switched, long-term memory against the capsular polysaccharides (Bai et al., 2013). In that study they also performed conditional ablation of CD1d on B cells and DCs to show that initial interactions with CD1d expressing DCs followed by NKT-B cognate interaction was required for this process (Bai et al., 2013). The exact mechanism by which NKT cells help the B cell response against carbohydrate antigens is unknown, but different subsets of NKT cells may be involved.

T-Dependent Antigens

When a protein Ag is administered with NKT-activating glycolipid α-GC, APCs like DCs and macrophages process protein antigens to peptides presented on class II MHC molecules, and take up α-GC simultaneously and present to the TCR on NKT cells via CD1d. This leads to priming of both classical Th cells and NKT cells that are activated and differentiate into follicular helpers, which further help the B cell response. Studies suggest that CD1d on DCs are involved with the initial activation of NKT cells, but CD1d on B cells is equally important for the NKT-enhanced humoral immune response. Adoptive transfer of $CD1d^{+/+}$ versus $CD1d^{-/-}$ B cells into μMT recipient mice that lack B cells showed that CD1d on B cells was essential for the NKT-enhanced Ab response (Lang et al., 2008). Vinuesa and colleagues used intravital microscopy to show that NKT and Ag-specific B cells interact directly for 4–50 min after immunization, indicating that the interaction between them occurs over a restricted time frame (Chang et al., 2012). Allan and coworkers (2011) showed that activation of human B cells leads to downregulation of CD1d where by CD1d is not detectable by day 5 after activation, also indicating a time-constrained NKT-B cell interaction.

We know CD40L on conventional Th cells is essential for the helper function of T cells. Bone marrow chimeras with $CD40L^{+/+}$ or $CD40L^{-/-}$ NKT cells were used to show that $CD40L^{-/-}$ NKT cells can provide help for Ab production in cooperation with $CD40L^{+/+}$ Th cells, and B cell helper function of NKT cells was not dependent on CD40L-CD40 between NKT and B cells (Shah et al., 2011).

SAP is an adaptor molecule essential for Tfh cells help to B cells. Detre et al. (2012) used mice that specifically lacked SAP expression in NKT cells, to show that SAP was required for cognate help to B cells by NKT cells but was not essential for the noncognate help.

Both cognate and noncognate help by NKT cells has been suggested for NKT-dependent B cell help. IL-4 and IFN-γ secreted upon NKT cell activation aids in the class switch of immunoglobulin to IgG1 and IgG2a, respectively (Devera et al., 2011). However, using bone marrow chimeras, where NKT cells were unable to express IL-4 and IFN-γ, it has been shown that NKT-derived IL-4 and IFN-γ are dispensable for the B cell help (Devera et al., 2011).

Immunization of mice lacking NKT cells with a combination of protein Ag and CD1d-binding ligand revealed that NKT cells also play a role in plasma cell homeostasis, inducing and/or maintaining the circulating Ab levels (Devera et al., 2008). BAFF and APRIL are two cytokines that are critical for PC survival. Shah et al. (2013) immunized bone marrow chimeric mice in which the entire hematopoietic compartment or NKT cells selectively lacked BAFF, APRIL, or both BAFF and APRIL with Ag and α-GC. Chimeras that lacked both BAFF and APRIL in the entire hematopoietic compartment or only in NKT cells had rapidly decaying Ab titers and reduced PC numbers in comparison with chimeras that lacked BAFF or APRIL alone (Shah et al., 2013). This showed that NKT cell-induced secretion of BAFF and APRIL plays an important role in the enhanced PC survival and maintenance of Ab titer stimulated by NKT cells (Shah et al., 2013).

We know that NKT cells differentiate into a subset called NKTfh cells when activated with α-GC (Chang et al., 2012). These NKTfh cells are known to migrate to the GCs and can facilitate rapid formation of GCs (Chang et al., 2012; Tonti et al., 2012). However, these GCs are short-lived and do not produce long-lived plasma cells and memory B cells against lipid antigens (Tonti et al., 2012). Cognate help from NKTfh cells can induce the formation of extra-follicular plasmablasts, Ig class switch, and moderate affinity maturation (Chang et al., 2012; King et al., 2012).

Lang and colleagues in a recent study showed that immunization with the C-terminal domain of *Clostridium difficile* toxin B (TcdB), accompanied by activation of iNKT cells with α-GC, had no direct effect on the numbers and phenotype of Th or Tfh cells, but NKT activation with α-GC transactivated Th cells to secrete IL-4 and IL-21. They also demonstrated, using conditional knockouts that lacked Bcl-6 in $CD4^+$ T cells, that Tfh and iNKTfh cells both contributed to B cell help. This work extends our understanding of the contributions by NKTfh cells to humoral immunity (Rampuria et al., 2016).

Overall evidence suggests that NKT cells help enhance the Ab response, generation of long-lived PCs and memory B cells, with both cognate interactions and noncognate help via soluble effector molecules.

Type II NKT Cells and Humoral Immunity

Recent studies have indicated that type II NKT cells also help in the induction of humoral immunity. Identification of Type II NKT cells has been difficult because of the absence of reagents like CD1d tetramers used for type I NKT identification that can be used to specifically classify and quantitate Type II NKT. Thus investigators have relied on the use of genetically deficient mouse models. Since CD1d-deficient ($CD1d^{-/-}$) mice lack both Type I and Type II NKT cells, and Jα18-deficient ($J\alpha18^{-/-}$) mice lack only Type I NKT cells, investigators have used a combination of these knock out and wild type B6 mice to allocate the contributions of Type I versus Type II NKT cells in the immune response.

Shah et al. immunized this combination of mouse models with keyhole limpet hemocyanin in the presence of Alum to demonstrate that Type II NKT cells are required for antigen-specific IgG1 and IgG3 production that is induced by Alum-adsorbed TD antigens. This study showed that Type II NKT cells transactivated Th cells to secrete IL-4 when mice are immunized with Alum (Shah et al., 2012). They also used CD1d blocking mAb to show that CD1d expression by APCs is required for the Type II NKT induced cytokine production and IgG1 production (Shah et al., 2012).

Another study investigated the role of type II NKT cells in Gaucher disease (GD) (Nair et al., 2015). GD occurs due to deficiency of acidic β-glucosidase enzyme, which in turn leads to buildup of lysosomal sphingolipids such as β-glucosylceramide (βGL1) and glucosylsphingosine (LGL1). This accumulation of lipids leads to inflammation and B cell activation. Nair et al. (2015) used βGL1 and LGL1 loaded CD1d tetramers and genomic profiling to demonstrate that type II NKT cells and not type I NKT cells are activated by these lipids. These type II NKT cells were found to constitutively display the Tfh phenotype. They showed this response was CD1d dependent and lead to the secretion of IL-5, IL-6, IL-10, IL-13, IL-17, granulocyte macrophage colony-stimulating factor, macrophage-derived chemokine, and macrophage inflammatory protein 1β (Nair et al., 2015). They demonstrated that type II NKT cells lead to enhanced Ab titers in vivo in a CD1d-dependent manner. They also showed that Type II NKT cells played a significant role in patients with the GD disease (Nair et al., 2015). Thus type II NKT cells which have been mostly described as immunosuppressive or immune-regulatory cells in certain environments can boost humoral immunity.

CONCLUDING REMARKS

We are now beginning to gain key mechanistic insights into how CD1d-restricted NKT cells can shape and direct the humoral immune response to T-dependent and T-independent antigens. The prototype glycolipid α-GC clearly has potent adjuvant properties that may be more effectively harnessed by synthesis of structural variants that can further direct the immune response (reviewed in Venkataswamy and Porcelli, 2010). In the coming few years, researchers may consider focusing on the contributions of NKTfh cells to B cell memory and to the mechanisms by which NKT cells can promote plasma cell longevity. Potentially CD1d-binding glycolipids based on α-GC could be incorporated into adjuvant platforms to enhance immune responses following vaccination.

Acknowledgments

The authors work referenced herein and ideas discussed were supported by National Institutes of Health grants AI078993 and AI125708 to M.L.L., an American Association of Immunologists (AAI) Careers in Immunology Fellowship to M.L.L. and P.R., and a Presbyterian Health Foundation (Oklahoma City) grant to M.L.L. We thank Gillian A. Lang (OUHSC) for assistance with preparing the manuscript. We apologize to those colleagues whose work we could not cite due to restrictions on bibliography length.

References

Akbari, O., Stock, P., Meyer, E., et al., 2003. Essential role of NKT cells producing IL-4 and IL-13 in the development of allergen-induced airway hyperreactivity. Nat. Med. 9, 582–588.

Allan, L.L., Stax, A.M., Zheng, D.J., et al., 2011. CD1d and CD1c expression in human B cells is regulated by activation and retinoic acid receptor signaling. J. Immunol. 186, 5261–5272.

Allen, C.D., Okada, T., and Cyster, J.G., 2007. Germinal-center organization and cellular dynamics. Immunity 27, 190–202.

Amprey, J.L., Im, J.S., Turco, S.J., et al., 2004. A subset of liver NK T cells is activated during *Leishmania donovani* infection by CD1d-bound lipophosphoglycan. J. Exp. Med. 200, 895–904.

An, D., Oh, S.F., Olszak, T., et al., 2014. Sphingolipids from a symbiotic microbe regulate homeostasis of host intestinal natural killer T cells. Cell 156, 123–133.

Apostolou, I., Takahama, Y., Belmant, C., et al., 1999. Murine natural killer T (NKT) cells contribute to the granulomatous reaction caused by mycobacterial cell walls. Proc. Natl. Acad. Sci. U.S.A. 96, 5141–5146.

Arase, H., Arase, N., Kobayashi, Y., et al., 1994. Cytotoxicity of fresh NK1.1+ T cell receptor alpha/beta+ thymocytes against a CD4 + 8+ thymocyte population associated with intact Fas antigen expression on the target. J. Exp. Med. 180, 423–432.

Arrenberg, P., Halder, R., and Kumar, V., 2009. Cross-regulation between distinct natural killer T cell subsets influences immune response to self and foreign antigens. J. Cell. Physiol. 218, 246–250.

Bai, L., Deng, S., Reboulet, R., et al., 2013. Natural killer T (NKT)-B-cell interactions promote prolonged antibody responses and long-term memory to pneumococcal capsular polysaccharides. Proc. Natl. Acad. Sci. U.S.A. 110, 16097–16102.

Ballas, Z.K., and Rasmussen, W., 1990. NK1.1+ thymocytes. Adult murine CD4-, CD8- thymocytes contain an NK1.1+, CD3+, CD5hi, CD44hi, TCR-V beta 8+ subset. J. Immunol. 145, 1039–1045.

Belperron, A.A., Dailey, C.M., and Bockenstedt, L.K., 2005. Infection-induced marginal zone B cell production of *Borrelia hermsii*-specific antibody is impaired in the absence of CD1d. J. Immunol. 174, 5681–5686.

Bendelac, A., Savage, P.B., and Teyton, L., 2007. The biology of NKT cells. Annu. Rev. Immunol. 25, 297–336.

Benlagha, K., Weiss, A., Beavis, A., et al., 2000. In vivo identification of glycolipid antigen-specific T cells using fluorescent CD1d tetramers. J. Exp. Med. 191, 1895–1903.

Blumberg, R.S., Gerdes, D., Chott, A., et al., 1995. Structure and function of the CD1 family of MHC-like cell surface proteins. Immunol. Rev. 147, 5–29.

Brigl, M., and Brenner, M.B., 2004. CD1: antigen presentation and T cell function. Annu. Rev. Immunol. 22, 817–890.

Brigl, M., Bry, L., Kent, S.C., et al., 2003. Mechanism of CD1d-restricted natural killer T cell activation during microbial infection. Nat. Immunol. 4, 1230–1237.

Brossay, L., Chioda, M., Burdin, N., et al., 1998. CD1d-mediated recognition of an alpha-galactosylceramide by natural killer T cells is highly conserved through mammalian evolution. J. Exp. Med. 188, 1521–1528.

Carnaud, C., Lee, D., Donnars, O., et al., 1999. Cutting edge: cross-talk between cells of the innate immune system: NKT cells rapidly activate NK cells. J. Immunol. 163, 4647–4650.

Casadevall, A., and Pirofski, L.A., 2012. A new synthesis for antibody-mediated immunity. Nat. Immunol. 13, 21–28.

Cernadas, M., Sugita, M., van der Wel, N., et al., 2003. Lysosomal localization of murine CD1d mediated by AP-3 is necessary for NK T cell development. J. Immunol. 171, 4149–4155.

Chackerian, A., Alt, J., Perera, V., et al., 2002. Activation of NKT cells protects mice from tuberculosis. Infect. Immun. 70, 6302–6309.

Chang, P.P., Barral, P., Fitch, J., et al., 2012. Identification of Bcl-6-dependent follicular helper NKT cells that provide cognate help for B cell responses. Nat. Immunol. 13, 35–43.

Chen, N., McCarthy, C., Drakesmith, H., et al., 2006. HIV-1 down-regulates the expression of CD1d via Nef. Eur. J. Immunol. 36, 278–286.

Chu, V.T., and Berek, C., 2013. The establishment of the plasma cell survival niche in the bone marrow. Immunol. Rev. 251, 177–188.

Crotty, S., 2011. Follicular helper CD4 T cells (TFH). Annu. Rev. Immunol. 29, 621–663.

Dellabona, P., Abrignani, S., and Casorati, G., 2014. iNKT-cell help to B cells: a cooperative job between innate and adaptive immune responses. Eur. J. Immunol. 44, 2230–2237.

Detre, C., Keszei, M., Garrido-Mesa, N., et al., 2012. SAP expression in invariant NKT cells is required for cognate help to support B-cell responses. Blood 120, 12212–12219.

Devera, T.S., Shah, H.B., Lang, G.A., et al., 2008. Glycolipid-activated NKT cells support the induction of persistent plasma cell responses and antibody titers. Eur. J. Immunol. 38, 1001–1011.

Devera, T.S., Aye, L.M., Lang, G.A., et al., 2010. CD1d-dependent B-cell help by NK-like T cells leads to enhanced and sustained production of *Bacillus anthracis* lethal toxin-neutralizing antibodies. Infect. Immun. 78, 1610–1617.

Devera, T.S., Joshi, S.K., Aye, L.M., et al., 2011. Regulation of anthrax toxin-specific antibody titers by natural killer T cell-derived IL-4 and IFNgamma. PLoS One 6, e23817.

Dougan, S.K., Salas, A., Rava, P., et al., 2005. Microsomal triglyceride transfer protein lipidation and control of CD1d on antigen-presenting cells. J. Exp. Med. 202, 529–539.

Duthie, M.S., Wleklinski-Lee, M., Smith, S., et al., 2002. During *Trypanosoma cruzi* infection CD1d-restricted NK T cells limit parasitemia and augment the antibody response to a glycophosphoinositol-modified surface protein. Infect. Immun. 70, 36–48.

Eberl, G., Lees, R., Smiley, S.T., et al., 1999. Tissue-specific segregation of CD1d-dependent and CD1d-independent NK T cells. J. Immunol. 162, 6410–6419.

Exley, M., Garcia, J., Wilson, S.B., et al., 2000. CD1d structure and regulation on human thymocytes, peripheral blood T cells, B cells and monocytes. Immunology 100, 37–47.

Exley, M.A., Bigley, N.J., Cheng, O., et al., 2003. Innate immune response to encephalomyocarditis virus infection mediated by CD1d. Immunology 110, 519–526.

Fischer, K., Scotet, E., Niemeyer, M., et al., 2004. Mycobacterial phosphatidylinositol mannoside is a natural antigen for CD1d-restricted T cells. Proc. Natl. Acad. Sci. U.S.A. 101, 10685–10690.

Fox, L.M., Cox, D.G., Lockridge, J.L., et al., 2009. Recognition of lyso-phospholipids by human natural killer T lymphocytes. PLoS Biol. 7, e1000228.

Gadola, S.D., Koch, M., Marles-Wright, J., et al., 2006. Structure and binding kinetics of three different human CD1d-alpha-galactosylceramide-specific T cell receptors. J. Exp. Med. 203, 699–710.

Galli, G., Pittoni, P., Tonti, E., et al., 2007. Invariant NKT cells sustain specific B cell responses and memory. Proc. Natl. Acad. Sci. U.S.A. 104, 3984–3989.

Gapin, L., 2010. iNKT cell autoreactivity: what is 'self' and how is it recognized? Nat. Rev. Immunol. 10, 272–277.

Gapin, L., Matsuda, J.L., Surh, C.D., et al., 2001. NKT cells derive from double-positive thymocytes that are positively selected by CD1d. Nat. Immunol. 2, 971–978.

Gausling, R., Trollmo, C., and Hafler, D.A., 2001. Decreases in interleukin-4 secretion by invariant CD4(-)CD8(-) V alpha 24J alpha Q T cells in peripheral blood of patients with relapsing-remitting multiple sclerosis. Clin. Immunol. 98, 11–17.

Godfrey, D.I., and Kronenberg, M., 2004. Going both ways: immune regulation via CD1d-dependent NKT cells. J. Clin. Invest. 114, 1379–1388.

Godfrey, D.I., MacDonald, H.R., Kronenberg, M., et al., 2004. NKT cells: what's in a name? Nat. Rev. Immunol. 4, 231–237.

Gonzalez-Aseguinolaza, G., de Oliveira, C., Tomaska, M., et al., 2000. alpha-galactosylceramide-activated Valpha 14 natural killer T cells mediate protection against murine malaria. Proc. Natl. Acad. Sci. U.S.A. 97, 8461–8466.

Good-Jacobson, K.L., and Tarlinton, D.M., 2012. Multiple routes to B-cell memory. Int. Immunol. 24, 403–408.

Grubor-Bauk, B., Arthur, J.L., and Mayrhofer, G., 2008. Importance of NKT cells in resistance to herpes simplex virus, fate of virus-infected neurons, and level of latency in mice. J. Virol. 82, 11073–11083.

Hammond, K.J., Poulton, L.D., Palmisano, L.J., et al., 1998. alpha/beta-T cell receptor (TCR) + CD4-CD8- (NKT) thymocytes prevent insulin-dependent diabetes mellitus in nonobese diabetic (NOD)/Lt mice by the influence of interleukin (IL)-4 and/or IL-10. J. Exp. Med. 187, 1047–1056.

Hong, S., Wilson, M.T., Serizawa, I., et al., 2001. The natural killer T-cell ligand alpha-galactosylceramide prevents autoimmune diabetes in non-obese diabetic mice. Nat. Med 7, 1052–1056.

Igarashi, K., Ochiai, K., Itoh-Nakadai, A., et al., 2014. Orchestration of plasma cell differentiation by Bach2 and its gene regulatory network. Immunol. Rev. 261, 116–125.

Iversen, M.B., Jensen, S.K., Hansen, A.L., et al., 2015. NKT cell activation by local alpha-galactosylceramide administration decreases susceptibility to HSV-2 infection. Immunobiology 220, 762–768.

Jahng, A.W., Maricic, I., Pedersen, B., et al., 2001. Activation of natural killer T cells potentiates or prevents experimental autoimmune encephalomyelitis. J. Exp. Med. 194, 1789–1799.

Jeannin, P., Magistrelli, G., Aubry, J.P., et al., 2000. Soluble CD86 is a costimulatory molecule for human T lymphocytes. Immunity 13, 303–312.

Kakimi, K., Guidotti, L.G., Koezuka, Y., et al., 2000. Natural killer T cell activation inhibits hepatitis B virus replication in vivo. J. Exp. Med. 192, 921–930.

Kawakami, K., Kinjo, Y., Uezu, K., et al., 2001. Monocyte chemoattractant protein-1-dependent increase of V alpha 14 NKT cells in lungs and their roles in Th1 response and host defense in cryptococcal infection. J. Immunol. 167, 6525–6532.

Kawano, T., Cui, J., Koezuka, Y., et al., 1997. CD1d-restricted and TCR-mediated activation of valpha14 NKT cells by glycosylceramides. Science 278, 1626–1629.

Kim, S., Lalani, S., Parekh, V.V., et al., 2008. Glycolipid ligands of invariant natural killer T cells as vaccine adjuvants. Expert Rev. Vaccines 7, 1519–1532.

King, I.L., Fortier, A., Tighe, M., et al., 2012. Invariant natural killer T cells direct B cell responses to cognate lipid antigen in an IL-21-dependent manner. Nat. Immunol. 13, 44–50.

Kinjo, Y., Wu, D., Kim, G., et al., 2005. Recognition of bacterial glycosphingolipids by natural killer T cells. Nature 434, 520–525.

Kobayashi, E., Motoki, K., Uchida, T., et al., 1995. KRN7000, a novel immunomodulator, and its antitumor activities. Oncol. Res. 7, 529–534.

Konieczny, B.T., Dai, Z., Elwood, E.T., et al., 1998. IFN-gamma is critical for long-term allograft survival induced by blocking the CD28 and CD40 ligand T cell costimulation pathways. J. Immunol. 160, 2059–2064.

Kumar, H., Belperron, A., Barthold, S.W., et al., 2000. Cutting edge: CD1d deficiency impairs murine host defense against the spirochete, *Borrelia burgdorferi*. J. Immunol. 165, 4797–4801.

Lang, G.A., Illarionov, P.A., Glatman-Freedman, A., et al., 2005. BCR targeting of biotin-{alpha}-galactosylceramide leads to enhanced presentation on CD1d and requires transport of BCR to CD1d-containing endocytic compartments. Int. Immunol. 17, 899–908.

Lang, G.A., Exley, M.A., and Lang, M.L., 2006. The CD1d-binding glycolipid alpha-galactosylceramide enhances humoral immunity to T-dependent and T-independent antigen in a CD1d-dependent manner. Immunology 119, 116–125.

Lang, G.A., Devera, T.S., and Lang, M.L., 2008. Requirement for CD1d expression by B cells to stimulate NKT cell-enhanced antibody production. Blood 111, 2158–2162.

Lang, G.A., Johnson, A.M., Devera, T.S., et al., 2011. Reduction of CD1d expression in vivo minimally affects NKT-enhanced antibody production but boosts B-cell memory. Int. Immunol. 23, 251–260.

Leadbetter, E.A., Brigl, M., Illarionov, P., et al., 2008. NK T cells provide lipid antigen-specific cognate help for B cells. Proc. Natl. Acad. Sci. U.S.A. 105, 8339–8344.

Legler, D.F., Loetscher, M., Roos, R.S., et al., 1998. B cell-attracting chemokine 1, a human CXC chemokine expressed in lymphoid tissues, selectively attracts B lymphocytes via BLR1/CXCR5. J. Exp. Med. 187, 655–660.

Lisbonne, M., Diem, S., de Castro Keller, A., et al., 2003. Cutting edge: invariant V alpha 14 NKT cells are required for allergen-induced airway inflammation and hyperreactivity in an experimental asthma model. J. Immunol. 171, 1637–1641.

Lotter, H., Gonzalez-Roldan, N., Lindner, B., et al., 2009. Natural killer T cells activated by a lipopeptidophosphoglycan from *Entamoeba histolytica* are critically important to control amebic liver abscess. PLoS Pathog. 5, e1000434.

Matsuda, J.L., Naidenko, O.V., Gapin, L., et al., 2000. Tracking the response of natural killer T cells to a glycolipid antigen using CD1d tetramers. J. Exp. Med. 192, 741–754.

Matsuda, J.L., Mallevaey, T., Scott-Browne, J., et al., 2008. CD1d-restricted iNKT cells, the 'Swiss-Army knife' of the immune system. Curr. Opin. Immunol. 20, 358–368.

Mattner, J., Debord, K.L., Ismail, N., et al., 2005. Exogenous and endogenous glycolipid antigens activate NKT cells during microbial infections. Nature 434, 525–529.

Morita, M., Motoki, K., Akimoto, K., et al., 1995. Structure-activity relationship of alpha-galactosylceramides against B16-bearing mice. J. Med. Chem. 38, 2176–2187.

Morshed, S.R., Mannoor, K., Halder, R.C., et al., 2002. Tissue-specific expansion of NKT and CD5 + B cells at the onset of autoimmune disease in (NZBxNZW)F1 mice. Eur. J. Immunol. 32, 2551–2561.

Motohashi, S., Nagato, K., Kunii, N., et al., 2009. A phase I-II study of alpha-galactosylceramide-pulsed IL-2/GM-CSF-cultured peripheral blood mononuclear cells in patients with advanced and recurrent non-small cell lung cancer. J. Immunol. 182, 2492–2501.

Nagarajan, N.A., and Kronenberg, M., 2007. Invariant NKT cells amplify the innate immune response to lipopolysaccharide. J. Immunol. 178, 2706–2713.

Nair, S., Boddupalli, C.S., Verma, R., et al., 2015. Type II NKT-TFH cells against Gaucher lipids regulate B-cell immunity and inflammation. Blood 125, 1256–1271.

Nieuwenhuis, E.E., Matsumoto, T., Exley, M., et al., 2002. CD1d-dependent macrophage-mediated clearance of *Pseudomonas aeruginosa* from lung. Nat. Med. 8, 588–593.

Nieuwenhuis, E.E., Matsumoto, T., Lindenbergh, D., et al., 2009. Cd1d-dependent regulation of bacterial colonization in the intestine of mice. J. Clin. Invest. 119, 1241–1250.

Olszak, T., An, D., Zeissig, S., et al., 2012. Microbial exposure during early life has persistent effects on natural killer T cell function. Science 336, 489–493.

Parekh, V.V., Wilson, M.T., Olivares-Villagomez, D., et al., 2005. Glycolipid antigen induces long-term natural killer T cell anergy in mice. J. Clin. Invest. 115, 2572–2583.

Porcelli, S., Yockey, C.E., Brenner, M.B., et al., 1993. Analysis of T cell antigen receptor (TCR) expression by human peripheral blood CD4-8- alpha/beta T cells demonstrates preferential use of several V beta genes and an invariant TCR alpha chain. J. Exp. Med. 178, 1–16.

Porcelli, S.A., and Modlin, R.L., 1999. The CD1 system: antigen-presenting molecules for T cell recognition of lipids and glycolipids. Annu. Rev. Immunol. 17, 297–329.

Rampuria, P., Lang, G.A., Devera, T.S., et al., 2016. Coordination between T helper cells, iNKT cells, and their follicular helper sub-sets in the humoral immune response against *Clostridium difficile* toxin B. J. Leukoc. Biol. http://dx.doi.org/10.1189/jlb.4A0616-271R.

Ravetch, J.V., and Bolland, S., 2001. IgG Fc receptors. Annu. Rev. Immunol. 19, 275–290.
Schmieg, J., Yang, G., Franck, R.W., et al., 2003. Superior protection against malaria and melanoma metastases by a C-glycoside analogue of the natural killer T cell ligand alpha-Galactosylceramide. J. Exp. Med. 198, 1631–1641.
Shah, H.B., Joshi, S.K., and Lang, M.L., 2011. CD40L-null NKT cells provide B cell help for specific antibody responses. Vaccine 29, 9132–9136.
Shah, H.B., Devera, T.S., Rampuria, P., et al., 2012. Type II NKT cells facilitate Alum-sensing and humoral immunity. J. Leukoc. Biol. 92, 883–893.
Shah, H.B., Joshi, S.K., Rampuria, P., et al., 2013. BAFF- and APRIL-dependent maintenance of antibody titers after immunization with T-dependent antigen and CD1d-binding ligand. J. Immunol. 191, 1154–1163.
Sharif, S., Arreaza, G.A., Zucker, P., et al., 2001. Activation of natural killer T cells by alpha-galactosylceramide treatment prevents the onset and recurrence of autoimmune Type 1 diabetes. Nat. Med. 7, 1057–1062.
Smyth, M.J., Wallace, M.E., Nutt, S.L., et al., 2005. Sequential activation of NKT cells and NK cells provides effective innate immunotherapy of cancer. J. Exp. Med. 201, 1973–1985.
Stetson, D.B., Mohrs, M., Reinhardt, R.L., et al., 2003. Constitutive cytokine mRNAs mark natural killer (NK) and NK T cells poised for rapid effector function. J. Exp. Med. 198, 1069–1076.
Takeda, K., and Dennert, G., 1993. The development of autoimmunity in C57BL/6 lpr mice correlates with the disappearance of natural killer type 1-positive cells: evidence for their suppressive action on bone marrow stem cell proliferation, B cell immunoglobulin secretion, and autoimmune symptoms. J. Exp. Med. 177, 155–164.
Tonti, E., Galli, G., Malzone, C., et al., 2009. NKT-cell help to B lymphocytes can occur independently of cognate interaction. Blood 113, 370–376.
Tonti, E., Fedeli, M., Napolitano, A., et al., 2012. Follicular helper NKT cells induce limited B cell responses and germinal center formation in the absence of CD4(+) T cell help. J. Immunol. 188, 3217–3222.
Trobonjaca, Z., Leithauser, F., Moller, P., et al., 2001. Activating immunity in the liver. I. Liver dendritic cells (but not hepatocytes) are potent activators of IFN-gamma release by liver NKT cells. J. Immunol. 167, 1413–1422.
van Dommelen, S.L., Tabarias, H.A., Smyth, M.J., et al., 2003. Activation of natural killer (NK) T cells during murine cytomegalovirus infection enhances the antiviral response mediated by NK cells. J. Virol. 77, 1877–1884.
Van Kaer, L., Parekh, V.V., Wu, L., et al., 2011. Invariant natural killer T cells: bridging innate and adaptive immunity. Cell Tissue Res. 343, 43–55.
van der Vliet, H.J., Molling, J.W., von Blomberg, B.M., et al., 2004. The immunoregulatory role of CD1d-restricted natural killer T cells in disease. Clin. Immunol. 112, 8–23.
Venkataswamy, M.M., and Porcelli, S.A., 2010. Lipid and glycolipid antigens of CD1d-restricted natural killer T cells. Semin. Immunol. 22, 68–78.
Wu, L., and Van Kaer, L., 2011. Natural killer T cells in health and disease. Front. Biosci. (Schol. Ed.) 3, 236–251.
Zeissig, S., and Blumberg, R.S., 2014. Commensal microbial regulation of natural killer T cells at the frontiers of the mucosal immune system. FEBS Lett. 588, 4188–4194.
Zeng, Z., Castano, A.R., Segelke, B.W., et al., 1997. Crystal structure of mouse CD1: an MHC-like fold with a large hydrophobic binding groove. Science 277, 339–345.
Zeng, D., Lewis, D., Dejbakhsh-Jones, S., et al., 1999. Bone marrow NK1.1(-) and NK1.1(+) T cells reciprocally regulate acute graft versus host disease. J. Exp. Med. 189, 1073–1081.

CHAPTER

6

Signaling by the Toll-Like Receptors Induces Autophagy Through Modification of Beclin 1: Molecular Mechanism

Neel R. Nabar, Chong-Shan Shi and John H. Kehrl

OUTLINE

Abstract

The innate immune system employs germline-encoded pattern recognition receptors (PRRs) that can detect pathogens and initiate the immune response. The Toll-like receptors (TLRs) are the best studied PRRs and they act by initiating an intracellular signaling cascade that is dependent on the adaptor proteins MyD88 and TRIF. This response activates the critical transcription factors NF-κB and IRF3, and functions to prime

M.A. Hayat (ed): Immunology, Volume 1.
DOI: http://dx.doi.org/10.1016/B978-0-12-809819-6.00006-X

© 2018 Elsevier Inc. All rights reserved.

both the innate and adaptive immune effector cells for pathogen clearance. Macrophages are one of the major effector cell types of innate immunity. They help clear extracellular infections by phagocytosis and eventual phagosome–lysosome fusion, and intracellular infections by enclosing microbes in autophagic vesicles for delivery to the lysosome. Initiation of autophagy requires phosphatidylinositol 3-phosphate [PtdIns(3)P], which is generated by the phosphatidylinositol 3-kinase class III (PtdIns3KC3/Vps34) complex. This chapter discusses the detailed molecular mechanisms through which TLR signaling primes macrophages for intracellular pathogen clearance by initiating autophagy. We focus on Beclin 1, a key component of the PtdIsn3KC3 complex, which is recruited to the TLR signalosome upon TLR stimulation and subsequently posttranslationally modified to disrupt its binding with its inhibitor, Bcl-2.

INTRODUCTION

The innate immune system is an evolutionarily conserved component of host defense that is considered the first line against infection. It plays an important role in initiating and perpetuating the inflammatory response and functions to augment and instruct the adaptive response through inflammatory signals. While pathogen recognition by adaptive immunity is characterized by receptor specificity through clonal selection from a diverse receptor pool generated by gene rearrangement, the innate immune system differs in that the response is less specific and is mediated by germline-encoded pattern recognition receptors (PRRs) (Janeway and Medzhitov, 2002).

PRRs recognize evolutionarily conserved and critical microbial patterns called pathogen-associated molecular patterns (PAMPs), as well as danger signals in the form of self-derived molecules from damaged cells called damage-associated molecular patterns. Mammals have several distinct PRR families based on protein homology that can be classified based on subcellular localization. There are three cytoplasmic receptors, namely the NOD-like receptors, RIG1-like receptors, and AIM2-like receptors, while Toll-like receptors (TLRs) and C-type lectin receptors are membrane bound and found either at the cell surface or on endosomal compartments (Akira et al., 2006; Takeuchi and Akira, 2010). Of the PRRs, the TLRs were the first to be identified and are the best characterized; TLR signaling results in the activation of NF-κB, interferon-regulatory factors (IRFs), and AP-1. The expression of proteins controlled by these transcription factors is essential for orchestrating the elimination of microbial infections.

Macroautophagy (referred to as autophagy herein) is an essential lysosome-dependent degradation pathway that breaks down cellular organelles and protein complexes too large for degradation by the proteasome. In this process, a double membrane vesicle termed the autophagosome envelops cytoplasmic components for delivery to the lysosome. Upon fusion of the outer autophagosome membrane to the lysosome, the vesicle is termed an autophagolysosome and the enclosed components along with the inner autophagosomal membrane are degraded (Mizushima et al., 2008). Autophagy is strongly induced by amino acid starvation in a mammalian-target of rapamycin dependent manner, and thus initially was considered a cellular adaptation to nutrient deprivation and stress. It has now been appreciated that autophagy is involved in diverse cellular processes, including programmed cell death, promotion and control of cancer, neurodegeneration, and the removal of intracellular bacteria, viruses, and protozoa (Levine et al., 2011). Genetic ablation of autophagy genes also increases the susceptibility to intracellular infection by pathogens, linking autophagy to innate immunity. The molecular mechanisms by which pathogen invasion induces autophagy is twofold: (1) nutrient deprivation due to competition by invading bacteria and viruses initiates the same signaling pathways as nutrient deprivation, and (2) PRR signaling directly interacts with the autophagy initiation machinery to induce the process (Levine et al., 2011; Deretic et al., 2013).

This chapter will focus on Beclin 1, which was named due to its initial discovery by Beth Levine's group as an interactor of the antiapoptotic B-cell lymphoma-2 (Bcl-2) protein (Beclin 1 stands for *Bcl*-2-interacting myos*in*-like coil-coiled protein) (Liang et al., 1998). Beclin 1 is a critical component of the autophagy initiation machinery; complementation of Beclin 1 into Beclin deficient cells as well as overexpression of Beclin 1 both initiate autophagy (Liang et al., 1999). Bcl-2 is a cell survival protein best known for its roles in inhibiting apoptosis (via interactions with the proapoptotic Bax and Bak) and promoting oncogenesis; however Bcl-2 is also an important inhibitor of autophagy due to its interaction with Beclin 1 (Pattingre et al., 2005). Using macrophages, the phagocytic effector cells of innate immunity, we will discuss the work that uncovered the molecular mechanisms through which TLR recognition of PAMPs modulates the Beclin 1-Bcl-2 interaction with downstream effects on autophagy (Pattingre et al., 2005; Shi and Kehrl, 2008, 2010).

TLR SIGNALING

TLR Localization, Structure, and PAMP Recognition

The TLR family is made up of 10 distinct members in humans (TLR1–10) and 12 in mice (TLR1–9, TLR11–13), with the mouse TLR10 homologue being a pseudogene due to the insertion of a stop codon. As TLRs are membrane bound, they can be found either at the cell surface or in endocytic compartments. TLR1, TLR2, TLR4, TLR5, TLR6, and TLR10 are located on the plasma membrane and thus primarily recognize microbial membrane components. TLR3, TLR7, TLR8, and TLR9 are found in the lumen of endocytic compartments, the location where uptaken microbial components are directed, where they recognize aberrant (microbial or autoimmune) nucleic acids (Kawasaki and Kawai, 2014).

Structurally, TLRs are type 1 integral transmembrane proteins with an N-terminal ectodomain of leucine-rich repeats (LRRs), a single transmembrane helix, and an intracellular Toll/IL-1 receptor (TIR) domain. Extracellularly, TLRs contain 19–25 LRRs that form a horseshoe shaped structure responsible for pathogen recognition. Upon PAMP binding TLRs typically dimerize with two extracellular domains sandwiching the ligand, causing the extracellular and intracellular domains to move closer together. As a result, the intracellular TIR domain (which is also found in many TLR adaptor proteins) initiates downstream signaling through homotypic interactions (Botos et al., 2011).

Despite all TLRs containing LRRs, the variation in length, number, and N-linked glycosylation of the LRRs change the ligand specificity. TLR2 forms heterodimers with TLR1 or TLR6 in a ligand specific manner to recognize a variety of PAMPs including lipopeptides, peptidoglycans, lipotechoic acid, zymosan, and mannan. TLR4 recognizes LPS, a hallmark component of gram-negative bacterial outer membranes. TLR5 recognizes bacterial flagellin, a protein found in nearly all flagellated bacteria. TLR10 senses the Influenza A virus or can work in conjunction with TLR2 to sense proteins from *Listeria*. Of the endocytic

TLRs, TLR3 recognizes viral dsRNA and siRNAs, including the synthetic polyinosinic-polycytidycyclic acid (poly(I:C)). TLR7 recognizes ssRNA from viruses, whereas human TLR8 recognizes both viral and bacterial RNA. Both also recognize imidazoquinoline derivatives, such as imiquimod and resiquimod (R-848). Finally, TLR9 recognizes unmethylated 2′-deoxyribo(cytidine-phosphate-guanosine) (CpG) motifs present in bacterial and viral, but not mammalian, DNA (Kawasaki and Kawai, 2014).

Engagement of a TLR results in a ligand specific and TIR domain dependent recruitment of one of several adaptor proteins. So far, 5 TIR-containing TLR adaptor proteins have been discovered: myeloid differentiation factor 88 (MyD88), TIR domain-containing adapter-inducing interferon-β (TRIF), MyD88 adapter-like (Mal/TIRAP), Trif-related adaptor molecule (TRAM), and sterile α- and armadillo motif-containing protein (SARM). MyD88 is used by all TLRs except TLR3 to initiate downstream activation of NF-κB and AP-1, and TIRAP is a bridge used by TLR2 and TLR4 to recruit MyD88. TRIF is recruited to TLR3 and TLR4 to activate NF-κB and IRF3, and TRAM is required for TLR4 dependent TRIF recruitment. SARM, whose function was most recently described, negatively regulates TRIF (O'Neill and Bowie, 2007). Together, MyD88 dependent and TRIF dependent signaling constitute the two major arms of the TLR response.

Myd88 Dependent Signaling and TRIF Dependent Signaling

Upon engagement of a MyD88 dependent TLR with its ligand, MyD88 is recruited directly to the receptor (or via TIRAP in the case of TLR2 and TLR4) and results in the formation of a large signaling complex called the "Myddosome." IRAK4 is recruited to this complex and activates IRAK1, resulting in IRAK1 autophosphorylation. IRAK1 then binds TRAF6, a RING-domain E3 ligase capable of K48 (degradative) and K63 (functional) ubiquitination. With the ubiquitin conjugating proteins UBC13 and UEV1a, TRAF6 promotes K63-linked self-ubiquitination as well as the ubiquitination and activation of the TAK1 complex. TAK1, an ubiquitin-dependent kinase of the MAPKKK family, exists in complex with its regulatory subunits TAB1, TAB2, and TAB3 and functions to directly activate two downstream signaling pathways, the NF-κB pathway and MAPKKK signaling (O'Neill and Bowie, 2007).

Basally, NF-κB exists in an inhibited state bound to IκBα. The IKK complex consists of the regulatory subunit NF-κB essential modulator (NEMO) and the catalytic subunits IKKα and IKKβ; it phosphorylates IκBα when activated, resulting in its proteasome mediated degradation and subsequent NF-κB nuclear translocation (Lawrence, 2009). Polyubiquitination by TRAF6 initiates this process by activating TAK1, which binds and activates the IKK complex in an ubiquitin-dependent manner via phosphorylation of IKKβ (Gohda et al., 2004). Simultaneously, TAK1 activates MAPK family members such as ERK1/2, p38, and JNK, culminating in the activation of AP-1.

Whereas the MyD88 dependent response induces proinflammatory cytokines production, the TRIF response also mediates the induction of interferon related genes. After ligand engagement, TLR4 use the bridge protein TRAM to recruit TRIF, which recruits the ubiquitin ligase TRAF3 (TRAF3 is also capable of binding MyD88). In MyD88 dependent signaling, TRAF3 is subject to degradative ubiquitination, whereas in TRIF dependent signaling TRAF3 is K63-linked ubiquitinated resulting in the recruitment of the kinases TBK1 and IKKi. These kinases together phosphorylate IRF3, resulting in the expression of type 1 IFN genes. Meanwhile, TRIF also functions to recruit TRAF6 and initiate NF-κB and MAPK signaling in a RIP1K dependent manner, albeit with different kinetics than MyD88 dependent signaling (Ullah et al., 2016).

Importantly, there is TLR signaling feedback loop mediated by the antiinflammatory protein A20. A20 is an ubiquitin-editing enzyme that has both deubiquitinating activity mediated by its ovarian tumor domain and E3 ubiquitin ligase activity mediated its zinc finger domain. As A20 is an NF-κB target gene, its levels rise rapidly after TLR signaling, at which point it feeds back to inhibit TLR signaling in a cell type dependent manner (Ma and Malynn, 2012). Specifically, A20 mediates this feedback inhibition either through cleavage of K63-linked polyubiquitin chains at the TRAF6-ubiquitin junction, by decreasing TRAF6 E3 ligase activity by preventing interaction with its E2 ligase Ubc13, or by targeting Ubc13 for degradation by the addition of K48-linked chains (Catrysse et al., 2014).

TLRs and Autophagy

Microtubule-associated protein 1-light chain 3 beta (LC3B-I) is an 18 kDa protein that is processed and conjugated to phosphatidylethanolamine upon autophagy induction to LC3B-II, which causes it to run at 16 kDa on an electrophoretic gel. LC3B-II is associated with the autophagosome membrane and is involved in autophagosome elongation. Thus, both visualization of LC3 puncta and an increase in LC3B-II (with the corresponding decrease in LC3B-I) are common markers of autophagy (Tanida et al., 2008). Nakagawa et al. (2004) first appreciated that autophagy was both induced by bacterial infection and protective against it. This group found that Group A *Streptococcus pyogenes* (GAS) infection of HeLa cells resulted in bacterial colocalization with LC3 and LAMP1 (a lysosomal marker) as well as increased conversion of LC3B-I to LC3B-II. In autophagy deficient cells ($Atg5^{-/-}$), the bacteria were better able survive and eventually escape from the cells, indicating that autophagy is protective against bacterial infection (Nakagawa et al., 2004). This was further demonstrated by Gutierezz et al. in an elegant study with *Mycobacterium tuberculosis*, a bacteria that thrives intracellularly in macrophages in part due to its ability to block autophagosome/lysosome fusion. Restoring autophagy in *M. tuberculosis* infected macrophages resulted in the suppression of bacterial survival (Gutierrez et al., 2004).

These studies led our group to hypothesize that TLR signaling may induce autophagy. Using the mouse murine macrophages line RAW264.1 (RAW) stably transfected with GFP-LC3, we found that TLR1, TLR3, TLR4, TLR5, and TLR7 stimulation all induced autophagosome formation upon examination with confocal microscopy (Shi and Kehrl, 2008). A similar study published around the same time also showed TLR induction of autophagy and its importance in the immune response (Delgado et al., 2008). We validated our results by looking at the conversation of endogenous LC3B-I to LC3B-II by western blot and by direct visualization of autophagosome double membrane formation by electron microscopy. As both TLR1 (MyD88 dependent) and TLR3 (TRIF dependent) signaling induced autophagosome formation—we went on to show that expression of either MyD88, TRIF, or TRAM could drive autophagosome formation. Knockdown of MyD88 or TRIF by shRNA and expression of MyD88 and TRIF dominant negative constructs both abrogated the induction of autophagy, demonstrating that TLR signaling induces autophagy in a MyD88 or TRIF dependent manner. These initial findings laid the groundwork for our understanding of TLR signaling regulation of Beclin 1, the protein through which TLR induction of autophagy occurs (Shi and Kehrl, 2008).

LINKING AUTOPHAGY TO TLRs: ROLE OF BECLIN 1

Role of Beclin 1 in the Initiation Complex

Mammalian autophagy is typically divided into six steps: initiation, nucleation, elongation, closure, maturation, and degradation. The first step, initiation of the autophagosome, requires generation of phosphatidylinositol 3-phosphate [PtdIns(3)P] by the activated phosphatidylinositol 3-kinase class III (PtdIns3KC3/Vps34) complex (Jaber et al., 2012). The coiled-coil protein Beclin 1 is a critical component of this complex and promotes its activity (and the initiation of autophagy) unless inhibited by protein–protein interactions. Structurally, Beclin 1 contains a BH3 domain, a coil-coiled domain (CCD), a nuclear export signal, and an evolutionarily conserved domain (ECD). These domains are critical for Beclin 1 protein–protein interactions, as Beclin 1 associates with PtdIns3KC3 via its ECD domain and with Bcl-2 via its BH3 domain. The BH3 domain and CCD domains also allow for Beclin 1 oligomerization, which is necessary for the formation of Beclin 1 containing protein complexes that stimulate PtdIns3KC3 activity and therefore autophagy (Kang et al., 2011).

Beclin 1 forms two known stable complexes, one of which (complex 1) is required for canonical autophagosome formation. Complex 1 consists of Beclin 1, PtdIns3KC3, Vps15, and Atg14L (also known as Barkor). Atg14L directs this complex to the phagophore assembly site, which allows the recruitment of other critical autophagy proteins and is required for autophagosome formation (Kang et al., 2011). It is widely held that the antiapoptotic protein Bcl-2 binds Beclin 1 and inhibits its ability to induce autophagy. Beth Levine's group showed the functional significance of this interaction in 2005, as overexpression of either wildtype or viral Bcl-2 inhibited Beclin 1 dependent autophagy by decreasing the interaction between Beclin 1 and PtdIsn3KC3. Moreover, expression of Beclin 1 mutants (L166A in the BH3 domain) that are unable to bind Bcl-2 promoted autophagy to a greater extent than WT Beclin 1. They then went on to show that the interaction between endogenous Beclin 1 and Bcl-2 is decreased upon starvation (a potent trigger for autophagy), indicating that the regulation of the interaction between Beclin 1 and Bcl-2 may be a physiological mechanism responsible for triggering autophagy (Pattingre et al., 2005). These observations laid the foundation for our group's investigation of whether TLR signaling also induced autophagy by regulating the Beclin 1/Bcl-2 interaction (Shi and Kehrl, 2008).

TLR Signaling Regulates the Beclin 1/Bcl-2 Interaction

TLR3 and TLR4 signaling not only induce autophagy in a MyD88 and TRIF dependent manner, but also recruit Beclin 1 to the TLR signalasome. To show this, we treated intact RAW cells with an agonistic TLR4 or control antibody followed by immunoprecipitation. Immunoblotting for associated proteins showed that Beclin 1 is recruited to the TLR4 signaling complex in a time dependent manner. This stimulation causes modification of Beclin 1, as its mobility on SDS-PAGE initially shifted slightly upwards (15 min) followed later by a dramatic upward shift. As expected, MyD88 and TRIF were recruited to the TLR complex, confirming initiation of TLR4 signaling (Shi and Kehrl, 2008).

The TLR signalasome consists of many proteins, so we performed coimmunoprecipitation studies to show that MyD88 and TRIF directly interact with Beclin 1 and that Beclin 1's recruitment was not simply mediated by other signalasome components. Coimmunoprecipitation experiments with overexpressed MyD88 and TRIF showed direct interaction with Beclin 1. MyD88 contains a death domain (DD) and a TIR domain (required for binding with TLRs); we mapped the Myd88/Beclin 1 interaction to the DD, which was consistent with our observation that expression of MyD88 DD initiated autophagy and that death domain-deficient MyD88 expression did not. Finally, to assay the functional importance of these interactions, we showed that Poly(I:C) and LPS (TLR3 and TLR4 ligands, respectively) increased Beclin 1 association with TRIF (TLR3 + TLR4) and MyD88 (TLR4 only). Taken together, these results indicated that following LPS stimulation, the TLR4 signalasome recruits Beclin 1 in an MyD88 DD dependent manner, and that TRIF recruits Beclin 1 after Poly(I:C) stimulation (Shi and Kehrl, 2008).

Given the physiological significance of the Bcl-2-Beclin 1 interaction in starvation-induced autophagy (Pattingre et al., 2005), we hypothesized that TLR signaling would decrease the Bcl-2/Beclin 1 interaction. Bcl-2 immunoprecipitates from LPS and Poly(I:C) treated RAW cells showed interaction of Bcl-2 with MyD88, TRIF, and Beclin 1, but that both treatment conditions drastically reduced the Bcl-2/Beclin 1 interaction compared to the control. This was supported by the observation that overexpression of MyD88 or MyD88 DD decreased the Bcl-2/Beclin 1 interaction, whereas overexpression of the MyD88 DD deficient mutant did not. Furthermore, overexpression of Bcl-2 decreased the Beclin 1 interaction with MyD88 and TRIF with a concurrent reduction in TLR induced autophagy. These experiments led us to the conclusion that TLR signaling, via MyD88 (DD dependent) and TRIF, target Beclin 1 in a manner competitive to the binding of the inhibitory Bcl-2 with opposing functional implications on downstream autophagy (Shi and Kehrl, 2008).

TRAF6 Directly Catalyzes K63-Linked Ubiquitination of Beclin 1 at Lys117 After TLR Signaling

As Beclin 1 immunoblot bands showed prominent posttranslational modification in TLR4 immunoprecipitates following TLR4 signaling, we set out to determine the specific modification. Immunoblotting with an antibody recognizing K63-linked ubiquitinated sites of Beclin 1 in Beclin 1 immunoprecipitates following LPS stimulation of RAW cells showed smeared patterns consistent with ubiquitination. TRAF6 is a key ubiquitin E3 ligase in the TLR pathway capable of K63-linked ubiquitination (Tseng et al., 2010), and coimmunoprecipitation studies confirmed a direct TRAF6/Beclin 1 interaction (Shi and Kehrl, 2010). To corroborate the importance of TRAF6 in Beclin 1 ubiquitination, we looked at IL-1R mediated signaling which also employs TRAF6 (Windheim et al., 2008) and confirmed that it too induces Beclin 1 ubiquitination. To test more directly if TRAF6 is involved in Beclin 1 ubiquitination, we expressed MyD88 and TRIF (which recruit TRAF6) as well as TRAF6 itself in RAW cells; these three stimuli all led to Beclin 1 ubiquitination even in the absence of LPS. Upon LPS stimulation after reducing TRAF6 expression by expressing a siRNA, the TRAF6 deficient cells had drastically less Beclin 1 ubiquitination. This, along with the observation that Bcl-2 overexpression decreased Beclin 1 ubiquitination in the presence of LPS, indicates that MyD88/TRIF and TRAF6 act upstream of Beclin 1 ubiquitination in an opposing manner to Bcl-2 (Shi and Kehrl, 2010).

We then found that TRAF6 directly interacts with and ubiquitinates Beclin 1. Beclin 1 has two sites (residues 54–58 and 297–301) that are canonical TRAF6 binding sites (P-X-E-X-X-aromatic/acidic). When the conserved glutamate residue at either of these sites on Beclin 1 is mutated to an alanine, both the Beclin 1/TRAF6 interaction and TRAF6 dependent Beclin 1 ubiquitination were abrogated. Finally, in vitro reconstitution using purified TRAF6, purified Beclin 1, and the essential cofactors, demonstrated that TRAF6 itself catalyzes Beclin 1 ubiquitination. In the BH3 domain of Beclin 1, there is a conserved lysine residue at amino acid 117 (right next to leucine 116, which is critical for Beclin 1's interaction with Bcl-2). Lysine-117 is the exact site of Beclin 1 ubiquitination, as modification of L117 to an arginine decreases Beclin 1 ubiquitination in LPS stimulated cells as well as the TRAF6/Beclin 1 in vitro reconstitution. This clearly illustrates that TLR signaling leads to Beclin 1 K63-linked ubiquitination by TRAF6. Experiments in cultured human monocytes showed this phenomenon is preserved in human cells (Shi and Kehrl, 2010).

A20 regulates TRAF6 enzymatic activity by direct deubiquitination (thereby reducing activation) and by K48-linked ubiquitination/degradation of essential TRAF6 coenzymes (Ma and Malynn, 2012; Catrysse et al., 2014). As such, knockdown of A20 increased Beclin 1 ubiquitination and expression of A20 decreased Beclin 1 ubiquitination as expected. However, experiments utilizing the A20 C103A mutant (which lacks deubiquitinase activity) and the A20 C624A/C627A (which lacks E3 ligase activity) showed unexpected results; the A20 624A/627A mutant had the same effect as wildtype A20 whereas the C103A mutant had no effect. As A20's regulatory effect on Beclin 1 ubiquitination was deubiquitinase activity dependent, we used in vitro studies using purified A20 (1-371, lacking ubiquitinase activity) in the absence of TRAF6 to show that A20 directly targets and deubiquitinates Beclin 1. In sum, A20 plays a regulatory role in the activation of autophagy by opposing the activity of TRAF6—both by limiting TRAF6 E3 ligase activity and by directly deubiquitinating Beclin 1 (Shi and Kehrl, 2010).

Beclin 1 Ubiquitination Amplifies TLR Induced Autophagy

Polyubiquitination, while originally simply considered a tag for proteasomal degradation, is an important posttranslational modification that influences the scaffolding ability of proteins (Nathan et al., 2013). This is because some proteins that undergo ubiquitination can also bind ubiquitin (11 families of ubiquitin binding domains have been identified). After the initial covalent attachment of ubiquitin to a substrate, additional ubiquitins can attach via one of 7 internal lysine residues (K6, K11, K27, K29, K33, K48, and K63) or by attachment of its carboxy-terminal carboxyl group to the amino-terminal amino group on the methionine of the substrate-linked ubiquitin (linear ubiquitination). Some ubiquitin binding proteins show specificity for the types of ubiquitin chains they link. The UBAN (ubiquitin binding in ABIN and NEMO) motif of NEMO specifically targets linear ubiquitin chains, whereas the TAK1 complex binds K63-linked chains based on recognition by the adaptor protein TAB2 (Komander and Rape, 2012). Beclin 1 shows specificity for K63-linked chains, and the K63-linked ubiquitination of Beclin 1 promotes the oligomerization and lipid kinase activity of PtdIns3KC3, as LPS stimulation of cell expressing K117R mutants of Beclin 1 neither have robust Beclin 1-Beclin 1 interactions nor do they produce as much PtdIns(3P) (Shi and Kehrl, 2010). Consistent with this, when K63-linked ubiquitination is inhibited either by addition of K63R ubiquitin chains (which cannot be polyubiquitinated) or by using Beclin 1 mutants not able to bind TRAF6, autophagosome formation is drastically decreased as assayed by LC3-GFP puncta and LC3-IB to LC3-IIB conversion.

CONCLUSION AND PERSPECTIVE

The involvement of Beclin 1 and Bcl-2 in TLR initiation of autophagy underscores the intimate and complex relationship between pathogen invasion, autophagy, and cell death. It is still unclear as to whether autophagy is protective against cell death or a cause of it (e.g., autophagic cell death). Since the initial study reporting the functional inhibition of Beclin 1 by Bcl-2 (Pattingre et al., 2005), a 2014 report from the Vaux group suggested that Bcl-2's effect on autophagy is mediated by its effect on the proapoptotic proteins Bax and Bak (Lindqvist et al., 2014). In these studies, there were no changes in autophagy initiation after Bcl-2 expression in $Bax^{-/-}$ $Bak^{-/-}$ cells. This group goes on to argue that many of the experiments that delineated the functional significance of Beclin 1 and Bcl-2 were either done using overexpression models or confounded by the initiation of apoptosis, which deletion of Bax and Bak inhibits. Regardless, the strength and quality of the genetic and biochemical data regarding the functional significance of the Beclin 1 and the Bcl-2 family members is undeniable (Pattingre et al., 2005; Maiuri et al., 2007), and more work will need to be done to understand the complex relationship between autophagy and cell death. In summary, autophagy is an evolutionary conserved process important in the clearance of intracellular pathogens. Innate immunity, which is considered the first response to pathogen invasion, functions to rapidly and nonspecifically detect pathogens to prime effector cells for pathogen clearance. Therefore, the initiation of autophagic processes by innate immune receptors is consistent with the notion that innate immunity can prime effector functions of immune cells.

Acknowledgments

This research was supported by the Intramural Research Program of NIAID, NIH. We thank Dr. Anthony Fauci for his continued support.

References

Akira, S., Uematsu, S., and Takeuchi, O., 2006. Pathogen recognition and innate immunity. Cell 124, 783–801.

Botos, I., Segal, M., and Davies, R., 2011. The structural biology of Toll-like receptors. Structure 19, 447–459.

Catrysse, L., Vereecke, L., Beyaert, R., et al., 2014. A20 in inflammation and autoimmunity. Trends Immunol. 35, 22–31.

Delgado, M.A., Elmaoued, R.A., Davis, A.S., et al., 2008. Toll-like receptors control autophagy. EMBO J. 27, 1110–1121.

Deretic, V., Saitoh, T., and Akira, S., 2013. Autophagy in infection, inflammation and immunity. Nat. Rev. Immunol. 13, 722–737.

Gohda, J., Matsumura, T., and Inoue, J., 2004. Cutting edge: TNFR-associated factor (TRAF) 6 is essential for MyD88-dependent pathway but not toll/IL-1 receptor domain-containing adaptor-inducing IFN-beta (TRIF)-dependent pathway in TLR signaling. J. Immunol. 173, 2913–2917.

Gutierrez, M.G., Master, S.S., Singh, S.B., et al., 2004. Autophagy is a defense mechanism inhibiting BCG and *Mycobacterium tuberculosis* survival in infected macrophages. Cell 119, 753–766.

Jaber, N., Dou, Z., Chen, J.-S., et al., 2012. Class III PI3K Vps34 plays an essential role in autophagy and in heart and liver function. Proc. Nat. Acad. Sci. 109, 2003–2008.

Janeway Jr., C.A., and Medzhitov, R., 2002. Innate immune recognition. Ann. Rev. Immunol. 20, 197–216.

Kang, R., Zeh, H.J., Lotze, M.T., et al., 2011. The Beclin 1 network regulates autophagy and apoptosis. Cell Death Differ. 18, 571–580.

Kawasaki, T., and Kawai, T., 2014. Toll-like receptor signaling pathways. Front. Immunol. 5, 461.

Komander, D., and Rape, M., 2012. The ubiquitin code. Ann. Rev. Biochem. 81, 203–229.

Lawrence, T., 2009. The nuclear factor NF-κB pathway in inflammation. Cold Spring Harbor Persp. Biol. 1, a001651.

Levine, B., Mizushima, N., and Virgin, H.W., 2011. Autophagy in immunity and inflammation. Nature 469, 323–335.

Liang, X.H., Kleeman, L.K., Jiang, H.H., et al., 1998. Protection against fatal Sindbis virus encephalitis by beclin, a novel Bcl-2-interacting protein. J. Virol. 72, 8586–8596.

Liang, X.H., Jackson, S., Seaman, M., et al., 1999. Induction of autophagy and inhibition of tumorigenesis by beclin 1. Nature 402, 672–676.

Lindqvist, L.M., Heinlein, M., Huang, D.C., et al., 2014. Prosurvival Bcl-2 family members affect autophagy only indirectly, by inhibiting Bax and Bak. Proc. Natl. Acad. Sci. U.S.A. 111, 8512–8517.

Ma, A., and Malynn, B.A., 2012. A20: linking a complex regulator of ubiquitylation to immunity and human disease. Nat. Rev. Immunol. 12, 774–785.

Maiuri, M.C., Criollo, A., Tasdemir, E., et al., 2007. BH3-Only proteins and BH3 mimetics induce autophagy by competitively disrupting the interaction between Beclin 1 and Bcl-2/Bcl-XL. Autophagy 3, 374–376.

Mizushima, N., Levine, B., Cuervo, A.M., et al., 2008. Autophagy fights disease through cellular self-digestion. Nature 451, 1069–1075.

Nakagawa, I., Amano, A., Mizushima, N., et al., 2004. Autophagy defends cells against invading group A *Streptococcus*. Science 306, 1037–1040.

Nathan, J.A., Kim, H.T., Ting, L., et al., 2013. Why do cellular proteins linked to K63-polyubiquitin chains not associate with proteasomes? EMBO J. 32, 552–565.

O'Neill, L.A.J., and Bowie, A.G., 2007. The family of five: TIR-domain-containing adaptors in Toll-like receptor signalling. Nat. Rev. Immunol. 7, 353–364.

Pattingre, S., Tassa, A., Qu, X., et al., 2005. Bcl-2 antiapoptotic proteins inhibit Beclin 1-dependent autophagy. Cell 122, 927–939.

Shi, C.S., and Kehrl, J.H., 2008. MyD88 and Trif target beclin 1 to trigger autophagy in macrophages. J. Biol. Chem. 283, 33175–33182.

Shi, C.S., and Kehrl, J.H., 2010. TRAF6 and A20 regulate lysine 63-linked ubiquitination of Beclin-1 to control TLR4-induced autophagy. Sci. Signal. 3 ra42.

Takeuchi, O., and Akira, S., 2010. Pattern recognition receptors and inflammation. Cell 140, 805–820.

Tanida, I., Ueno, T., and Kominami, E., 2008. LC3 and autophagy. Methods Mol. Biol. 445, 77–88.

Tseng, P.-H., Matsuzawa, A., Zhang, W., et al., 2010. Different modes of ubiquitination of the adaptor TRAF3 selectively activate the expression of type I interferons and proinflammatory cytokines. Nat. Immunol. 11, 70–75.

Ullah, M.O., Sweet, M.J., Mansell, A., et al., 2016. TRIF-dependent TLR signaling, its functions in host defense and inflammation, and its potential as a therapeutic target. J. Leukoc. Biol. 100, 27–45.

Windheim, M., Stafford, M., Peggie, M., et al., 2008. Interleukin-1 (IL-1) induces the Lys63-linked polyubiquitination of IL-1 receptor-associated kinase 1 to facilitate NEMO binding and the activation of IkappaBalpha kinase. Mol. Cell. Biol. 28, 1783–1791.

CHAPTER

7

Molecular Mechanisms Underlying the Action of Ipilimumab Against Metastatic Melanoma

Lucia Festino, Vito Vanella, Martina Strudel and Paolo A. Ascierto

OUTLINE

Abstract

Cytotoxic T lymphocyte-associated antigen (CTLA-4) is a coinhibitory transmembrane protein which binds to the B-71 and B7-2 ligands on antigen presenting cells (APCs) with an affinity approximately 20-fold higher than that of CD28. While CD28 has a costimulatory function and helps facilitate and maintain a T cell response, CTLA-4 plays a role in blocking T cell activation. Understanding the precise mechanism of CTLA-4 activity in vivo, and by extension, the mechanism of antitumor immune activity mediated by CTLA-4 blockade, is an area of active investigation, and two hypotheses to explain CTLA-4 activity—distinct but not mutually exclusive—involve intrinsic and extrinsic cellular mechanisms. Several models have been proposed to explain the intrinsic inhibitory function of CTLA-4, including recruitment of inhibitory proteins to T cell synapsis, nonligand binding CTLA-4 splice variant, ligand competition to prevent

M.A. Hayat (ed): Immunology, Volume 1.
DOI: http://dx.doi.org/10.1016/B978-0-12-809819-6.00007-1

© 2018 Elsevier Inc. All rights reserved.

CD28 signaling, inhibition of lipid-raft and microcluster formation and inhibition of the T cell stop signal. Proposed models for extrinsic mechanisms include transendocytosis of CD80/CD86, reverse signaling through CD80 and CD86 on APCs, modulation of inhibitory cytokine production, blocking CD80 capture and the cell-mediated cytotoxicity of regulatory T cells. The identification of CTLA-4 has led to the clinical development of CTLA-4 blocking antibodies that are able to stimulate potent antitumor immunity. Clinical studies have confirmed that the CTLA-4 inhibitor ipilimumab can induce durable responses and improve overall survival in patients with advanced melanoma.

CTLA-4

Cytotoxic T lymphocyte-associated antigen (CTLA-4), also known as CD152, is a coinhibitory molecule involved in regulating T cell activation. The normal function of the immune system involves several complex and balanced interactions, involving both stimulatory and inhibitory components, which helps protect the host from exogenous pathogens and, at same time, prevent autoimmunity. T cell activation is regulated by stimulatory and inhibitory signals (Sharpe, 2009). The first activating signal is provided by engagement of the T cell receptor (TCR) with a cognate peptide-bound major histocompatibility complex. A second costimulatory signal is provided by engagement of a coreceptor. B-71 (CD80) and B7-2 (CD86) are the major costimulatory ligands on antigen presenting cells (APCs). The canonical coreceptor, CD28, binds to members of the B7 family present on APCs. APCs and especially dendritic cells always express moderate levels of B7-2; these levels increase after their activation, enhancing their function. Sufficient levels of CD28:B7-1/2 binding lead to proliferation of T cells, increased T cell survival, and differentiation through the production of growth cytokines such as interleukin-2 (IL-2), increased energy metabolism, and upregulation of cell survival genes. Brunet et al. (1987) described CTLA-4, a 223-amino acid protein belonging to the immunoglobulin superfamily that is mainly expressed in activated lymphocytes and coinduced with T cell-mediated cytotoxicity. The human homolog of the gene was subsequently cloned in 1988 (Brunet et al., 1988). Studies with soluble fusion protein indicate that CTLA-4 binds both B7 family members with an affinity ~20 fold higher than that of CD28 (Chambers et al., 2001; Collins et al., 2002). While CD28 and CTLA-4 share the same ligand-binding motifs, CD28 binds its ligands monovalently and CTLA-4 binds bivalently. Multiple groups have demonstrated that CTLA-4 expression can prevent CD28 ligand binding, possibly through its recruitment to the immunological synapse. CD28 facilitates and maintains a T cell response while CTLA-4, triggering a negative signal, blocks T cells activation (Krummel and Allison, 1995).

One question is how T cells are activated given that the affinity of CTLA-4 is much higher than that of CD28. Unlike CD28, which is constitutively expressed in most populations of T cells at rest and following activation, CTLA-4 is not expressed on the surface of resting T cells, but is maintained in the endosome compartment. Only upon antigen stimulation is CTLA-4 surface expression induced, through both de novo gene expression and cycling of existing CTLA-4 protein to the cell surface. When CTLA-4 engages its ligands, transductional cascades are triggered leading to inactivation of T cells (Teft et al., 2006). Understanding the precise mechanism of CTLA-4 activity in vivo, and by extension, the mechanism of antitumor immune activity mediated by CTLA-4 blockade, is an area of active investigation, and will help to better understand the mechanism of action of anti-CTLA-4 antibodies. Two hypotheses to explain CTLA-4 activity, distinct but not mutually exclusive, have gained experimental support, these being cell intrinsic and cell-extrinsic mechanisms.

Intrinsic Mechanisms

Both clinical and preclinical data suggest that CTLA-4 blockade results in direct activation of $CD4^+$ and $CD8^+$ effector cells. The CTLA-4 intrinsic inhibitory function consists of a direct inhibitory effect on CTLA-4-expressing cells, which can be reversed by the use of inhibitory anti-CTLA-4 antibodies (Fig. 7.1). Several models have been proposed to explain the intrinsic inhibitory function of CTLA-4.

Recruitment of Inhibitory Proteins to T Cell Synapsis

In this model, CTLA-4 generates inhibitory signals that interfere with proximal signaling by either TCR or CD28 (Fig. 7.1A). CTLA-4 can recruit and activate several phosphatases leading to dephosphorylation of key proteins involved in TCR and CD28 signaling. For example, it has been reported that the cytoplasmic domain of CTLA-4 interacts and recruits the tyrosine phosphatase SHP-2 and serine/threonine phosphatase PP2A. A low level of protein phosphorylation in the early steps of the TCR pathway could result in a reduced response to its stimulation. Furthermore, mutations of the CTLA-4 domain that alter the binding capacity with these phosphatases affect the ability of CTLA-4 to inhibit T cell activation (Baroja et al., 2002).

Nonligand Binding CTLA-4 Splice Variant

Another model emerged from data on a splice variant of CTLA-4 that lacks the region encoded by exon 2; this variant is a membrane-anchored cytoplasmic molecule without the extracellular ligand-binding domain which can generate a constitutive inhibitory signal (Fig. 7.1B). The cytoplasmic domain of CTLA-4 (ctCTLA-4) has been found to be 100% conserved among different species, due to its criticality for CTLA-4 function (Ling et al., 1999). What is clear is that this domain has the capacity to inhibit T cell activation in vitro and in vivo. Lower expression of this variant is associated with increased incidence of autoimmune diabetes in mice and its inhibition can prevent arthritic symptoms while overexpression results in inhibition of T cell responses. However, activated T cells continue to proliferate though expressing substantial levels of CTLA-4 cytoplasmic domain. Thus, the natural expression of this CTLA-4 ligand does not appear to be able to inhibit T cell function alone.

Ligand Competition Prevents CD28 Signal

In the third model, CTLA-4 does not signal itself but rather prevents CD28 from generating positive signals by acting as a high-affinity competitor for ligands. CD80 and CD86 bind to CTLA-4 with 50–100-fold greater avidity than to CD28 (Fig. 7.1C). This finding led to the idea that CTLA-4 acts by outcompeting CD28 for its ligand. In support of this model, overexpression of a truncated form of CTLA-4 that lacks a cytoplasmic domain and is therefore unable to induce intracellular signals, can partially delay the onset of disease (Masteller et al., 2000).

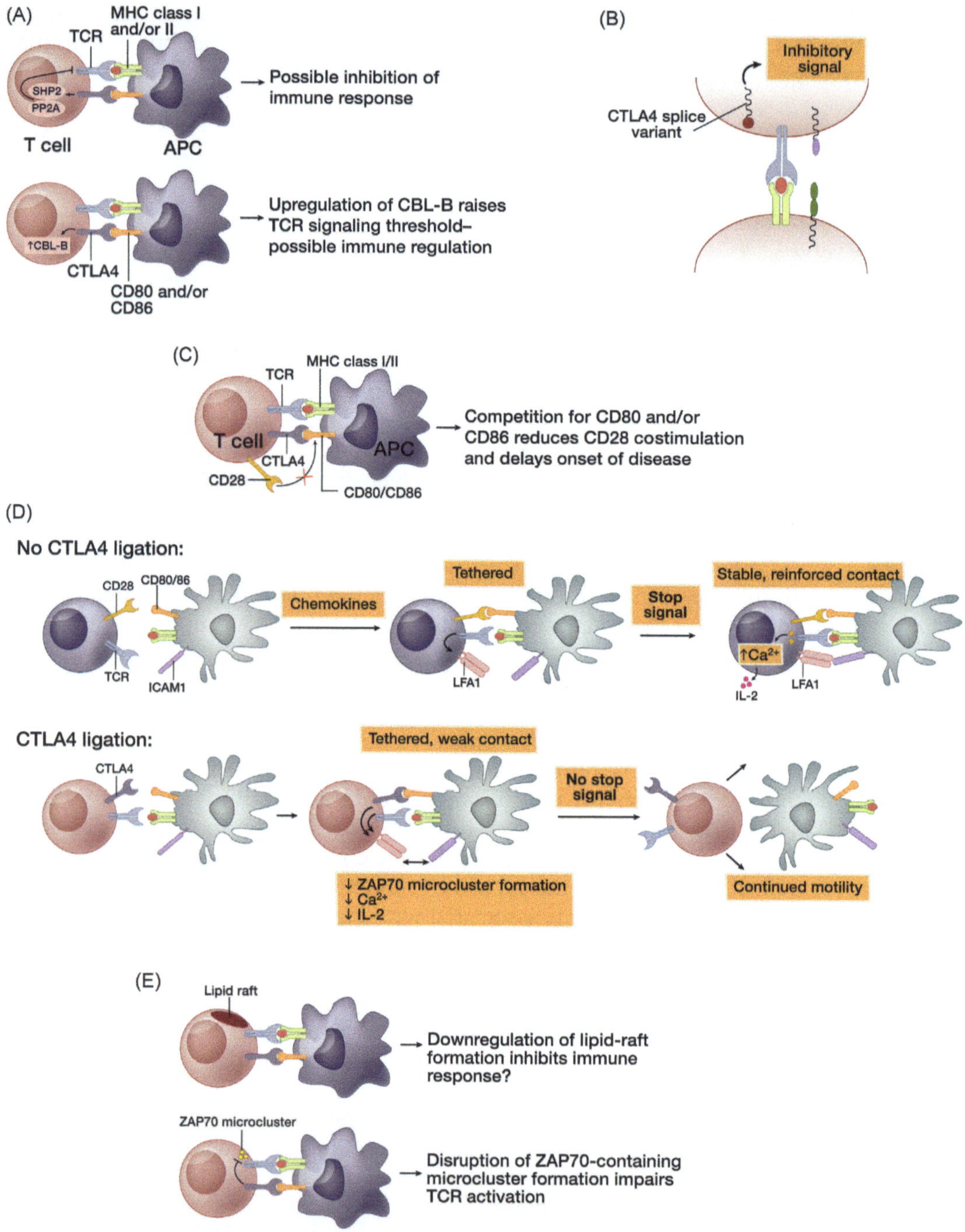

FIGURE 7.1 Mechanism of action of ipilimumab: intrinsic mechanism: (A) recruitment of inhibitory proteins to T cell synapsis; (B) nonligand binding CTLA-4 splice variant; (C) ligand competition prevents CD28 signal; (D) inhibition of T cell stop signal; (E) inhibition of lipid-raft and microcluster formation.

Inhibition of Lipid-Raft and Microcluster Formation

Another model is based on the opposing effects of CD28 and CTLA-4 on the expression of lipid rafts. This model proposes that these receptors differentially regulate T-cell activation by increasing or decreasing the formation of lipid rafts and modulating the availability of key components in these lipid rafts that are useful for TCR signaling (Fig. 7.1E). TCR ligation induces the formation of lipid rafts and microclusters acting as platforms for signaling in T cells. These lipid rafts are enriched in cholesterol and signaling proteins. CTLA-4 has been reported to block lipid-raft formation on the surface of naïve T cells (Martin et al., 2001). In contrast, CD28 potentiates lipid-raft formation following TCR ligation. A similar model is based on microcluster inhibition. Microclusters of the TCR complex are composed of the kinase ζ-chain-associated protein kinase of 70 kDa (ZAP70) and adaptors, such as LAT and SH2-domain-containing leukocyte protein of 76 kDa (SLP76), and are the sites of tyrosine phosphorylation (Yokosuka et al., 2005). Their formation occurs within a few seconds following TCR ligation. It has been demonstrated that coligation of TCR and CTLA-4 can effectively block the formation of ZAP70 containing microclusters in T cells (Schneider et al., 2007). Impaired microcluster formation could explain the reduced phosphorylation of TCR substrates that has been observed without involving PTPases.

Inhibition of T Cell Stop Signal

TCR ligation leads to the reduction of T-cell motility, an event termed the "stop signal." Without the stop signal, T cells continue to move and are unable to form a stable interface with the APC and thereby an effective immunological synapse (Fig. 7.1D). It has been shown that CTLA-4 ligation can effectively reverse or override the TCR-induced stop signal, such that T cells continue to move as if they never encountered an antigen (Schneider et al., 2006; Rudd, 2008). This function of CTLA-4 might be mediated by a completely different mechanism based on an ability to engage positive signaling pathways. In view of these results, the model of CTLA-4 as a negative signaling coreceptor may be in part correct since it is possible that CTLA-4 can function by engaging a positive event that produces a negative outcome in terms of T cell activation.

Extrinsic Mechanisms

Beyond direct inhibition of T cell activation, several other extrinsic mechanisms have been proposed to explain the role of CTLA-4 in regulating the immune system.

Transendocytosis of CD80/CD86

In an important study, Qureshi et al. (2011) demonstrated that CTLA-4 facilitates the transendocytosis of CD80/CD86 from APCs (Fig. 7.2A). In both human and murine T cells, CTLA-4 selectively removed its ligands from the surface of adjacent cells which were then targeted for degradation in endocytic vesicles inside CTLA-4-expressing cells. Based on this mechanism, blockade of CTLA-4 with monoclonal antibodies preserved the expression of CD80/CD86 on APCs and enhanced costimulatory function of T cells.

Reverse Signaling Through CD80 and CD86 on APCs

Another potential explanation for the ability of CTLA-4 to function in a cell-extrinsic manner is that it can bind to its ligands on APCs and initiate a "reverse signal" (Fig. 7.2B).

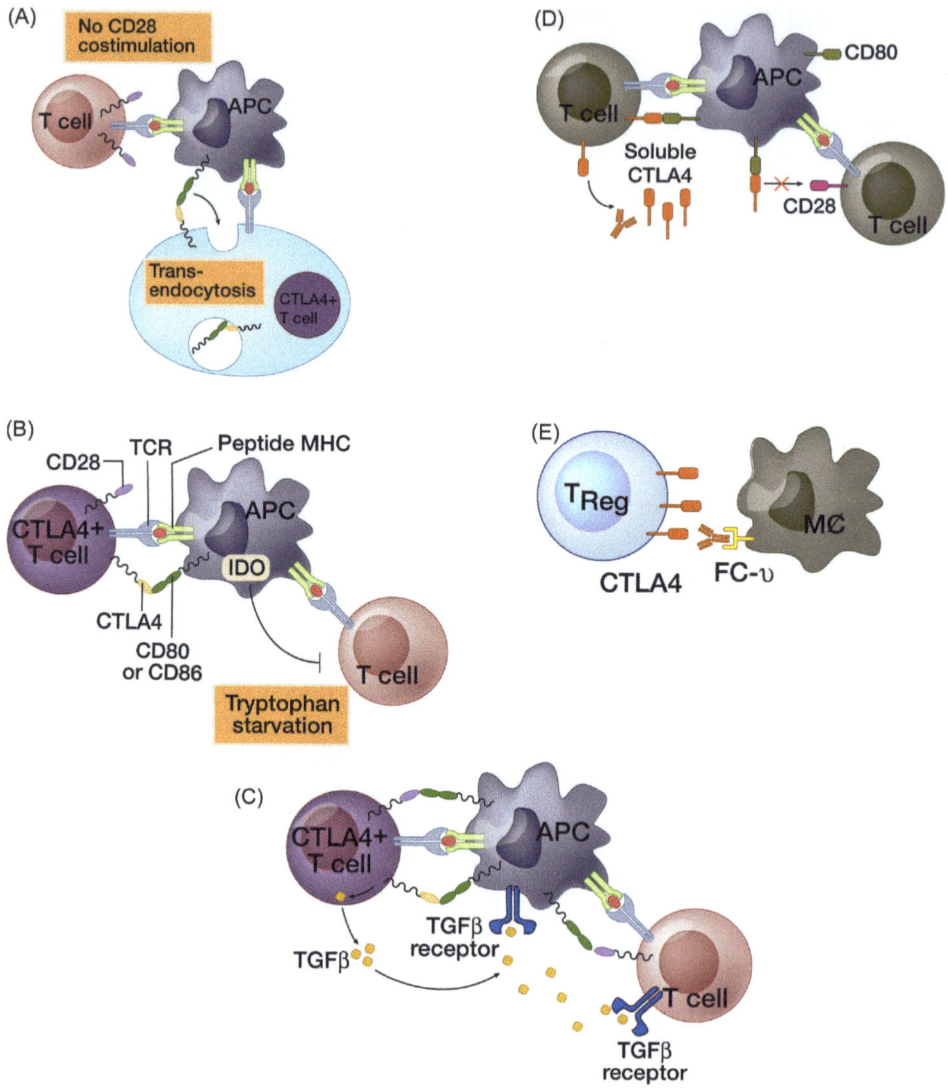

FIGURE 7.2 Mechanism of action of ipilimumab: extrinsic mechanism: (A) transendocytosis of CD80/CD86; (B) reverse signaling through CD80 and CD86 on APCs; (C) modulation of inhibitory cytokine production; (D) blocking CD80 capture; (E) Treg depletion by ADCC.

CTLA-4 has been demonstrated to activate indoleamine 2,3-dioxygenase (IDO) activity in APCs that express CD80 and CD86 (Fallarino et al., 2003). IDO catabolizes the amino acid tryptophan, which is essential for T cell proliferation, leading to localized tryptophan depletion and the consequent inhibition of T cell proliferation. T cells activated in tryptophan-deficient conditions are able to synthesize protein, and to start the cell cycle, but the progression of the cell cycle is stopped in the G1 phase. T cells blocked in G1 do not resume the cell cycle after exposure to tryptophan, but need a second activation of the T receptor in the presence of tryptophan. These data have shown a new mechanism of regulatory T cells by APCs (Munn et al., 1999). Anti-CTLA-4 antibody, blocking this reverse signal, can inhibit the activity of IDO, favoring T cell activation. How CTLA-4 induces IDO activation is unclear. One study demonstrated that the ability of CTLA-4-Ig to activate IDO is mediated by interferon-γ (IFNγ) (Grohmann et al., 2002), although another study has refuted this hypothesis (Mellor et al., 2004). Therefore, although IDO itself plays an important role in immune regulation, it seems unlikely that IDO activation is a mechanism of CTLA-4 function in all situations.

Modulation of Inhibitory Cytokine Production

CTLA-4 signals stimulate the production of regulatory cytokines (such as transforming growth factor-β (TGF-β) and IL10) (Fig. 7.2C). TGF-β elicits several cellular responses depending on cell type, state of differentiation, culture conditions, and presence of other cytokines (Massagué, 1990). TGF-β is a potent inhibitor of the T cell-mediated immune response both in vitro (Kehrl et al., 1986) and in vivo (Barral-Netto et al., 1992), although it can also exert upregulatory effects on T cell proliferative responses under some conditions. TGF-β is synthesized by different types of cells, including T cells, but the pathway involved in T cell production of TGF-β is not clear. The evidence that mice deficient in the TGF-β1 gene develop massive lymphoproliferative disorders that lead to organ failure and death after 2–3 weeks, similar to the syndrome of CTLA-$4^{-/-}$ mice, led to the hypothesis that the CTLA-4 pathway may be involved in the production of TGF-β. Subsequently, it was demonstrated that the activation of CTLA-4 leads to the production of TGF-β by $CD4^+$ T cells. In this way, CTLA-4, through the production of TGF-β, may counterbalance the production of IL-2 induced through the costimulation of CD28 (Chen et al., 1998).

Some studies have since shown that immunosuppression induced by CTLA-4 is independent of TGF-β, since it also occurred in mice that do not express TGF-β. It was thus suggested that IL10 may be as important as TGF-β for this purpose (Jovasevic et al., 2004).

Blocking CD80 Capture

Another mechanism by which CTLA-4 performs its immunosuppressive action is to bind CD80 and CD86, thereby reducing the availability of these ligands to bind to CD28 (Fig. 7.2D). Furthermore, a soluble form of CTLA-4 was identified, obtained by alternative splicing, without the transmembrane portion (sCTLA-4). sCTLA-4 secretion can be increased during immune responses and has potent inhibitory properties. Regulatory T cells (Tregs) were demonstrated to be a prominent source of sCTLA-4, which contributed to suppression in vitro when their numbers were limited. Its function could be to bind B7, reducing the concentration of free receptor interacting with CD28. In this way, sCTLA-4 may act as an important immune regulator, responsible for at least some of the inhibitory effects previously ascribed to the membrane-bound isoform. These results suggest that the immune system exploits the different CTLA-4 isoforms for either intrinsic or extrinsic regulation of T-cell activity (Ward et al., 2013).

Cell-Mediated Cytotoxicity of Regulatory T Cells

One of the most studied theories to explain the extrinsic mode of action of CTLA-4 is the ability to modulate the activity of Tregs. Tregs comprise several subsets of immunosuppressive cells that play critical roles in maintaining immune homeostasis and self-tolerance. They are involved in autoimmunity control, infection, graft-versus-host disease, inflammation, fetal-maternal tolerance, and tumor immunity. In particular, thymic-derived Tregs are essential for preserving self-tolerance and preventing autoimmunity, while peripheral Tregs maintain peripheral tolerance at mucosal interfaces and in response to external antigens (Sakaguchi et al., 2008). Both subsets of Tregs are defined by expression of the Forkhead Box P3 (FoxP3) transcription factor, a "master regulator" of the suppressive lineage, and the IL-2 receptor α chain (CD25). In cancers, Tregs are able to suppress antitumor immune responses and contribute to immune evasion and cancer progression (Elkord et al., 2010).

Tregs have been extensively characterized in the peripheral blood and immune infiltrates of different cancers (Nishikawa and Sakaguchi, 2014). It has been demonstrated that a higher Treg:T effector cell (Teff) ratio within tumor tissue is associated with a worse prognosis in many cancers. For example, transient Treg depletion can induce regression of metastatic lesions in advanced stage melanoma patients (Rasku et al., 2008). It has also been shown that CTLA-4 is expressed by Treg cells and plays a crucial role in the control of their immunosuppressive functions (Walker, 2013). The demonstration that Tregs express high levels of CTLA-4 has suggested that anti-CTLA-4 antibodies may impact directly on the Treg compartment, either through a reduction in their number or by modifying their suppressive activity (Read et al., 2000). However, several studies have demonstrated that anti-CTLA-4 antibodies trigger an expansion of Treg cells in secondary lymphoid organs and blood (Quezada et al., 2011). In practice, the use of anti-CTLA-4 antibodies appears to result in an increase in the number of both Teff cells and Tregs in blood and lymphoid organs, while in the tumor microenvironment there is an increase in Teffs associated with a reduction in Tregs. Several studies have been conducted to define the mechanism underlying the antitumor activity of α-CTLA-4 by focusing on factors controlling the selective increase in the Teff:Treg cell ratio within the tumor.

In recent years, it has been suggested that anti-CTLA-4 antibodies can lead to a reduction of Treg cells via an antibody-dependent cell-mediated cytotoxicity (ADCC) mechanism (Fig. 7.2E). In particular, it was shown that FcgRIV-expressing macrophages in the tumor microenvironment can recognize and phagocyte Treg cells opsonized by anti-CTLA-4 antibodies (Simpson et al., 2013). In vitro, the anti-CTLA-4 antibody, ipilimumab, leads to the depletion of Tregs via an ADCC mechanism, selectively mediated by FcgRIIIA (CD16)-expressing, nonclassical monocytes. In contrast, classical monocytes, which do not express FcgRIIIA, are unable to deplete Tregs in an ADCC assay. In a study of 15 patients with melanoma, responding patients displayed significantly higher baseline peripheral frequencies of nonclassical monocytes than nonresponder patients. These results highlighted the contribution of the tumor stroma into the final outcome of antibody-based immunomodulatory therapies and suggest nonclassical monocytes as a potential biomarker of response (Romano et al., 2015).

ANTI-CTLA-4 ANTIBODIES

The interaction between cancer and the immune system is complex and multifaceted. While evidence of an antitumor immune response is detectable in many cancer patients, tumors develop multiple strategies to evade immune detection and destruction (Dunn et al., 2004). Immunotherapies aim to generate or potentiate antitumor immunity to gain clinical benefit. Advances in defining the mechanisms that regulate immune responses have provided new targets for therapeutic intervention. The identification of CTLA-4, a coinhibitory molecular expressed on T cells, led to the clinical development of CTLA-4 blocking antibodies that are able to stimulate potent antitumor immunity (Korman et al., 2006). Clinical studies have confirmed that the CTLA-4 inhibitor ipilimumab can induce durable responses and improve overall survival (OS) in patients with advanced melanoma (Wolchok et al., 2010). Two phase III studies demonstrated the efficacy of two different dosages of ipilimumab (3 and 10 mg) in metastatic melanoma. In the first phase III trial, ipilimumab 3 mg/kg with or without glycoprotein 100 peptide (gp100) vaccine was compared with gp100 vaccine monotherapy in patients with unresectable stage III or stage IV melanoma. Ipilimumab monotherapy improved OS compared with gp100 vaccine monotherapy (10.1 vs 6.4 months). However, more immune-related adverse events were observed with ipilimumab treatment. The combination of ipilimumab and gp100 did not improve OS compared with ipilimumab alone (Hodi et al., 2010). In another randomized phase III trial, ipilimumab 10 mg/kg combined with dacarbazine 850 mg/m^2 demonstrated a modest but significant improvement in OS compared with placebo plus dacarbazine (11.2 vs 9.1 months). A higher incidence of adverse events was recorded in the ipilimumab-dacarbazine group compared with placebo-dacarbazine, including elevation of alanine aminotransferase levels (29.1% vs 4.4% of patients), elevation of aspartate aminotransferase levels (26.7% vs 3.2%), diarrhea (32.8% vs 15.9%), pruritus (29.6% vs 8.8%), and rash (24.7% vs 6.8%). Grade 3 or 4 adverse events occurred in 56.3% of patients treated with ipilimumab plus dacarbazine, as compared to 27.5% treated with dacarbazine alone ($p < .001$) (Robert et al., 2011). On the basis of these two phase III studies, ipilimumab was approved for the treatment of unresectable or metastatic melanoma at the dosage of 3 mg/kg in 2011 by the US Food and Drug Administration (FDA).

Subsequently, the CA 184-169 trial directly compared ipilimumab 3 versus 10 mg/kg in 727 patients with unresectable stage III/IV melanoma who may have received chemotherapy but had not previously received a v-raf murine sarcoma viral oncogene homologe B1 (BRAF) inhibitor or checkpoint inhibitor. Ipilimumab 10 mg/kg every 3 weeks led to better OS than 3 mg/kg every 3 weeks (median 15.7 vs 11.5 months; hazard ratio (HR) 0.84; $p = .04$). Three-year survival rates were 31% versus 23%, respectively. However, the higher dose was associated with higher toxicity rates, with treatment-related adverse events leading to discontinuation in about 31% in the 10 mg/kg arm versus 19% in the 3 mg/kg arm (Ascierto et al., 2016). In 2015, a large pooled analysis of long-term survival data including 1861 patients from 10 prospective and 2 retrospective studies of ipilimumab, and 2 phase III trials with an additional 2985 patients from an expanded access program, showed a median OS of 9.5 months (95% confidence interval (CI), 9.0–10.0), with a plateau at 21% in the survival curve beginning around 3 years, with follow-up of up to 10 years (Schadendorf et al., 2015).

Recently ipilimumab at a dose of 10 mg/kg was shown to be effective when used as adjuvant treatment in patients with melanoma after complete resection of stage III disease (Eggermont et al., 2016). At a median follow-up of 5.3 years, the 5-year recurrence-free survival rate was 40.8% in the ipilimumab group compared with 30.3% in the placebo group (HR for recurrence or death, 0.76; 95% CI 0.64–0.89; $p < .001$). OS at 5 years was 65.4% with ipilimumab and 54.4% with placebo (HR for death, 0.72; 95.1% CI 0.58–0.88; $p = 0.001$).

Following these results, the FDA expanded the approved use of ipilimumab to include adjuvant therapy for patients with stage III melanoma.

With regard to the other anti-CTLA-4 antibody, tremelimumab, treatment with 15 mg/kg every 90 days was defined at the most appropriate regimen in phase I /II studies (Camacho et al., 2009; Kirkwood et al., 2010) and then compared to standard of care chemotherapy (dacarbazine or temozolomide) in a phase III trial of 655 patients with advanced melanoma. However, the primary endpoint was not achieved, with median OS of 12.6 versus 10.7 months with chemotherapy ($p = .127$) (Ribas et al., 2013). Long-term survival data were recently published and are very similar to ipilimumab data, with a 5-year survival rate of 20% (95% CI 13–26%), and 10 and 12.5-year survival rates of 16% (95% CI 9–23%) (Eroglu et al., 2015). Patient selection, the use of a suboptimal schedule, and the possibility that patients in the chemotherapy arm received compassionate use ipilimumab after disease progression, were considered as possible explanations for the failure of this trial. In addition, the study did not use immune-related response criteria (Ascierto, 2013).

In recent years, the combination of ipilimumab with nivolumab, an anti-PD1 antibody, has been shown to be superior to ipilimumab monotherapy in terms of both response and survival. The CheckMate 069 trial compared ipilimumab plus nivolumab versus ipilimumab alone in patients with advanced melanoma. The objective response rate and progression-free survival (PFS) among patients with advanced melanoma who had not previously received treatment were significantly greater with nivolumab combined with ipilimumab, or nivolumab alone, than with ipilimumab monotherapy (Postow et al., 2015). Recently, survival data from this trial were published confirming that the combination of first-line nivolumab plus ipilimumab might lead to improved outcomes compared with first-line ipilimumab alone. In fact, 2-year OS was 63.8% (95% CI 53.3–72.6) for those treated with nivolumab plus ipilimumab and 53.6% (95% CI 38.1–66.8) for those treated with ipilimumab alone; median OS had not been reached in either group (HR 0.74, 95% CI 0.43–1.26; $p = .26$) (Hodi et al., 2016). In the phase III CheckMate 067 study, 945 untreated patients with metastatic melanoma were randomized to receive nivolumab monotherapy, nivolumab plus ipilimumab, or ipilimumab monotherapy. Nivolumab alone or combined with ipilimumab resulted in significantly longer PFS than ipilimumab alone (6.9 and 11.5 months vs 2.9 months). The overall response rate (ORR) in the three arms was respectively 43.8%, 57.6%, and 19% (Larkin et al., 2015). On the basis of these results, the FDA approved the treatment of BRAF wild type unresectable melanoma with nivolumab plus ipilimumab.

CONCLUSIONS

Although our understanding of CTLA-4 activity has improved and several models have been proposed for its mechanism of action on T cell inhibition, the exact processes involved and their exact contributions remain unclear. However, this continues to be an active focus of research. Importantly, better understanding of the antitumor immune effect mediated by CTLA-4 inhibition should provide insight into the effects of anti-CTLA-4 antibodies, such as ipilimumab. Such knowledge may provide greater understanding of the place of ipilimumab and other CTLA-4 checkpoint blockers in the rapidly evolving therapeutic armamentarium against melanoma and other cancers, in particular how these agents may be best used in combination with other treatments.

References

Ascierto, P.A., 2013. Is there still a role for tremelimumab in the treatment of cancer? Transl. Cancer Res. 2, 48–50.

Ascierto, P.A., Del Vecchio, M., Robert, C., et al., 2016. Overall survival (OS) and safety results from a phase 3 trial of ipilimumab (IPI) at 3 mg/kg vs 10 mg/kg in patients with metastatic melanoma (MEL). Ann. Oncol. 27 (Suppl. 6) ABST 1106O.

Baroja, M.L., Vijayakrishnan, L., Bettelli, E., et al., 2002. Inhibition of CTLA-4 function by the regulatory subunit of serine/threonine phosphatase 2A. J. Immunol. 168, 5070–5078.

Barral-Netto, M., Barral, A., Brownell, C.E., et al., 1992. Transforming growth factor-beta in leishmanial infection: a parasite escape mechanism. Science 257, 545–548.

Brunet, J.F., Denizot, F., Luciani, M.F., et al., 1987. A new member of the immunoglobulin superfamily CTLA-4. Nature 328, 267–270.

Brunet, J.F., Denizot, F., and Golstein, P., 1988. A differential molecular biology search for genes preferentially expressed in functional T lymphocytes: the CTLA genes. Immunol. Rev. 103, 21–36.

Camacho, L.H., Antonia, S., Sosman, J., et al., 2009. Phase I/II trial of tremelimumab in patients with metastatic melanoma. J. Clin. Oncol. 27, 1075–1081.

Chambers, C.A., Kuhns, M.S., Egen, J.G., et al., 2001. CTLA-4-mediated inhibition in regulation of T cell responses: mechanisms and manipulation in tumor immunotherapy. Ann. Rev. Immunol. 19, 565–594.

Chen, W., Jin, W., and Wahl, S.M., 1998. Engagement of cytotoxic T lymphocyte-associated antigen 4 (CTLA-4) induces transforming growth factor β (TGF-β) production by murine $CD4^+$ T cells. J. Exp. Med. 188, 1849–1857.

Collins, A.V., Brodie, D.W., Gilbert, R.J., et al., 2002. The interaction properties of costimulatory molecules revisited. Immunity 17, 201–210.

Dunn, G.P., Old, L.J., and Schreiber, R.D., 2004. The immunobiology of cancer immunosurveillance and immunoediting. Immunity 21, 137–148.

Eggermont, A.M., Chiarion-Sileni, V., Grob, J.J., et al., 2016. Prolonged survival in stage III melanoma with ipilimumab adjuvant therapy. N. Engl. J. Med. 375, 1845–1855.

Elkord, E., Alcantar-Orozco, E.M., Dovedi, S.J., et al., 2010. T regulatory cells in cancer: recent advances and therapeutic potential. Expert Opin. Biol. Ther. 10, 1573–1586.

Eroglu, Z., Kim, D.W., Wang, X., et al., 2015. Long term survival with cytotoxic T lymphocyte-associated antigen 4 blockade using tremelimumab. Eur. J. Cancer 51, 2689–2697.

Fallarino, F., Grohmann, U., Hwang, K.W., et al., 2003. Modulation of tryptophan catabolism by regulatory T cells. Nat. Immunol. 4, 1206–1212.

Grohmann, U., Orabona, C., Fallarino, F., et al., 2002. CTLA-4-Ig regulates tryptophan catabolism in vivo. Nat. Immunol. 3, 1097–1101.

Hodi, F.S., O'Day, S.J., McDermott, D.F., et al., 2010. Improved survival with ipilimumab in patients with metastatic melanoma. N. Engl. J. Med. 363, 711–723.

Hodi, F.S., Chesney, J., Pavlick, A.C., et al., 2016. Combined nivolumab and ipilimumab versus ipilimumab alone in patients with advanced melanoma: 2-year overall survival outcomes in a multicentre, randomised, controlled, phase 2 trial. Lancet Oncol. 17, 1558–1568.

Jovasevic, V.M., Gorelik, L., Bluestone, J.A., et al., 2004. Importance of IL-10 for CTLA-4-mediated inhibition of tumor-eradicating immunity. J. Immunol. 172, 1449–1454.

Kehrl, J.H., Wakefield, L.M., Roberts, A.B., et al., 1986. Production of transforming growth factor beta by human T lymphocytes and its potential role in the regulation of T cell growth. J. Exp. Med. 163, 1037–1050.

Kirkwood, J.M., Lorigan, P., Hersey, P., et al., 2010. Phase II trial of tremelimumab (CP-675,206) in patients with advanced refractory or relapsed melanoma. Clin. Cancer Res. 16, 1042–1048.

Korman, A.J., Peggs, K.S., and Allison, J.P., 2006. Checkpoint blockade in cancer immunotherapy. Adv. Immunol. 90, 297–339.

Krummel, M.F., and Allison, J.P., 1995. CD28 and CTLA-4 have opposing effects on the response of T cells to stimulation. J. Exp. Med. 182, 459–465.

Larkin, J., Chiarion-Sileni, V., Gonzalez, R., et al., 2015. Combined nivolumab and ipilimumab or monotherapy in untreated melanoma. N. Engl. J. Med. 373, 23–34.

Ling, V., Wu, P.W., Finnerty, H.F., et al., 1999. Complete sequence determination of the mouse and human CTLA-4 gene loci: cross-species DNA sequence similarity beyond exon borders. Genomics 60, 341–355.

Martin, M., Schneider, H., Azouz, A., et al., 2001. Cytotoxic T lymphocyte antigen 4 and CD28 modulate cell surface raft expression in their regulation of T cell function. J. Exp. Med. 194, 1675–1681.
Massagué, J., 1990. The transforming growth factor-beta family. Annu. Rev. Cell Biol. 6, 597–641.
Masteller, E.L., Chuang, E., Mullen, A.C., et al., 2000. Structural analysis of CTLA-4 function in vivo. J. Immunol. 164, 5319–5327.
Mellor, A.L., Chandler, P., Baban, B., et al., 2004. Specific subsets of murine dendritic cells acquire potent T cell regulatory functions following CTLA-4-mediated induction of indoleamine 2,3 dioxygenase. Int. Immunol. 16, 1391–1401.
Munn, D.H., Shafizadeh, E., Attwood, J.T., et al., 1999. Inhibition of T cell proliferation by macrophage tryptophan catabolism. J. Exp. Med. 189, 1363–1372.
Nishikawa, H., and Sakaguchi, S., 2014. Regulatory T cells in cancer immunotherapy. Curr. Opin. Immunol. 27, 1–7.
Postow, M.A., Chesney, J., Pavlick, A.C., et al., 2015. Nivolumab and ipilimumab versus ipilimumab in untreated melanoma. N. Engl. J. Med. 372, 2006–2017.
Quezada, S.A., Peggs, K., Simpson, T.R., et al., 2011. Shifting the equilibrium in cancer immunoediting: from tumor tolerance to eradication. Immunol. Rev 241, 104–118.
Qureshi, O.S., Zheng, Y., Nakamura, K., et al., 2011. Trans-endocytosis of CD80 and CD86: a molecular basis for the cell-extrinsic function of CTLA-4. Science 332, 600–603.
Rasku, M.A., Clem, A.L., Telang, S., et al., 2008. Transient T cell depletion causes regression of melanoma metastases. J. Transl. Med. 6, 12.
Read, S., Malmström, V., and Powrie, F., 2000. Cytotoxic T lymphocyte-associated antigen 4 plays an essential role in the function of CD25(+)CD4(+) regulatory cells that control intestinal inflammation. J. Exp. Med. 192, 295–302.
Ribas, A., Kefford, R., Marshall, M.A., et al., 2013. Phase III randomized clinical trial comparing tremelimumab with standard-of-care chemotherapy in patients with advanced melanoma. J. Clin. Oncol. 31, 616–622.
Robert, C., Thomas, L., Bondarenko, I., et al., 2011. Ipilimumab plus dacarbazine for previously untreated metastatic melanoma. N. Engl. J. Med. 364, 2517–2526.
Romano, E., Kusio-Kobialka, M., Foukas, P.G., et al., 2015. Ipilimumab-dependent cell-mediated cytotoxicity of regulatory T cells ex vivo by nonclassical monocytes in melanoma patients. Proc. Natl. Acad. Sci. U.S.A. 112, 6140–6145.
Rudd, C.E., 2008. The reverse stop-signal model for CTLA-4 function. Nat. Rev. Immunol. 8, 153–160.
Sakaguchi, S., Yamaguchi, T., Nomura, T., et al., 2008. Regulatory T cells and immune tolerance. Cell 133, 775–787.
Schadendorf, D., Hodi, F.S., Robert, C., et al., 2015. Pooled analysis of long-term survival data from phase II and phase III trials of ipilimumab in unresectable or metastatic melanoma. J. Clin. Oncol. 33, 1889–1894.
Schneider, H., Downey, J., Smith, A., et al., 2006. Reversal of the TCR stop signal by CTLA-4. Science 313, 1972–1975.
Schneider, H., Smith, X., Liu, H., et al., 2007. CTLA-4 expression disrupts ZAP-70 microcluster formation, T-cell/APC conjugation and calcium mobilization. Eur. J. Immunol. 38, 40–47.
Sharpe, A.H., 2009. Mechanisms of costimulation. Immunol. Rev. 229, 5–11.
Simpson, T.R., Li, F., Montalvo-Ortiz, W., et al., 2013. Fc-dependent depletion of tumor-infiltrating regulatory T cells co-defines the efficacy of anti-CTLA-4 therapy against melanoma. J. Exp. Med. 210, 1695–1710.
Teft, W.A., Kirchhof, M.G., and Madrenas, J., 2006. A molecular perspective of CTLA-4 function. Ann. Rev. Immunol. 24, 65–97.
Walker, L.S., 2013. Treg and CTLA-4: two intertwining pathways to immune tolerance. J. Autoimmun. 45, 49–57.
Ward, F.J., Dahal, L.N., Wijesekera, S.K., et al., 2013. The soluble isoform of CTLA-4 as a regulator of T-cell responses. Eur. J. Immunol. 43, 1274–1285.
Wolchok, J.D., Neyns, B., Linette, G., et al., 2010. Ipilimumab monotherapy in patients with pretreated advanced melanoma: a randomised, double-blind, multicentre, phase 2, dose-ranging study. Lancet Oncol. 11, 155–164.
Yokosuka, T., Sakata-Sogawa, K., Kobayashi, W., et al., 2005. Newly generated T cell receptor microclusters initiate and sustain T cell activation by recruitment of Zap70 and SLP-76. Nat. Immunol. 6, 1253–1262.

CHAPTER

8

To Help or To Harm: Dynamic Roles of CD4$^+$ T Helper Cells in Solid Tumor Microenvironments

Shannon M. Liudahl and Lisa M. Coussens

OUTLINE

Abstract

In addition to their functional relevance in pathogenic infection and autoimmune disease, leukocytes also play critical roles in regulating cancer progression, and immunotherapies designed to enhance antitumor immune responses are becoming mainstays in clinical oncology. Complex interactions between different leukocyte subtypes recruited to neoplastic microenvironments can result in either tumor promotional or tumor suppressive programming. Therefore, understanding the mechanisms regulating the balance between these states is critical for developing improved immune-based cancer therapeutics. Effector and memory CD8$^+$ T cells are well established as key mediators of tumor cytotoxicity, but the roles of CD4$^+$ T cells in tumor microenvironments have not been as extensively explored. In this chapter, we review current understandings of the diverse roles played by CD4$^+$ T helper cell subsets in either promoting or suppressing progression of solid tumors.

M.A. Hayat (ed): *Immunology*, Volume 1.
DOI: http://dx.doi.org/10.1016/B978-0-12-809819-6.00008-3

© 2018 Elsevier Inc. All rights reserved.

INTRODUCTION

Immune cell recruitment to and persistence within solid tumors is one of the fundamental hallmarks of cancer (Hanahan and Weinberg, 2011). Tissue resident and infiltrating leukocytes engage in crosstalk with neoplastic cells, stromal cells, and other immune cells within tumor microenvironments (TMEs) throughout all stages of neoplastic progression to ultimately shape tumor development and responses to anticancer therapy. Dynamic interactions between leukocytes and other cell types within evolving TMEs influence overall immune complexity and immune effector function, and based on tumor stage and/or anatomic location, can give rise to either pro- or antitumor immune driven programs (Hanahan and Coussens, 2012; Palucka and Coussens, 2016).

Historically, tumor immunology has largely focused on cytotoxic $CD8^+$ T lymphocyte (CTL) recruitment and effector status within TMEs due to their cytotoxic activity against neoplastic cells. Indeed, $CD8^+$ T cell abundance within several solid tumor types is now recognized as a robust prognostic indicator of clinical outcomes (Fridman et al., 2012; Galon et al., 2014). Moreover, $CD8^+$ T cells receive significant attention based on their promise as a potent cell-based therapy, particularly in adoptive cell transfer strategies (Restifo et al., 2012). Preclinical and clinical studies aiming to reveal and therapeutically manipulate regulatory checkpoint programs controlling $CD8^+$ T cell effector activity have also become a central focus of immuno-oncology (for a recent review, see Topalian et al., 2015).

CTLs, however, do not represent the entirety of T cell activity within TMEs. $CD4^+$ T cells also infiltrate neoplastic tissues and have central roles in shaping pro- and antitumor immune responses. Naïve $CD4^+$ T cells can differentiate into either T regulatory cells (Tregs) or one of several $CD4^+$ T helper (Th) subsets depending on the particular cytokine milieu in the local microenvironment at the time of T cell receptor engagement by cognate antigen bound to major histocompatibility complex (MHC) class II on antigen presenting cells (APCs). Once differentiated into a particular subset, Th cells engage in a variety of activities necessary for productive immune responses by way of direct cell–cell interactions and production of soluble factors. $CD4^+$ T cell help is critical for optimal $CD8^+$ T cell priming and memory responses (Bevan, 2004). Th cells also provide essential help to B cells during immunoglobulin (Ig) production and can induce activation and enhanced phagocytic function of myeloid cells during infection and other disease states (Zhu and Paul, 2008; Crotty, 2015). Such Th functions are important during cancer progression. Notably, recent studies have also indicated roles for $CD4^+$ Th cells in direct tumor cytotoxicity (Quezada et al., 2010; Xie et al., 2010). However, due to the extensive heterogeneity of Th populations, it is unsurprising that not all Th subsets confer host protection during tumorigenesis and can instead foster neoplastic growth and survival, as will be discussed in the following sections.

$CD4^+$ Tregs are also significant in TMEs, given their well-described immunosuppressive functions that align them as critical regulators of tumor immune evasion and escape (Facciabene et al., 2012). Treg expression of antiinflammatory soluble factors, including IL-10 and transforming growth factor (TGF)-β, generally positions them as potent inhibitors of productive antitumor immune programs in a wide range of tumors, as has been reviewed extensively elsewhere (Mougiakakos et al., 2010; Ruffell et al., 2010; Fridman et al., 2012; Josefowicz et al., 2012; Adeegbe and Nishikawa, 2013). $CD4^+$ T helper cells and their roles in tumor immunity, on the other hand, have received comparatively less attention. This chapter thus focuses specifically on $CD4^+$ Th cells and their functions in neoplastic progression. We review molecular mediators required for differentiation of each of the major recognized $CD4^+$ Th populations (summarized in Fig. 8.1), and provide a comprehensive overview of how, once committed to a given lineage, activated T helper cells regulate tumor immunity within different solid TMEs. Developing a deeper understanding of and appreciation for the dynamic and sometimes contrasting roles of Th subsets in TMEs will illuminate new strategies by which we may harness particular Th cells and Th-mediated pathways for efficacious cancer therapies.

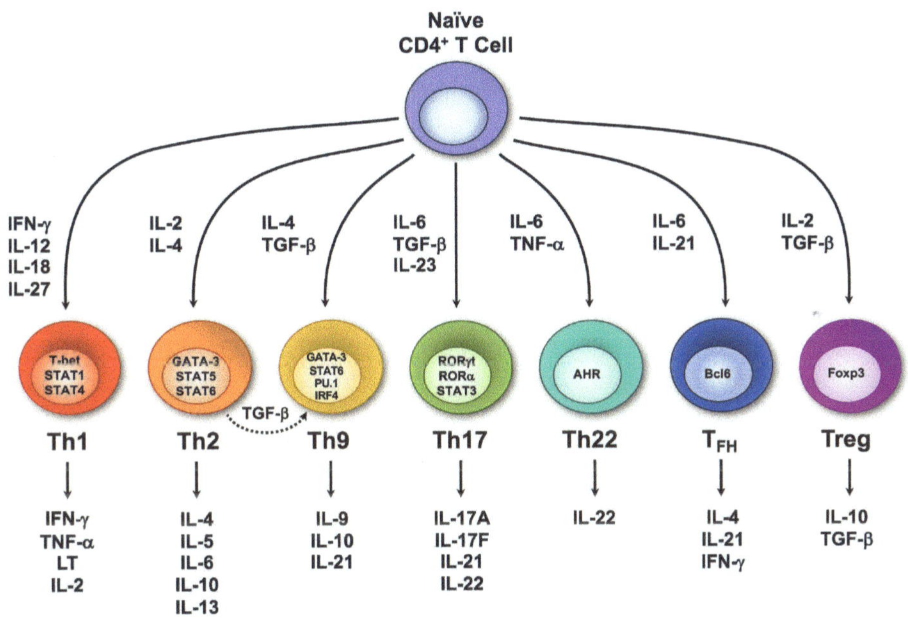

FIGURE 8.1 **Differentiation of effector $CD4^+$ T helper lineages is directed by specific cytokine milieus.** Naïve $CD4^+$ T cells activated by antigen presenting cells (APCs) presenting cognate antigen on MHC class II molecules can differentiate into T regulatory cells (Tregs) or one of at least six known T helper subsets. Specific differentiation-inducing cytokines (top) and required lineage-specific transcription factors (depicted inside cells) promote effector cell lineage commitment. Once committed to a given lineage, Th cells produce characteristic sets of effector cytokines (bottom). Importantly, differentiation of Th cells is not rigidly fixed, and some populations have considerable plasticity—e.g., mature Th2 cells exposed to TGF-β may be converted to a Th9 phenotype (dotted line). Th17 cells are also known for their significant phenotypic plasticity (omitted here for clarity), as is described in detail in the text.

T HELPER SUBSETS AND THEIR DYNAMIC FUNCTIONS IN CANCER

Th1 and Th2 Cells

A Finely Tuned Balance of Environmental Cues Dictates Th1 Versus Th2 Cell Differentiation Fate

T helper 1 (Th1) and T helper 2 (Th2) cells were the first two classes of $CD4^+$ Th cells to be recognized as phenotypically and functionally distinct populations within the $CD4^+$ T cell pool. Differential effector cytokine profiles of Th1 and Th2 cell clones were first described by Mosmann and Coffman in studies reporting that upon antigen recognition and activation, Th1 cells upregulate immunostimulatory cytokines interleukin (IL)-2, interferon (IFN)-γ, lymphotoxin (LT), and tumor necrosis factor (TNF)-α, whereas Th2 cells produce IL-4, IL-5, IL-6, IL-10, and IL-13 (Mosmann et al., 1986; Mosmann and Coffman, 1989). These unique cytokine profiles shape the effector functions of these cell populations, and decades of characterization of the dichotomous Th1 and Th2 responses has led to the accepted model wherein Th1 cells regulate cell-mediated immunity against intracellular pathogens, while Th2 cells are the predominant orchestrators of humoral immunity and response to extracellular pathogens (O'Garra, 1998). Similar to their divergent functions in response to infection, Th1 and Th2 cells generally exhibit highly polarized roles in the context of tumors, which will be discussed in detail below.

Pathways regulating differentiation of naïve T cells into either Th1 or Th2 lineages are mutually exclusive, and commitment to either requires cooperative function of a series of specific cytokines, polypeptide growth factors, and transcription factors. IL-12 is the primary cytokine regulating Th1 development, acting along with its cofactor, IL-18, to induce IFN-γ production (Trinchieri, 1995; Magram et al., 1996; Szabo et al., 1997; Takeda et al., 1998). IL-27 was subsequently identified to function in synergy with IL-12 to promote IFN-γ expression (Pflanz et al., 2002; Owaki et al., 2005). Collectively, this set of cytokines drives preferential Th1 development via activation of *s*ignal *t*ransducer and *a*ctivator of *t*ranscription (STAT) protein family members STAT1 and STAT4. Signaling through these STAT proteins results in transcription of *T-bet*, the primary transcription factor required for Th1 differentiation and upregulation of IFN-γ and IL-12 receptor beta 2 (IL-12Rβ2) (Szabo et al., 2000, 2002; Mullen et al., 2001; Afkarian et al., 2002). IFN-γ itself also engages in positive feedback that supports its continued production by directly stimulating T-bet expression in a STAT1-dependent manner (Bradley et al., 1996; Lighvani et al., 2001; Afkarian et al., 2002). Importantly, T-bet expression initiates and maintains Th1 programming in part through inhibition of Th2 polarization. T-bet suppresses the Th2 transcription factor GATA-3, and transduction of T-bet into mature Th2-polarized cells in vitro leads to concurrent upregulation of Th1 effector cytokines and suppression of critical Th2 effectors (Szabo et al., 2000; Usui et al., 2006).

Th2 development also requires a unique set of cytokines and transcription factors that function to simultaneously promote Th2 commitment and suppress Th1 developmental pathways. As is the case with IFN-γ acting as both a Th1 product and mediator of Th1 differentiation, the Th2 effector cytokine IL-4 likewise operates in an autocrine loop to drive Th2 differentiation (Noben-Trauth et al., 2002). IL-4 engagement of IL-4 receptor α and γ subunits on naïve $CD4^+$ T cells triggers activation of STAT6 and subsequent induction of the Th2-specific transcription factor GATA-3 and transcription of *Il4* (Kaplan et al., 1996; Shimoda et al., 1996; Takeda et al., 1996; Zhu et al., 2001). GATA-3 is required for Th2 differentiation and maintenance, as well as Th1 inhibition (Zheng and Flavell, 1997; Pai et al., 2004; Zhu et al., 2006). IL-2 also plays an important role in Th2 differentiation via signal transduction through STAT5 that stabilizes access to the *Il4* gene locus (Cote-Sierra et al., 2004).

Th1 Cells Are Classical T Helper Regulators of Antitumor Immunity, While Th2 Cells Exhibit Functional Duality in the TME

At present, Th1 cells are the only identified $CD4^+$ T helper population to be uniformly correlated with improved disease outcome across a wide spectrum of solid tumors. This is due in part to their robust production of IFN-γ and its pleiotropic downstream effects, including antitumor myeloid cell programming, enhancement of CTL maturation and function, and inhibition of angiogenesis (Ruffell et al., 2010; Fridman et al., 2012). Early tumor immunology studies using murine models of carcinogen-induced sarcoma, mastocytoma, mammary carcinoma, and lung adenocarcinoma pointed toward antitumor properties of Th1 cells by demonstrating that deletion of either IFN-γ, IFN-γ receptor, or STAT1 promoted tumor growth (Fallarino and Gajewski, 1999; Shankaran et al., 2001; Street et al., 2002). Of course, antitumor effects of IFN-γ and STAT1 signaling cannot be solely attributed to Th1 cells, given that $CD8^+$ T and natural killer (NK) cells also rely on STAT1 and produce IFN-γ. However, more direct support for a Th1 role in IFN-γ-mediated antitumor immunity has come from studies in which delivery of recombinant IL-12 to mice bearing sarcomas or colon carcinomas suppressed tumor growth in an IFN-γ-dependent manner that required both $CD8^+$ and $CD4^+$ T cells (Nastala et al., 1994). The authors of this study suggested that $CD4^+$ T cell help, presumably from Th1 cells, was necessary for optimal CTL activity and tumor suppression. In addition, studies in a transgenic murine model of islet cell carcinoma revealed that IFN-γ and TNF-α specifically derived from Th1 cells were required for cytokine-induced cancer cell senescence, thereby reflecting another mechanism through which Th1-polarized cells contribute to antitumor immunity (Braumuller et al., 2013).

The presence of Th1 cells and/or Th1 cytokines within TMEs has now been associated with improved patient prognosis in a range of human cancers. In gastric cancer, $T\text{-bet}^+$ T cell infiltration was associated with improved relapse-free survival (Hennequin et al., 2016). High gene expression levels of *T-bet* and other Th1-associated genes has also been significantly correlated with extended disease-free survival in colorectal carcinoma (Camus et al., 2009; Tosolini et al., 2011), improved prognosis in hepatocellular carcinoma (HCC) (Gao et al., 2011), and increased overall survival in patients with advanced ovarian carcinomas (Kusuda et al., 2005). A recent prospective epidemiological study also reported that high circulating levels of Th1 cytokines IFN-γ and TNF-α correlate with reduced risk of developing prostate carcinoma (Bhavsar et al., 2014).

In contrast to the consistent antitumor nature of Th1 cells and Th1 cytokines in TMEs, the role of Th2 cells in cancer may have somewhat greater nuance and contextual dependence. It has been proposed that Th2 cells guide antitumor immunity via recruitment of specific populations of innate immune cells to TMEs. Implantable murine mammary adenocarcinomas, melanomas, and sarcomas fail to grow in models wherein syngeneic tumor cell lines were engineered to exogenously express IL-4 prior to subcutaneous tumor inoculation

(Tepper et al., 1989). Tumor inhibition in these and other transplantable murine tumor models was later reported to be dependent on IL-4-mediated eosinophil recruitment that led to significant tumor cell cytotoxicity (Tepper et al., 1992; Mattes et al., 2003). Others have also reported that IL-4 and IL-13 inhibit proliferation of breast carcinoma cells via signaling through IL-4/IL-13 receptors expressed on neoplastic epithelium (Nagai and Toi, 2000; Gooch et al., 2002). Conversely, $CD4^+$ T cells expressing IL-4, IL-13, and IL-10 also promote pulmonary metastasis in the MMTV-PyMT murine model of mammary adenocarcinoma by activating protumorigenic programs in tumor-associated macrophages (TAMs) (DeNardo et al., 2009). Depletion of either $CD4^+$ T cells or IL-4 in this context significantly reduced numbers of circulating tumor cells and metastatic outgrowth. $CD4^+$ T cell depletion or neutralization of either IL-4 or IL-13 also conferred improved therapeutic responses to radiation therapy and paclitaxel chemotherapy in a syngeneic orthotopic model of murine mammary carcinoma (Shiao et al., 2015).

Th2 T cell polarization has also been implicated as a poor prognostic predictor in several human malignancies. In early stage resectable human pancreatic ductal adenocarcinoma, high GATA-3/T-bet ratios were found to be predictive of decreased patient survival (De Monte et al., 2011). In this setting, cancer-associated fibroblast crosstalk with dendritic cells (DCs) promoted Th2-priming of naïve T cells in tumor-draining lymph nodes (De Monte et al., 2011), paralleling DC-mediated induction of Th2 cells observed in a murine model of pancreatic neoplasia (Ochi et al., 2012). Others have reported significant Th2 polarization in the microenvironment of multiple human gynecologic cancers, and high expression of Th2 gene signatures was found to be an independent negative prognostic indicator for patient survival in ovarian cancer specifically (Sheu et al., 2001; Kusuda et al., 2005). Systemic Th2 dominance was also observed in human stage IV metastatic melanomas, which contrasted with Th1-skewed blood samples from healthy controls and patients with fully resected nonmetastatic melanomas (Nevala et al., 2009). In sum, there is increasing agreement in both preclinical and human studies that Th2 cells predominantly engage in and regulate tumor promotional mechanisms, often via activation of and crosstalk with tumor infiltrating myeloid cells.

Th17 Cells

Th17 Cells Are a Functionally Distinct T Cell Lineage

For several years following the discovery of Th1 and Th2 subsets, the prevailing data indicated that T helper cells fell within one of the two lineages described above. This concept was challenged when a unique $IFN\text{-}\gamma^- IL\text{-}17^+$ $CD4^+$ T cell population was discovered and found to be associated with pathogenesis of some autoimmune disorders (Murphy et al., 2003; Langrish et al., 2005). Subsequent research elegantly demonstrated that $IL\text{-}17^+$ T helper cells represent a population developmentally and functionally distinct from Th1 and Th2 lineages (Harrington et al., 2005; Park et al., 2005), and they were designated Th17 cells based on their characteristic IL-17 expression. The initial paradigm-altering studies characterizing this population revealed that Th17 cells employ differentiation programs mutually exclusive to those of Th1 and Th2 cells. Transcription factors and cytokine signals necessary for differentiation of Th1 (STAT1, STAT4, T-bet, IL-12) and Th2 cells (STAT6) are dispensable for Th17 differentiation, and IFN-γ and IL-4 were found to be potent inhibitors of Th17 development (Harrington et al., 2005; Park et al., 2005). It was later revealed that Th17 cell differentiation is regulated by STAT3 signaling upstream of the Th17 master transcription factors retinoic-acid-related orphan receptors gamma t (RORγt) and alpha (RORα) (Ivanov et al., 2006; Zhou et al., 2007; Yang et al., 2008). Furthermore, Th17 polarization from a naïve T cell state requires TGF-β and IL-6 within the cytokine milieu, as well as IL-23 for sustained differentiation and survival (Bettelli et al., 2006; Ivanov et al., 2006; Veldhoen et al., 2006).

Once differentiated, Th17 cells exert proinflammatory effector functions through secretion of IL-17A, IL-17F, IL-21, and IL-22 (Korn et al., 2009). IL-17 receptor (IL-17RA) is systemically expressed by leukocytes, epithelial cells, and vascular endothelium, enabling Th17 cells to engage in crosstalk with a variety of other cell types during immune responses. IL-17A and IL-17F signaling induces production of granulopoietic factors, such as granulocyte-colony stimulating factor and stem cell factor, as well as a series of neutrophil-attracting chemokines, including CXCL1, CXCL2, and CXCL5 (Ouyang et al., 2008). Outcomes of Th17-mediated immunity vary considerably depending upon the specific type of inflammatory environment in which they are activated. While Th17 cells are key players in promoting autoimmune diseases, they are also critical for protective host defense against pathogens at mucosal sites (Dubin and Kolls, 2008).

Th17 Cells Have Opposing Roles in Tumor Immunity

Much is now understood regarding the role of Th17 cells in several disease states, but their functional significance in cancer remains controversial. Both tumor promotional and tumor inhibitory functions for Th17 cells have been reported in human and murine studies across a variety of cancer types, indicating that Th17-mediated tumor regulation may be tumor site specific (Wilke et al., 2011). In human disease, high levels of Th17 cells correlate with improved outcomes in prostate carcinoma (Sfanos et al., 2008), non-small cell lung cancer (NSCLC) (Ye et al., 2010), breast cancer (Yang et al., 2012), and ovarian cancer (Kryczek et al., 2009a). In human ovarian carcinomas, neoplastic cells, cancer-associated fibroblasts, and tumor-associated APCs promote generation of Th17 cells via production of Th17-stimulating cytokines, including IL-1β, IL-6, and TGF-β (Miyahara et al., 2008). Kryczek et al. (2009a) profiled 200 human ovarian tumors and found that Th17 cells were more abundant within tumors than in tumor-draining lymph nodes or in peripheral blood of either cancer patients or healthy blood donors, indicating specific recruitment to TMEs. High percentages of Th17 cells within these ovarian tumors positively correlated with high frequencies of intratumoral IFN-γ^+ CD8$^+$ T cells and NK cells, and negatively correlated with Treg frequency. The link between Th17 cells and effector CD8$^+$ T cells was reported to be at least partially due to IL-17-mediated induction of CXCL9 and CXCL10 by neoplastic cells and TAMs, which stimulates trafficking of CD8$^+$ T cells expressing cognate chemokine receptors (Kryczek et al., 2009a). The ability of Th17 cells to recruit other antitumor immune populations such as CD8$^+$ T cells to TMEs positions them as important mediators of protective tumor immunity, and, indeed, IL-17 levels within ovarian cancer patient ascites positively associates with increased survival.

Similar findings indicative of antitumor functions for Th17 cells have been reported in several murine models of solid tumor development. When in vitro polarized Th17 cells were adoptively transferred into mice bearing pulmonary melanoma, Martin-Orozco and colleagues observed enhanced recruitment of activated $CD4^+$ and $CD8^+$ T cells and DCs to lungs and tumor-draining lymph nodes (Martin-Orozco et al., 2009). The authors determined that, when compared to adoptively transferred Th1 cells, Th17 cells engaged in increased tumor-specific antigen loading onto DCs within tumor-draining lymph nodes, ultimately leading to improved $CD8^+$ T cell priming. They further demonstrated that Th17 cells possessed antitumor activities superior to those of Th1 cells in models of subcutaneous melanoma and fibrosarcoma, which supported related studies reporting that Th17 transfer results in improved overall survival in murine melanoma (Muranski et al., 2008). These antitumor properties of Th17 cells were found to be driven, at least in part, through Th17 production of IL-17, which was necessary for induction of chemotactic signals needed for leukocyte recruitment to tumors and lymph nodes. IL-17-deficient mice challenged with pulmonary melanoma exhibited increased tumor burden as compared to IL-17-proficient controls that corresponded to reduced total immune infiltration and decreased T cell activation (Muranski et al., 2008). IL-17 deficiency has been found to similarly promote growth and metastasis of murine MC38 colon carcinomas (Kryczek et al., 2009b). High levels of Th17 cells in a murine model of pancreatic ductal adenocarcinoma also correlates with longer survival and increased $CD8^+$ T cell trafficking to tumors, which was hypothesized to occur via Th17-mediated immune recruitment (Gnerlich et al., 2010).

On the other hand, many other investigators have reported tumor promotional roles for Th17 cells and IL-17 during solid tumor progression. In subcutaneous B16 melanoma and MB49 bladder carcinoma, tumor growth was significantly attenuated in IL-17-deficient mice compared to controls (Wang et al., 2009). IL-17 production in this context was found to stimulate STAT3 signaling in neoplastic cells, resulting in activation of downstream pathways promoting neoplastic cell proliferation, induction of antiapoptotic programs, and tumor angiogenesis (Numasaki et al., 2003; Wang et al., 2009). Th17 cell activation in a bacteria-induced model of colitis also promoted colon tumorigenesis, and while the underlying molecular mechanisms have not been elucidated, the authors suggested that protumorigenic functions of Th17 cells in this setting could be related to their role in recruiting tumor supportive granulocytes (Wu et al., 2009). Expression of Th17-associated genes as well as high in situ $IL\text{-}17^+$ cell density have also been linked to poor prognosis in human colorectal cancer (CRC) (Tosolini et al., 2011).

Growth of subcutaneous melanoma and prostate carcinoma are reportedly impaired in IL-17R-deficient mice, which was attributed to reduced recruitment of $CD11b^+$ $Gr\text{-}1^+$ suppressive myeloid cells to tumors and spleen as compared to IL-17R-proficient controls (He et al., 2010). Similar observations were reported in a murine model of lung adenocarcinoma, wherein Th17 cells were found to foster tumorigenesis through recruitment of tumor promotional myeloid cells and IL-17-mediated promotion of angiogenesis (Chang et al., 2014). Finally, in murine hepatoma, tumor-associated monocytes and macrophages induced expansion of Th17 cells, which ultimately correlated with increased tumor growth (Kuang et al., 2009). This finding complemented the earlier observation that enrichment of Th17 cells in human HCC positively associated with poor patient survival (Zhang et al., 2009).

Taken together, research to date has generated a highly contradictory picture of how Th17 cells mediate tumor progression and supports the notion that distinct TMEs can differentially influence the function of Th17 cells recruited to them. Beyond this, it is also critical to avoid overly conflating the functions of Th17 cells and IL-17. Although Th17 cells are a major source of IL-17 in many settings, numerous other immune cell populations also produce this cytokine and are therefore additional contributors to the pro- or antitumor effects ascribed to IL-17 signaling. For example, IL-17A produced by gamma delta T cells and neutrophils in numerous types of cancer has been associated with tumor progression (Ma et al., 2014; Rei et al., 2014; Wu et al., 2014; Coffelt et al., 2015; Housseau et al., 2016). A recent systematic review of Th17 and IL-17 prognostic associations in human cancers determined that high IL-17 levels in serum correlated with poor prognosis in 50% of studies examined, and that density of IL-17$^+$ cells in tumor tissue negatively correlated with survival in the majority of cases (Punt et al., 2015). In contrast, frequency of Th17 T cells within tumors was more often associated with improved patient prognosis, thus highlighting the importance of determining the cellular source of IL-17 on a case-by-case basis.

Phenotypic Plasticity Adds Increased Dimension to Th17 Function in Disease

Distinguishing a clear and consistent role for Th17 cells in cancer has been further complicated by the apparent plasticity of this lineage. Given the appropriate contextual signals, differentiated Th17 cells exhibit functional characteristics of other T helper subsets and Tregs. Likewise, under specific stimulatory conditions, some other CD4$^+$ lineages can upregulate Th17 signature transcription factors and cytokines (Guery and Hugues, 2015). Since other CD4$^+$ T cell subsets, particularly Th1 and Tregs, generally have opposing roles in tumor immunity in and of themselves, the ability for Th17 cells to shift between these opposing phenotypes offers some explanation for how they may have tumor promotional properties in certain settings and antitumor roles in others.

TGF-β and IL-12 levels strongly regulate Th17 phenotype by influencing Th17 to Th1 reprogramming. Differentiated Th17 cells recruited to environments with low or absent TGF-β and high IL-12 decrease expression of IL-17 and Th17 transcription factors while upregulating expression of IFN-γ and Th1 regulatory proteins STAT4 and T-bet (Lee et al., 2009; Nistala et al., 2010). Th17/Th2 intermediate effector cells have also been observed. Peripheral blood from healthy human donors contains low levels of Th17/Th2 intermediate cells, but in patients suffering from chronic allergic asthma, there is a marked enrichment of T cells that concurrently express Th2 and Th17 transcription factors as well as IL-4, IL-5, IL-13, and IL-17 (Cosmi et al., 2010; Wang et al., 2010). Th17/Th2 intermediates are the most abundant CD4$^+$ T cell population in a mouse model of chronic asthma and were responsible for promoting disease-related inflammation (Wang et al., 2010).

Potential for Th17 conversion to Th17/Treg intermediates is poorly understood. Th17-derived Th17/Treg cells have been reported in a model of heart allograft, wherein mesenchymal stem cell transfer in combination with immune suppressive therapeutics induced Th17/Treg formation whose presence correlated with improved graft tolerance (Obermajer et al., 2014). Further investigations are necessary to determine if Th17 cells transition to Th17/Treg intermediates in the context of disease, including cancer. However, it is now well established that differentiated Tregs can assume Th17 properties under the necessary environmental conditions in a range of inflammatory disorders. Human Tregs upregulate RORγt and IL-17 in the presence of exogenous IL-2 or IL-15, and these intermediate cells retain immunosuppressive effector functions in vitro (Koenen et al., 2008). This phenotypic switch is further enhanced when IL-1β, IL-23, or TGF-β are also present (Voo et al., 2009; Valmori et al., 2010). Other studies reported that IL-6 can also regulate Treg to Th17 plasticity;

exogenous IL-6 induces Foxp3$^+$ murine Tregs to express IL-17 (Xu et al., 2007), and recent research in a mouse model of arthritis revealed that fibroblast-secreted IL-6 induces Treg/Th17 intermediates similar to Treg/Th17 cells found in synovial fluid of humans with rheumatoid arthritis (Komatsu et al., 2014).

The role of Treg/Th17 intermediates in cancer is starting to be investigated. IL-17$^+$ Foxp3$^+$ CD4$^+$ T cells are found in human chronic ulcerative colitis and associated colon carcinomas where they have been described as inflammatory Tregs that induce cancer-promoting chronic inflammation and suppress antitumor immune responses (Kryczek et al., 2011). APCs play a critical role in inducing differentiation of these intermediate cells by producing the IL-2 and TGF-β needed to stimulate a Treg/Th17 intermediate phenotype. While Treg/Th17 intermediates were abundant in colitis and colon carcinoma samples, this cell population was not similarly detected in several other human cancers (ovarian, melanoma, and renal cell carcinoma), indicating cancer-specific roles for this population, perhaps in malignancies with well-defined pathologic transitions from chronic inflammation to invasive cancer (Kryczek et al., 2011).

Th9 Cells

Th9 cells are among the most recently described subsets of CD4$^+$ T helper cells. They are characterized by robust production of the cytokine IL-9, as well as IL-10 and IL-21 (Kaplan et al., 2015). Like Th2 cells, Th9 cells rely on IL-4, STAT6, and GATA-3 expression for their differentiation but acquire distinct cytokine expression patterns as they develop. Notably, although Th2 cells were long believed to be strictly limited in plasticity following their initial differentiation, it has been revealed that differentiated Th2 cells can be "reprogrammed" to an IL-9 producing Th9 phenotype when stimulated with TGF-β (Veldhoen et al., 2008). In addition to Th2-to-Th9 reprogramming, it is also accepted that IL-4 and TGF-β signaling can drive naïve CD4$^+$ T cells toward direct Th9 differentiation upon cell activation (Dardalhon et al., 2008). The transcription factors PU.1 and interferon response factor 4 (IRF4), in addition to GATA-3, are critical for Th9 induction from naïve CD4$^+$ precursors (Kaplan, 2013).

Since their discovery, Th9 cell function has been evaluated primarily in the context of allergic inflammation and autoimmunity where they have been found to have proinflammatory functions that exacerbate disease (Kaplan, 2013). In contrast, beneficial antitumor roles for Th9 cells are also emerging. In one of the earliest reports to elucidate Th9 function in the context of cancer, it was reported that adoptive transfer of differentiated Th9 cells significantly suppressed tumor growth in a B16-F10 murine model of melanoma via mechanisms dependent on IL-9 production (Purwar et al., 2012). While the authors of this study went on to determine that Th9 cells directly induce melanoma cell death in vitro, they concluded that Th9 crosstalk with other immune cells, namely mast cells, was the primary mechanism by which Th9 cells exert antitumor effects in their in vivo system. Melanoma growth was restored in the absence of mast cells, and taken together with the knowledge that IL-9 was produced predominately by T cells, and that IL-9 promotes mast cell proliferation and recruitment (Sehra et al., 2015), the authors suggested that Th9 cells could have downstream antitumor effects by mediating cytokine-driven myeloid cell activity (Purwar et al., 2012).

Others have corroborated the antitumor properties of Th9 using a murine model of pulmonary metastatic melanoma, and have indicated that in addition to potential effects on mast cells, Th9 cells also influence DC recruitment and downstream $CD8^+$ T cell cytotoxic function (Lu et al., 2012). Th9-dependent upregulation of the chemokine CCL20 by lung epithelial cells was found to be critical for recruitment of chemokine receptor CCR6-positive DCs to tumor sites and subsequent DC trafficking to tumor-draining lymph nodes where they could in turn prime $CD8^+$ T cells to engage in antitumor responses.

It remains to be determined whether Th9 cells similarly regulate antitumor immunity in other tumor types, and whether their presence and function in human disease parallels that observed thus far in murine models. It is certainly possible that Th9 cells instead engage in tumor promotion in some tumor contexts, especially given that IL-9 has previously been implicated in enhancement of Treg immunosuppressive activity (Eller et al., 2011) and impaired Th1 responses (Wu et al., 2008). Furthermore, since Th9 cells may play a significant role in mediating mast cell activation and recruitment, it must be considered that mast cells themselves have dynamic roles in tumor responses, as has been reported in multiple solid tumor models (Coussens et al., 1999; de Visser et al., 2005; Andreu et al., 2010; Masso-Valles et al., 2015). As T cell-based immunotherapies continue to be developed for clinical applications in cancer, it will be critical to build a more comprehensive understanding of how Th9 cells and the cytokines they produce either help or hinder tumor eradication.

Th22 Cells

Like Th9 cells, Th22 cells have only gained recognition as a distinct $CD4^+$ T cell lineage within the past decade. Following the convention of other recently discovered T helper subsets, the Th22 cell lineage was so-named based on characteristic production of IL-22 (Duhen et al., 2009). Expression of IL-22 is not restricted to the Th22 T cell lineage, as Th17 cells and NK cells are also capable of IL-22 production. However, Th22 T cells are unique in their expression of IL-22 in the absence of IL-17 and IFN-γ (Duhen et al., 2009; Eyerich et al., 2009; Trifari et al., 2009). This cell population was first observed in healthy human peripheral blood where, in addition to production of IL-22, Th22 cells were characterized by surface expression of the chemokine receptors CCR4, CCR6, and CCR10 that are associated with cutaneous T cell homing (Duhen et al., 2009). These initial studies further demonstrated that human naïve T cells differentiate into Th22 in the presence of IL-6 and TNF-α, and that plasmacytoid DCs in inflammatory microenvironments serve as a primary source of these Th22 differentiation factors (Duhen et al., 2009). Signaling downstream of the transcription factor aryl hydrocarbon receptor is thought to be necessary for Th22 production of IL-22 (Trifari et al., 2009).

The major known functions of IL-22 are to regulate innate defense programs (e.g., antibacterial protein production), induce cytokine and chemokine production, inhibit differentiation of keratinocytes and some epithelial cells in the gastrointestinal and respiratory tracts, and stimulate tissue remodeling. Effects of IL-22 signaling have been evaluated in several chronic inflammatory diseases. IL-22 expression has been demonstrated to have protective effects in models of asthma, inflammatory bowel disease, pancreatitis, hepatitis, and bacterial infection by virtue of its ability to promote antimicrobial responses, inhibit epithelial cell damage, and stimulate epithelial repair (Sabat et al., 2014). Conversely, in the context of cutaneous inflammatory disorders, such as psoriasis, IL-22 has been found to be a significant mediator of inflammation and is required for disease progression (Ma et al., 2008). Recent research evaluating Th22 and IL-22 regulation of solid tumor progression has revealed that while this cell population may be beneficial during chronic inflammation in certain tissues within the gastrointestinal tract (Sabat et al., 2014), this role can shift to one of disease promotion once invasive neoplasms have developed. Th22 cells are reportedly enriched in human CRC (Huang et al., 2015) and gastric cancer (Zhuang et al., 2012). It has been reported that IL-22 production by T cells promotes stemness of colon carcinoma cells via activation of STAT3 signaling and downstream epigenetic modifications of core stem cell gene promoters, such as *NANOG* and *SOX2* (Kryczek et al., 2014). Additional studies have found that Th22 cells are also positively associated with tumor stage in HCC, and that intratumoral IL-22 expression inversely correlates with disease-free survival and overall survival (Kuang et al., 2014). It should be noted, however, that some of these pioneering studies evaluating Th22 function in cancer have focused at least partially on "nonclassical" subsets of Th22 cells that, in addition to IL-22, are also reported to express IL-17 and/or IFN-γ (Kuang et al., 2014). This raises questions of the capacity for Th22 plasticity and the context(s) in which a "classical" Th22 may take on phenotypic properties of other Th populations, or vice versa. Further, it could be argued that perhaps these "nonclassical" Th22 cells represent the same cell population that others have referred to as Th17/Th1 intermediate cells. Additional studies will be required to more robustly characterize the relationship between Th22 and other T helper cells in the context of different TMEs and to determine the functional significance of Th22 plasticity in disease.

T Follicular Helper Cells

The primary function of T follicular helper (T_{FH}) cells is to provide help to B cells within B cell follicles of secondary lymphoid tissues. T_{FH} cellular interactions with B cells are essential for germinal center formation, B cell isotype switching, and B cell differentiation into plasma and memory B cells (Qi, 2016). T_{FH} were first described as a unique subset of T helper cells based on surface expression of CXCR5, the chemokine receptor for CXCL13, that is critical for B cell homing to follicles (Breitfeld et al., 2000; Schaerli et al., 2000; Kim et al., 2001). T_{FH} expression of CXCR5 also enables their own chemotactic recruitment to B cell follicles wherein they engage in direct and indirect interactions with B cells. Differentiation of naïve T cells into mature T_{FH} cells is regulated by IL-6 and IL-21, which together drive downstream expression of the necessary T_{FH} transcription factor Bcl6 (Nurieva et al., 2008, 2009; Vogelzang et al., 2008; Eddahri et al., 2009; Johnston et al., 2009). T_{FH} cells recruited to follicles do not produce a "signature" cytokine or set of cytokines as most other T helper subsets, but they are instead capable of secreting an array of factors including IL-21, IFN-γ, and IL-4 in a context-dependent manner (Crotty, 2011). In addition, T_{FH} cells recruited to follicles express the costimulatory molecules ICOS and CD40L, which play important roles in B cell differentiation and Ig production (Breitfeld et al., 2000; Schaerli et al., 2000).

T_{FH} cells have been described to have pathogenic functions in several autoimmune diseases and hematologic malignancies (Ueno et al., 2015), and the role of T_{FH} cells in solid tumors is also beginning to come into focus. In contrast to their observed tumor promotional roles in cancers of the blood, initial T_{FH} research in human solid tumors has correlated T_{FH} presence with improved patient prognosis. It was recently demonstrated that tertiary lymphoid structures rich in T_{FH} cells are detectable in regions directly adjacent to human breast tumors, particularly in TMEs characterized by extensive intratumoral immune infiltration (Gu-Trantien et al., 2013). Furthermore, T_{FH}-specific gene signatures detected in these immune-rich breast tumors positively associate with patient disease-free survival as well as pathologic complete response to neoadjuvant chemotherapy. Gu-Trantien and colleagues hypothesized that in parallel to their role in promoting germinal center development in secondary lymphoid tissues, T_{FH} cells are critical for germinal center formation within tertiary lymphoid structures, thereby contributing to generation and maintenance of these lymphocyte-dense regions that have capacity to participate in long-term antitumor immune responses (Gu-Trantien et al., 2013).

In human CRC, upregulation of T_{FH} associated genes and genes related to B cell costimulation also correlate with prolonged patient survival (Bindea et al., 2013). Loss of the chemokine CXCL13 in this human CRC patient cohort was associated with both lower densities of B and T_{FH} cells at tumor invasive margins, and increased risk of disease relapse. T_{FH} cell infiltration in human NSCLC has been similarly associated with improved survival outcomes (Ma et al., 2016). Collectively, these findings implicate T_{FH} in durable antitumor responses, although detailed cellular mechanisms underlying this biology have not yet been determined.

Given the strong interactions between T_{FH} cells and B cells, it will be interesting to determine if and how T_{FH} cells impact tumor responses in other solid malignancies, especially since B cells and humoral immunity are now recognized to foster tumor progression in some cancer contexts, including squamous cell carcinoma, pancreatic ductal adenocarcinoma, and prostate carcinoma (de Visser et al., 2005; Ammirante et al., 2010; Affara et al., 2014; Shalapour et al., 2015; Gunderson et al., 2016; Lee et al., 2016; Pylayeva-Gupta et al., 2016). It is also important to consider that, in addition to the surface proteins mentioned previously, differentiated T_{FH} also express high levels of the immunoinhibitory checkpoint receptor programmed cell death protein 1 (PD-1) (Crotty, 2011). PD-1 binding to its ligand PD-L1 on B cells, myeloid cells, and/or neoplastic cells can lead to reduced T cell proliferation, effector activity, and survival (Topalian et al., 2015). The data indicate that sustained T cell–B cell interactions and long-lasting T cell receptor stimulation within germinal centers give rise to high expression of PD-1 on T_{FH} cells that could serve to limit their proliferation and survival. However, studies have indicated that although PD-1/PD-L1 binding between T_{FH} and B cells in germinal centers can restrict the absolute number of T_{FH} cells, loss of PD-1 leads to reduced levels of T_{FH} cytokines and impairs germinal center formation, corresponding with a reduction in B cell survival (Good-Jacobson et al., 2010). As PD-1 and PD-L1 inhibitors are now being used clinically to treat some malignancies and are being evaluated in a broader set of clinical trials in many tumor types, it will be necessary to assess how blockade of this pathway impacts T_{FH} cell contributions to tumor immunity on a tumor type-specific basis.

CONCLUSIONS

Based on the constantly expanding knowledge surrounding the types and functions of CD4$^+$ T cell subsets within TMEs, it is increasingly clear that one size does not fit all when assessing how different lineages of T helper cells regulate tumor immunity (Fig. 8.2). While it is true that some Th populations appear to have strong tumor promotional capabilities in certain settings, it is also evident that T helper cells as a whole should not be underestimated in their ability to direct significant antitumor responses given the

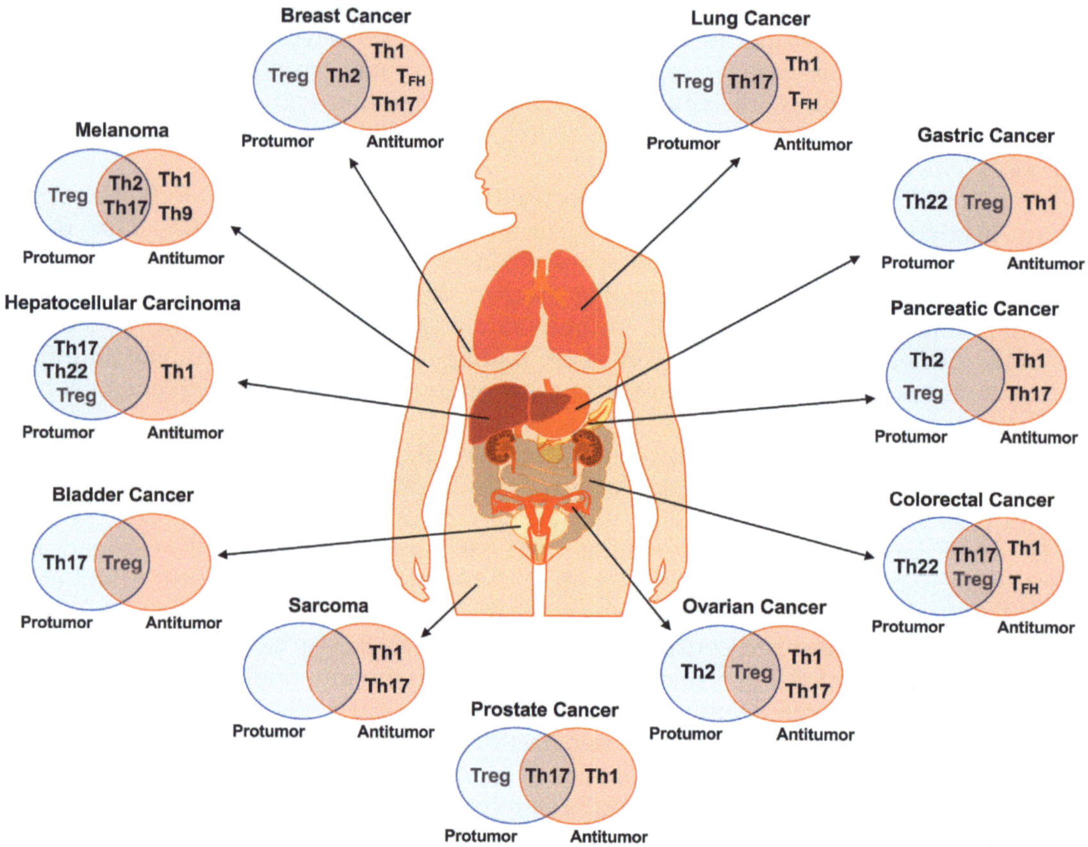

FIGURE 8.2 **CD4$^+$ T cells are important regulators of tumor promotion and/or tumor suppression in solid tumors.** T helper subsets are involved in critical immunoregulatory processes across a range of cancers and exhibit dynamic functions across tumor types. Th populations associated with protumor processes (blue circles) and antitumor processes (red circles) in various human cancers, as well as in murine models that recapitulate aspects of human disease, are summarized by general tumor type/anatomic location. Th subsets implicated in both pro- and antitumor mechanisms in a given tumor type are depicted in centers of Venn diagrams. Though not a T helper population, CD4$^+$ T regulatory cells (Treg) are also included, as they are an additional CD4$^+$ T cell population known to play significant roles in malignancy. Their depicted functions are based on research summarized in reviews by the Galon and Johansson laboratories, which we encourage readers to visit for more detailed information about Treg roles and potential plasticity during neoplastic progression (Mougiakakos et al., 2010; Fridman et al., 2012).

appropriate environmental context, and could perhaps be better leveraged in adoptive cell transfer, cancer vaccines, and/or other immunotherapeutic platforms to help eradicate tumors. Future research aimed at better understanding Th cell subset functionality, plasticity, and tumor antigen recognition in cancer is highly warranted, as it will surely illuminate new avenues by which we might take advantage of Th biology to therapeutically tackle cancers.

References

Adeegbe, D.O., and Nishikawa, H., 2013. Natural and induced T regulatory cells in cancer. Front. Immunol. 4, 190.

Affara, N.I., Ruffell, B., Medler, T.R., et al., 2014. B cells regulate macrophage phenotype and response to chemotherapy in squamous carcinomas. Cancer Cell 25, 809–821.

Afkarian, M., Sedy, J.R., Yang, J., et al., 2002. T-bet is a STAT1-induced regulator of IL-12R expression in naive $CD4^+$ T cells. Nat. Immunol. 3, 549–557.

Ammirante, M., Luo, J.L., Grivennikov, S., et al., 2010. B-cell-derived lymphotoxin promotes castration-resistant prostate cancer. Nature 464, 302–306.

Andreu, P., Johansson, M., Affara, N.I., et al., 2010. FcRgamma activation regulates inflammation-associated squamous carcinogenesis. Cancer Cell 17, 121–134.

Bettelli, E., Carrier, Y., Gao, W., et al., 2006. Reciprocal developmental pathways for the generation of pathogenic effector TH17 and regulatory T cells. Nature 441, 235–238.

Bevan, M.J., 2004. Helping the CD8(+) T-cell response. Nat. Rev. Immunol. 4, 595–602.

Bhavsar, N.A., Bream, J.H., Meeker, A.K., et al., 2014. A peripheral circulating TH1 cytokine profile is inversely associated with prostate cancer risk in CLUE II. Cancer Epidemiol. Biomarkers Prev. 23, 2561–2567.

Bindea, G., Mlecnik, B., Tosolini, M., et al., 2013. Spatiotemporal dynamics of intratumoral immune cells reveal the immune landscape in human cancer. Immunity 39, 782–795.

Bradley, L.M., Dalton, D.K., and Croft, M., 1996. A direct role for IFN-gamma in regulation of Th1 cell development. J. Immunol. 157, 1350–1358.

Braumuller, H., Wieder, T., Brenner, E., et al., 2013. T-helper-1-cell cytokines drive cancer into senescence. Nature 494, 361–365.

Breitfeld, D., Ohl, L., Kremmer, E., et al., 2000. Follicular B helper T cells express CXC chemokine receptor 5, localize to B cell follicles, and support immunoglobulin production. J. Exp. Med. 192, 1545–1552.

Camus, M., Tosolini, M., Mlecnik, B., et al., 2009. Coordination of intratumoral immune reaction and human colorectal cancer recurrence. Cancer Res. 69, 2685–2693.

Chang, S.H., Mirabolfathinejad, S.G., Katta, H., et al., 2014. T helper 17 cells play a critical pathogenic role in lung cancer. Proc. Natl. Acad. Sci. U.S.A. 111, 5664–5669.

Coffelt, S.B., Kersten, K., Doornebal, C.W., et al., 2015. IL-17-producing gammadelta T cells and neutrophils conspire to promote breast cancer metastasis. Nature 522, 345–348.

Cosmi, L., Maggi, L., Santarlasci, V., et al., 2010. Identification of a novel subset of human circulating memory CD4(+) T cells that produce both IL-17A and IL-4. J. Allergy Clin. Immunol. 125 222–230 e221–224.

Cote-Sierra, J., Foucras, G., Guo, L., et al., 2004. Interleukin 2 plays a central role in Th2 differentiation. Proc. Natl. Acad. Sci. U.S.A. 101, 3880–3885.

Coussens, L.M., Raymond, W.W., Bergers, G., et al., 1999. Inflammatory mast cells up-regulate angiogenesis during squamous epithelial carcinogenesis. Genes Dev. 13, 1382–1397.

Crotty, S., 2011. Follicular helper $CD4^+$ T cells (TFH). Annu. Rev. Immunol. 29, 621–663.

Crotty, S., 2015. A brief history of T cell help to B cells. Nat. Rev. Immunol. 15, 185–189.

Dardalhon, V., Awasthi, A., Kwon, H., et al., 2008. IL-4 inhibits TGF-beta-induced Foxp3+ T cells and, together with TGF-beta, generates IL-9+ IL-10+ Foxp3(−) effector T cells. Nat. Immunol. 9, 1347–1355.

De Monte, L., Reni, M., Tassi, E., et al., 2011. Intratumor T helper type 2 cell infiltrate correlates with cancer-associated fibroblast thymic stromal lymphopoietin production and reduced survival in pancreatic cancer. J. Exp. Med. 208, 469–478.

de Visser, K.E., Korets, L.V., and Coussens, L.M., 2005. De novo carcinogenesis promoted by chronic inflammation is B lymphocyte dependent. Cancer Cell 7, 411–423.

DeNardo, D.G., Barreto, J.B., Andreu, P., et al., 2009. CD4(+) T cells regulate pulmonary metastasis of mammary carcinomas by enhancing protumor properties of macrophages. Cancer Cell 16, 91–102.

Dubin, P.J., and Kolls, J.K., 2008. Th17 cytokines and mucosal immunity. Immunol. Rev. 226, 160–171.

Duhen, T., Geiger, R., Jarrossay, D., et al., 2009. Production of interleukin 22 but not interleukin 17 by a subset of human skin-homing memory T cells. Nat. Immunol. 10, 857–863.

Eddahri, F., Denanglaire, S., Bureau, F., et al., 2009. Interleukin-6/STAT3 signaling regulates the ability of naive T cells to acquire B-cell help capacities. Blood 113, 2426–2433.

Eller, K., Wolf, D., Huber, J.M., et al., 2011. IL-9 production by regulatory T cells recruits mast cells that are essential for regulatory T cell-induced immune suppression. J. Immunol. 186, 83–91.

Eyerich, S., Eyerich, K., Pennino, D., et al., 2009. Th22 cells represent a distinct human T cell subset involved in epidermal immunity and remodeling. J. Clin. Invest. 119, 3573–3585.

Facciabene, A., Motz, G.T., and Coukos, G., 2012. T-regulatory cells: key players in tumor immune escape and angiogenesis. Cancer Res. 72, 2162–2171.

Fallarino, F., and Gajewski, T.F., 1999. Cutting edge: differentiation of antitumor CTL in vivo requires host expression of Stat1. J. Immunol. 163, 4109–4113.

Fridman, W.H., Pages, F., Sautes-Fridman, C., et al., 2012. The immune contexture in human tumours: impact on clinical outcome. Nat. Rev. Cancer 12, 298–306.

Galon, J., Mlecnik, B., Bindea, G., et al., 2014. Towards the introduction of the 'Immunoscore' in the classification of malignant tumours. J. Pathol. 232, 199–209.

Gao, Q., Wang, X.Y., Qiu, S.J., et al., 2011. Tumor stroma reaction-related gene signature predicts clinical outcome in human hepatocellular carcinoma. Cancer Sci. 102, 1522–1531.

Gnerlich, J.L., Mitchem, J.B., Weir, J.S., et al., 2010. Induction of Th17 cells in the tumor microenvironment improves survival in a murine model of pancreatic cancer. J. Immunol. 185, 4063–4071.

Gooch, J.L., Christy, B., and Yee, D., 2002. STAT6 mediates interleukin-4 growth inhibition in human breast cancer cells. Neoplasia 4, 324–331.

Good-Jacobson, K.L., Szumilas, C.G., Chen, L., et al., 2010. PD-1 regulates germinal center B cell survival and the formation and affinity of long-lived plasma cells. Nat. Immunol. 11, 535–542.

Gu-Trantien, C., Loi, S., Garaud, S., et al., 2013. CD4(+) follicular helper T cell infiltration predicts breast cancer survival. J. Clin. Invest. 123, 2873–2892.

Guery, L., and Hugues, S., 2015. Th17 cell plasticity and functions in cancer immunity. Biomed. Res. Int. 2015, 314620.

Gunderson, A.J., Kaneda, M.M., Tsujikawa, T., et al., 2016. Bruton tyrosine kinase-dependent immune cell cross-talk drives pancreas cancer. Cancer Discov. 6, 270–285.

Hanahan, D., and Coussens, L.M., 2012. Accessories to the crime: functions of cells recruited to the tumor microenvironment. Cancer Cell 21, 309–322.

Hanahan, D., and Weinberg, R.A., 2011. Hallmarks of cancer: the next generation. Cell 144, 646–674.

Harrington, L.E., Hatton, R.D., Mangan, P.R., et al., 2005. Interleukin 17-producing $CD4^+$ effector T cells develop via a lineage distinct from the T helper type 1 and 2 lineages. Nat. Immunol. 6, 1123–1132.

He, D., Li, H., Yusuf, N., et al., 2010. IL-17 promotes tumor development through the induction of tumor promoting microenvironments at tumor sites and myeloid-derived suppressor cells. J. Immunol. 184, 2281–2288.

Hennequin, A., Derangere, V., Boidot, R., et al., 2016. Tumor infiltration by Tbet+ effector T cells and CD20+ B cells is associated with survival in gastric cancer patients. Oncoimmunology 5, e1054598.

Housseau, F., Wu, S., Wick, E.C., et al., 2016. Redundant innate and adaptive sources of IL17 production drive colon tumorigenesis. Cancer Res. 76, 2115–2124.

Huang, Y.H., Cao, Y.F., Jiang, Z.Y., et al., 2015. Th22 cell accumulation is associated with colorectal cancer development. World J. Gastroenterol. 21, 4216–4224.

Ivanov, I.I., McKenzie, B.S., Zhou, L., et al., 2006. The orphan nuclear receptor RORgammat directs the differentiation program of proinflammatory IL-17+ T helper cells. Cell 126, 1121–1133.

Johnston, R.J., Poholek, A.C., DiToro, D., et al., 2009. Bcl6 and Blimp-1 are reciprocal and antagonistic regulators of T follicular helper cell differentiation. Science 325, 1006–1010.

Josefowicz, S.Z., Lu, L.F., and Rudensky, A.Y., 2012. Regulatory T cells: mechanisms of differentiation and function. Annu. Rev. Immunol. 30, 531–564.

Kaplan, M.H., 2013. Th9 cells: differentiation and disease. Immunol. Rev. 252, 104–115.

Kaplan, M.H., Schindler, U., Smiley, S.T., et al., 1996. Stat6 is required for mediating responses to IL-4 and for development of Th2 cells. Immunity 4, 313–319.

Kaplan, M.H., Hufford, M.M., and Olson, M.R., 2015. The development and in vivo function of T helper 9 cells. Nat. Rev. Immunol. 15, 295–307.

Kim, C.H., Rott, L.S., Clark-Lewis, I., et al., 2001. Subspecialization of CXCR5+ T cells: B helper activity is focused in a germinal center-localized subset of CXCR5+ T cells. J. Exp. Med. 193, 1373–1381.

Koenen, H.J., Smeets, R.L., Vink, P.M., et al., 2008. Human CD25highFoxp3pos regulatory T cells differentiate into IL-17-producing cells. Blood 112, 2340–2352.

Komatsu, N., Okamoto, K., Sawa, S., et al., 2014. Pathogenic conversion of Foxp3+ T cells into TH17 cells in autoimmune arthritis. Nat. Med. 20, 62–68.

Korn, T., Bettelli, E., Oukka, M., et al., 2009. IL-17 and Th17 Cells. Annu. Rev. Immunol. 27, 485–517.

Kryczek, I., Banerjee, M., Cheng, P., et al., 2009a. Phenotype, distribution, generation, and functional and clinical relevance of Th17 cells in the human tumor environments. Blood 114, 1141–1149.

Kryczek, I., Wei, S., Szeliga, W., et al., 2009b. Endogenous IL-17 contributes to reduced tumor growth and metastasis. Blood 114, 357–359.

Kryczek, I., Wu, K., Zhao, E., et al., 2011. IL-17+ regulatory T cells in the microenvironments of chronic inflammation and cancer. J. Immunol. 186, 4388–4395.

Kryczek, I., Lin, Y., Nagarsheth, N., et al., 2014. IL-22(+)CD4(+) T cells promote colorectal cancer stemness via STAT3 transcription factor activation and induction of the methyltransferase DOT1L. Immunity 40, 772–784.

Kuang, D.M., Zhao, Q., Peng, C., et al., 2009. Activated monocytes in peritumoral stroma of hepatocellular carcinoma foster immune privilege and disease progression through PD-L1. J. Exp. Med. 206, 1327–1337.

Kuang, D.M., Xiao, X., Zhao, Q., et al., 2014. B7-H1-expressing antigen-presenting cells mediate polarization of protumorigenic Th22 subsets. J. Clin. Invest. 124, 4657–4667.

Kusuda, T., Shigemasa, K., Arihiro, K., et al., 2005. Relative expression levels of Th1 and Th2 cytokine mRNA are independent prognostic factors in patients with ovarian cancer. Oncol. Rep. 13, 1153–1158.

Langrish, C.L., Chen, Y., Blumenschein, W.M., et al., 2005. IL-23 drives a pathogenic T cell population that induces autoimmune inflammation. J. Exp. Med. 201, 233–240.

Lee, K.E., Spata, M., Bayne, L.J., et al., 2016. Hif1a deletion reveals pro-neoplastic function of B cells in pancreatic neoplasia. Cancer Discov. 6, 256–269.

Lee, Y.K., Turner, H., Maynard, C.L., et al., 2009. Late developmental plasticity in the T helper 17 lineage. Immunity 30, 92–107.

Lighvani, A.A., Frucht, D.M., Jankovic, D., et al., 2001. T-bet is rapidly induced by interferon-gamma in lymphoid and myeloid cells. Proc. Natl. Acad. Sci. U.S.A. 98, 15137–15142.

Lu, Y., Hong, S., Li, H., et al., 2012. Th9 cells promote antitumor immune responses in vivo. J. Clin. Invest. 122, 4160–4171.

Ma, H.L., Liang, S., Li, J., et al., 2008. IL-22 is required for Th17 cell-mediated pathology in a mouse model of psoriasis-like skin inflammation. J. Clin. Invest. 118, 597–607.

Ma, Q.Y., Huang, D.Y., Zhang, H.J., et al., 2016. Function of follicular helper T cell is impaired and correlates with survival time in non-small cell lung cancer. Int. Immunopharmacol. 41, 1–7.

Ma, S., Cheng, Q., Cai, Y., et al., 2014. IL-17A produced by gammadelta T cells promotes tumor growth in hepatocellular carcinoma. Cancer Res. 74, 1969–1982.

Magram, J., Connaughton, S.E., Warrier, R.R., et al., 1996. IL-12-deficient mice are defective in IFN gamma production and type 1 cytokine responses. Immunity 4, 471–481.

Martin-Orozco, N., Muranski, P., Chung, Y., et al., 2009. T helper 17 cells promote cytotoxic T cell activation in tumor immunity. Immunity 31, 787–798.

Masso-Valles, D., Jauset, T., Serrano, E., et al., 2015. Ibrutinib exerts potent antifibrotic and antitumor activities in mouse models of pancreatic adenocarcinoma. Cancer Res. 75, 1675–1681.

Mattes, J., Hulett, M., Xie, W., et al., 2003. Immunotherapy of cytotoxic T cell-resistant tumors by T helper 2 cells: an eotaxin and STAT6-dependent process. J. Exp. Med. 197, 387–393.

Miyahara, Y., Odunsi, K., Chen, W., et al., 2008. Generation and regulation of human $CD4^+$ IL-17-producing T cells in ovarian cancer. Proc. Natl. Acad. Sci. U.S.A. 105, 15505–15510.

Mosmann, T.R., and Coffman, R.L., 1989. TH1 and TH2 cells: different patterns of lymphokine secretion lead to different functional properties. Annu. Rev. Immunol. 7, 145–173.

Mosmann, T.R., Cherwinski, H., Bond, M.W., et al., 1986. Two types of murine helper T cell clone. I. Definition according to profiles of lymphokine activities and secreted proteins. J. Immunol. 136, 2348–2357.

Mougiakakos, D., Choudhury, A., Lladser, A., et al., 2010. Regulatory T cells in cancer. Adv. Cancer Res. 107, 57–117.

Mullen, A.C., High, F.A., Hutchins, A.S., et al., 2001. Role of T-bet in commitment of TH1 cells before IL-12-dependent selection. Science 292, 1907–1910.

Muranski, P., Boni, A., Antony, P.A., et al., 2008. Tumor-specific Th17-polarized cells eradicate large established melanoma. Blood 112, 362–373.

Murphy, C.A., Langrish, C.L., Chen, Y., et al., 2003. Divergent pro- and antiinflammatory roles for IL-23 and IL-12 in joint autoimmune inflammation. J. Exp. Med. 198, 1951–1957.

Nagai, S., and Toi, M., 2000. Interleukin-4 and breast cancer. Breast Cancer 7, 181–186.

Nastala, C.L., Edington, H.D., McKinney, T.G., et al., 1994. Recombinant IL-12 administration induces tumor regression in association with IFN-gamma production. J. Immunol. 153, 1697–1706.

Nevala, W.K., Vachon, C.M., Leontovich, A.A., et al., 2009. Evidence of systemic Th2-driven chronic inflammation in patients with metastatic melanoma. Clin. Cancer Res. 15, 1931–1939.

Nistala, K., Adams, S., Cambrook, H., et al., 2010. Th17 plasticity in human autoimmune arthritis is driven by the inflammatory environment. Proc. Natl. Acad. Sci. U.S.A. 107, 14751–14756.

Noben-Trauth, N., Hu-Li, J., and Paul, W.E., 2002. IL-4 secreted from individual naive $CD4^+$ T cells acts in an autocrine manner to induce Th2 differentiation. Eur. J. Immunol. 32, 1428–1433.

Numasaki, M., Fukushi, J., Ono, M., et al., 2003. Interleukin-17 promotes angiogenesis and tumor growth. Blood 101, 2620–2627.

Nurieva, R.I., Chung, Y., Hwang, D., et al., 2008. Generation of T follicular helper cells is mediated by interleukin-21 but independent of T helper 1, 2, or 17 cell lineages. Immunity 29, 138–149.

Nurieva, R.I., Chung, Y., Martinez, G.J., et al., 2009. Bcl6 mediates the development of T follicular helper cells. Science 325, 1001–1005.

O'Garra, A., 1998. Cytokines induce the development of functionally heterogeneous T helper cell subsets. Immunity 8, 275–283.

Obermajer, N., Popp, F.C., Soeder, Y., et al., 2014. Conversion of Th17 into IL-17A(neg) regulatory T cells: a novel mechanism in prolonged allograft survival promoted by mesenchymal stem cell-supported minimized immunosuppressive therapy. J. Immunol. 193, 4988–4999.

Ochi, A., Nguyen, A.H., Bedrosian, A.S., et al., 2012. MyD88 inhibition amplifies dendritic cell capacity to promote pancreatic carcinogenesis via Th2 cells. J. Exp. Med. 209, 1671–1687.

Ouyang, W., Kolls, J.K., and Zheng, Y., 2008. The biological functions of T helper 17 cell effector cytokines in inflammation. Immunity 28, 454–467.

Owaki, T., Asakawa, M., Morishima, N., et al., 2005. A role for IL-27 in early regulation of Th1 differentiation. J. Immunol. 175, 2191–2200.

Pai, S.Y., Truitt, M.L., and Ho, I.C., 2004. GATA-3 deficiency abrogates the development and maintenance of T helper type 2 cells. Proc. Natl. Acad. Sci. U.S.A. 101, 1993–1998.

Palucka, A.K., and Coussens, L.M., 2016. The basis of oncoimmunology. Cell. 164, 1233–1247.

Park, H., Li, Z., Yang, X.O., et al., 2005. A distinct lineage of $CD4^+$ T cells regulates tissue inflammation by producing interleukin 17. Nat. Immunol. 6, 1133–1141.

Pflanz, S., Timans, J.C., Cheung, J., et al., 2002. IL-27, a heterodimeric cytokine composed of EBI3 and p28 protein, induces proliferation of naive $CD4^+$ T cells. Immunity 16, 779–790.

Punt, S., Langenhoff, J.M., Putter, H., et al., 2015. The correlations between IL-17 vs. Th17 cells and cancer patient survival: a systematic review. Oncoimmunology 4, e984547.

Purwar, R., Schlapbach, C., Xiao, S., et al., 2012. Robust tumor immunity to melanoma mediated by interleukin-9-producing T cells. Nat. Med. 18, 1248–1253.

Pylayeva-Gupta, Y., Das, S., Handler, J.S., et al., 2016. IL35-producing B cells promote the development of pancreatic neoplasia. Cancer Discov. 6, 247–255.

Qi, H., 2016. T follicular helper cells in space-time. Nat. Rev. Immunol. 16, 612–625.

Quezada, S.A., Simpson, T.R., Peggs, K.S., et al., 2010. Tumor-reactive CD4(+) T cells develop cytotoxic activity and eradicate large established melanoma after transfer into lymphopenic hosts. J. Exp. Med. 207, 637–650.

Rei, M., Goncalves-Sousa, N., Lanca, T., et al., 2014. Murine CD27(−) Vgamma6(+) gammadelta T cells producing IL-17A promote ovarian cancer growth via mobilization of protumor small peritoneal macrophages. Proc. Natl. Acad. Sci. U.S.A. 111, E3562–3570.

Restifo, N.P., Dudley, M.E., and Rosenberg, S.A., 2012. Adoptive immunotherapy for cancer: harnessing the T cell response. Nat. Rev. Immunol. 12, 269–281.

Ruffell, B., DeNardo, D.G., Affara, N.I., et al., 2010. Lymphocytes in cancer development: polarization towards pro-tumor immunity. Cytokine Growth Factor Rev. 21, 3–10.

Sabat, R., Ouyang, W., and Wolk, K., 2014. Therapeutic opportunities of the IL-22-IL-22R1 system. Nat. Rev. Drug Discov. 13, 21–38.

Schaerli, P., Willimann, K., Lang, A.B., et al., 2000. CXC chemokine receptor 5 expression defines follicular homing T cells with B cell helper function. J. Exp. Med. 192, 1553–1562.

Sehra, S., Yao, W., Nguyen, E.T., et al., 2015. TH9 cells are required for tissue mast cell accumulation during allergic inflammation. J. Allergy Clin. Immunol. 136 433–440 e431.

Sfanos, K.S., Bruno, T.C., Maris, C.H., et al., 2008. Phenotypic analysis of prostate-infiltrating lymphocytes reveals TH17 and Treg skewing. Clin. Cancer Res. 14, 3254–3261.

Shalapour, S., Font-Burgada, J., Di Caro, G., et al., 2015. Immunosuppressive plasma cells impede T-cell-dependent immunogenic chemotherapy. Nature 521, 94–98.

Shankaran, V., Ikeda, H., Bruce, A.T., et al., 2001. IFNgamma and lymphocytes prevent primary tumour development and shape tumour immunogenicity. Nature 410, 1107–1111.

Sheu, B.C., Lin, R.H., Lien, H.C., et al., 2001. Predominant Th2/Tc2 polarity of tumor-infiltrating lymphocytes in human cervical cancer. J. Immunol. 167, 2972–2978.

Shiao, S.L., Ruffell, B., DeNardo, D.G., et al., 2015. Th2-polarized $CD4^+$ T cells and macrophages limit efficacy of radiotherapy. Cancer Immunol. Res. 3, 518–525.

Shimoda, K., van Deursen, J., Sangster, M.Y., et al., 1996. Lack of IL-4-induced Th2 response and IgE class switching in mice with disrupted Stat6 gene. Nature 380, 630–633.

Street, S.E., Trapani, J.A., MacGregor, D., et al., 2002. Suppression of lymphoma and epithelial malignancies effected by interferon gamma. J. Exp. Med. 196, 129–134.

Szabo, S.J., Dighe, A.S., Gubler, U., et al., 1997. Regulation of the interleukin (IL)-12R beta 2 subunit expression in developing T helper 1 (Th1) and Th2 cells. J. Exp. Med. 185, 817–824.

Szabo, S.J., Kim, S.T., Costa, G.L., et al., 2000. A novel transcription factor, T-bet, directs Th1 lineage commitment. Cell 100, 655–669.

Szabo, S.J., Sullivan, B.M., Stemmann, C., et al., 2002. Distinct effects of T-bet in TH1 lineage commitment and IFN-gamma production in CD4 and CD8 T cells. Science 295, 338–342.

Takeda, K., Tanaka, T., Shi, W., et al., 1996. Essential role of Stat6 in IL-4 signalling. Nature 380, 627–630.

Takeda, K., Tsutsui, H., Yoshimoto, T., et al., 1998. Defective NK cell activity and Th1 response in IL-18-deficient mice. Immunity 8, 383–390.

Tepper, R.I., Pattengale, P.K., and Leder, P., 1989. Murine interleukin-4 displays potent anti-tumor activity in vivo. Cell. 57, 503–512.

Tepper, R.I., Coffman, R.L., and Leder, P., 1992. An eosinophil-dependent mechanism for the antitumor effect of interleukin-4. Science 257, 548–551.

Topalian, S.L., Drake, C.G., and Pardoll, D.M., 2015. Immune checkpoint blockade: a common denominator approach to cancer therapy. Cancer Cell 27, 450–461.

Tosolini, M., Kirilovsky, A., Mlecnik, B., et al., 2011. Clinical impact of different classes of infiltrating T cytotoxic and helper cells (Th1, th2, treg, th17) in patients with colorectal cancer. Cancer Res. 71, 1263–1271.

Trifari, S., Kaplan, C.D., Tran, E.H., et al., 2009. Identification of a human helper T cell population that has abundant production of interleukin 22 and is distinct from T(H)-17, T(H)1 and T(H)2 cells. Nat. Immunol. 10, 864–871.

Trinchieri, G., 1995. Interleukin-12: a proinflammatory cytokine with immunoregulatory functions that bridge innate resistance and antigen-specific adaptive immunity. Annu. Rev. Immunol. 13, 251–276.

Ueno, H., Banchereau, J., and Vinuesa, C.G., 2015. Pathophysiology of T follicular helper cells in humans and mice. Nat. Immunol. 16, 142–152.

Usui, T., Preiss, J.C., Kanno, Y., et al., 2006. T-bet regulates Th1 responses through essential effects on GATA-3 function rather than on IFNG gene acetylation and transcription. J. Exp. Med. 203, 755–766.

Valmori, D., Raffin, C., Raimbaud, I., et al., 2010. Human RORgammat+ TH17 cells preferentially differentiate from naive FOXP3+ Treg in the presence of lineage-specific polarizing factors. Proc. Natl. Acad. Sci. U.S.A. 107, 19402–19407.

Veldhoen, M., Hocking, R.J., Atkins, C.J., et al., 2006. TGFbeta in the context of an inflammatory cytokine milieu supports de novo differentiation of IL-17-producing T cells. Immunity 24, 179–189.

Veldhoen, M., Uyttenhove, C., van Snick, J., et al., 2008. Transforming growth factor-beta 'reprograms' the differentiation of T helper 2 cells and promotes an interleukin 9-producing subset. Nat. Immunol. 9, 1341–1346.

Vogelzang, A., McGuire, H.M., Yu, D., et al., 2008. A fundamental role for interleukin-21 in the generation of T follicular helper cells. Immunity 29, 127–137.

Voo, K.S., Wang, Y.H., Santori, F.R., et al., 2009. Identification of IL-17-producing FOXP3+ regulatory T cells in humans. Proc. Natl. Acad. Sci. U.S.A. 106, 4793–4798.

Wang, L., Yi, T., Kortylewski, M., et al., 2009. IL-17 can promote tumor growth through an IL-6-Stat3 signaling pathway. J. Exp. Med. 206, 1457–1464.

Wang, Y.H., Voo, K.S., Liu, B., et al., 2010. A novel subset of CD4(+) T(H)2 memory/effector cells that produce inflammatory IL-17 cytokine and promote the exacerbation of chronic allergic asthma. J. Exp. Med. 207, 2479–2491.

Wilke, C.M., Kryczek, I., Wei, S., et al., 2011. Th17 cells in cancer: help or hindrance? Carcinogenesis 32, 643–649.

Wu, B., Huang, C., Kato-Maeda, M., et al., 2008. IL-9 is associated with an impaired Th1 immune response in patients with tuberculosis. Clin. Immunol. 126, 202–210.

Wu, P., Wu, D., Ni, C., et al., 2014. gammadeltaT17 cells promote the accumulation and expansion of myeloid-derived suppressor cells in human colorectal cancer. Immunity 40, 785–800.

Wu, S., Rhee, K.J., Albesiano, E., et al., 2009. A human colonic commensal promotes colon tumorigenesis via activation of T helper type 17 T cell responses. Nat. Med. 15, 1016–1022.

Xie, Y., Akpinarli, A., Maris, C., et al., 2010. Naive tumor-specific CD4(+) T cells differentiated in vivo eradicate established melanoma. J. Exp. Med. 207, 651–667.

Xu, L., Kitani, A., Fuss, I., et al., 2007. Cutting edge: regulatory T cells induce $CD4^+$ CD25-Foxp3- T cells or are self-induced to become Th17 cells in the absence of exogenous TGF-beta. J. Immunol. 178, 6725–6729.

Yang, L., Qi, Y., Hu, J., et al., 2012. Expression of Th17 cells in breast cancer tissue and its association with clinical parameters. Cell. Biochem. Biophys. 62, 153–159.

Yang, X.O., Pappu, B.P., Nurieva, R., et al., 2008. T helper 17 lineage differentiation is programmed by orphan nuclear receptors ROR alpha and ROR gamma. Immunity 28, 29–39.

Ye, Z.J., Zhou, Q., Gu, Y.Y., et al., 2010. Generation and differentiation of IL-17-producing $CD4^+$ T cells in malignant pleural effusion. J. Immunol. 185, 6348–6354.

Zhang, J.P., Yan, J., Xu, J., et al., 2009. Increased intratumoral IL-17-producing cells correlate with poor survival in hepatocellular carcinoma patients. J. Hepatol. 50, 980–989.

Zheng, W., and Flavell, R.A., 1997. The transcription factor GATA-3 is necessary and sufficient for Th2 cytokine gene expression in CD4 T cells. Cell 89, 587–596.

Zhou, L., Ivanov, I.I., Spolski, R., et al., 2007. IL-6 programs T(H)-17 cell differentiation by promoting sequential engagement of the IL-21 and IL-23 pathways. Nat. Immunol. 8, 967–974.

Zhu, J., and Paul, W.E., 2008. CD4 T cells: fates, functions, and faults. Blood 112, 1557–1569.

Zhu, J., Guo, L., Watson, C.J., et al., 2001. Stat6 is necessary and sufficient for IL-4's role in Th2 differentiation and cell expansion. J. Immunol. 166, 7276–7281.

Zhu, J., Yamane, H., Cote-Sierra, J., et al., 2006. GATA-3 promotes Th2 responses through three different mechanisms: induction of Th2 cytokine production, selective growth of Th2 cells and inhibition of Th1 cell-specific factors. Cell. Res. 16, 3–10.

Zhuang, Y., Peng, L.S., Zhao, Y.L., et al., 2012. Increased intratumoral IL-22-producing CD4(+) T cells and Th22 cells correlate with gastric cancer progression and predict poor patient survival. Cancer Immunol. Immunother. 61, 1965–1975.

CHAPTER

9

Development and Function of Natural Killer Cells and Its Importance in Cancer Immunotherapy

Sourav Paul and Girdhari Lal

OUTLINE

Abstract

Natural killer (NK) cells are a group of innate immune cells. They show spontaneous cytolytic activity against cells under stress or infected with a virus. NK cells show a wide tissue distribution and a variable phenotype from tissue to tissue. Unlike adaptive immune cells, the cytolytic function of NK cells does not require clonal expansion. The function of NK cells is regulated by a cumulative signal received from the activating and inhibitory receptors. Mostly, expression of inhibitory receptors is controlled by genetic factors and provides tolerance to the self-tissues, whereas activation of receptor expression is controlled by environmental factors, and helps to mount the effector immune response against virus infection and tumor.

M.A. Hayat (ed): *Immunology, Volume 1.*
DOI: http://dx.doi.org/10.1016/B978-0-12-809819-6.00009-5

© 2018 Elsevier Inc. All rights reserved.

NK cells isolated from the peripheral blood show antitumor cytolytic activity, which has led to several clinical trials using NK cells as adoptive cellular therapy which, however, have showed very limited success. With the advancement of or knowledge about NK cell biology, a combination of modified NK cells or NK cells in combination with other therapeutics showed better clinical output. In this review, we discuss the development of NK cells in humans and mice, various NK cell subsets and tissue distribution, an interaction of NK cells with other innate immune cells and adaptive immune cells, and the mechanism of NK cell-mediated antitumor immunity. We further discuss the use of NK cells to control tumor growth and suggested possible ways to improve the NK cell-based therapy.

INTRODUCTION

The innate immune system provides the first line of defense against invading pathogens. There are several innate immune cells in the body such as γδ T cells, macrophages, dendritic cells (DCs), natural killer (NK) cells, and NKT cells. The NK cell has the ability to lyse tumor or virus-infected cells. NK cells are characterized as $CD3^{-}CD56^{+}$ cells in humans and $CD3^{-}NK1.1^{+}$ cells in some strain of mice (C57BL/6, FVB/N and NZB). NK1.1 is not expressed on the AKR, BALB/c, CBA/J, C3H, DBA, NOD, SJH and 129 strain of mice but in these mice NK cells can be stained with CD49b (DX5). These cells represent 5–15% of lymphocytes in peripheral blood of humans and 2–5% of mouse splenic lymphocytes. In steady state, mouse splenic NK cells are predominantly localized in the red pulp (Gregoire et al., 2007). NK cells are also distributed among various lymphoid organs such as thymus, spleen, lymph nodes, and bone marrow, as well as in other tissues such as liver, lung, uterus, placenta, gut, kidney, and skin. NK cells constitute 5–10% of mouse hepatic lymphocytes whereas in humans they represent 30–50% of total hepatic lymphocytes. Hepatic NK cells are preferentially located in hepatic sinusoids and contribute to liver tolerance under physiological conditions. Hepatic NK cells are known to promote Treg by potentiating the function of DCs (Jinushi et al., 2007). NK cells represent around 10% of total lymphocytes in the normal healthy lung (Gregoire et al., 2007). Lung-resident NK cells play an important role in host defense against bacterial infection. For example, they improve the phagocytic ability of alveolar macrophages following *Staphylococcus aureus* infection (Small et al., 2008). In normal healthy human skin, about 10% of total lymphocytes are $CD3^{-}CD56^{+}$ NK cells (Ebert et al., 2006). Gut-associated NK cells are localized in the lamina propria of the intestine, Peyer's patches and mesenteric lymph nodes, and protect against viral and other pathogens. Uterine NK cells are abundantly present in the decidua and they represent about 70% of total lymphocytes during the first trimester of pregnancy (King et al., 1991). Uterine NK cells play a pivotal role in tissue remodeling and help in establishing tolerance towards fetal allograft.

Based on the expression of CD11b and CD27, NK cells can be further differentiated into four subsets: $CD27^{-}CD11b^{-}$, $CD27^{+}CD11b^{-}$, $CD27^{+}CD11b^{+}$, and $CD27^{-}CD11b^{+}$ (Hayakawa and Smyth, 2006; Paul et al., 2016). The $CD27^{-}CD11b^{+}$ subset is the terminally mature NK cell (mNK) subset and expresses inhibitory Ly49C/I receptors and killer cell lectin-like receptor subfamily G member 1 (KLRG1) but lacks expression of CD69, c-kit, and CD127. This subset is predominantly found in peripheral tissues such as liver and lung and has a limited turnover rate. $CD27^{+}CD11b^{+}$ subset represents intermediate or less mature phenotypes and expresses CD69, c-kit, and CD127. $CD27^{+}CD11b^{+}$ NK cells display higher cytotoxicity, increased cytokine production, and are also known to control function of DCs in mice as compared to the $CD27^-CD11b^+$ subset (Hayakawa and Smyth, 2006). $CD27^-CD11b^-$ and $CD27^+CD11b^-$ NK cells display immature phenotypes and have the potential to further differentiate into mNK subsets. $CD27^+CD11b^-$ and $CD27^+CD11b^+$ subsets are dominant subsets in bone marrow and lymph nodes. The double negative $CD27^-CD11b^-$ subset is the precursor of $CD27^+CD11b^-$ NK cells and represents the most immature NK (iNK) cells (Chiossone et al., 2009). Various types of tumor such as fibrosarcoma, gastrointestinal sarcoma, renal cell sarcoma, lung cancer, and breast tumor were reported to have a high infiltration of NK cells (Hayakawa et al., 2011; Ishigami et al., 2000). Fibrosarcoma and melanoma tumor microenvironments harbor the $CD27^+CD11b^+$ subset of NK cells (Hayakawa et al., 2011; Paul et al., 2016), while $CD27^-CD11b^-$NK cells predominantly migrate in lung tumors (Jin et al., 2013). Using mass cytometry, it has been shown that human NK cells are much more heterogeneous than was previously thought (Horowitz et al., 2013; Newell and Cheng, 2016). Horowitz et at investigated the 35 parameters simultaneously in 5 sets of monozygotic twins as well as in 12 unrelated donors and showed that about 30,000 phenotypically diverse NK cell populations can exist in a given individual (Horowitz et al., 2013). By comparing the twin versus unrelated donors, they showed that genetic factors primarily determine the expression of inhibitory receptors, whereas environment controls the expression of activating receptors. This implies that NK cell self-tolerance is mostly contributed by the inhibitory receptors, whereas response to virus-infected cells or tumors are guided by the activating receptors (Horowitz et al., 2013). The details of the role of inhibitory and activating receptors in distinguishing the phenotypic and functional heterogeneity are discussed in the following sections.

DEVELOPMENT AND MATURATION OF NK CELLS

Development of NK cells predominantly takes place in bone marrow but can also happen in the liver and thymus. The various developmental stages of mouse and human NK cells are schematically represented in Fig. 9.1. Common lymphoid precursor (CLP) is the first stage of NK cell development and is characterized by expression of IL-7Rα (CD122), c-kit (CD117), Sca-1 and Flt-3 (CD135), and the absence of a lineage-specific marker (Kondo et al., 1997). CLPs subsequently develop into an intermediate stage called pre-NKP before developing into NK cell precursors (NKP). Pre-NKP is defined as lineage negative, $CD244^+$c-kit$^-$Flt3$^-$IL-7Rα$^+$CD122$^-$ (Carotta et al., 2011; Fathman et al., 2011). NKP is associated with expression of the IL-2Rβ subunit (CD122), coexpression of CD27 and CD244, and lack of NK1.1 and CD49b (Fathman et al., 2011; Rosmaraki et al., 2001). NKP cells give rise to iNK cells, which are characterized by the loss of IL-7Rα expression and acquisition of the NK1.1 molecule (Rosmaraki et al., 2001). iNK cells further develop into mNK cells through the acquisition of CD49b, CD11b, CD43, and Ly49 receptors (Kim et al., 2002). As NK cells develop from immature to mature stage, they undergo an "education" process whereby they acquire receptors specific for self-major histocompatibility complex (MHC) class I molecules. The engagement of inhibitory receptors with the self-MHC ligands during the education process leads to the development of fully functional mNK cells. These educated mNK cells are fully responsive to activating receptor signaling and display cytotoxicity against target cells and secrete IFN-γ (Orr and Lanier, 2010). The mNK cells migrate to the peripheral tissues.

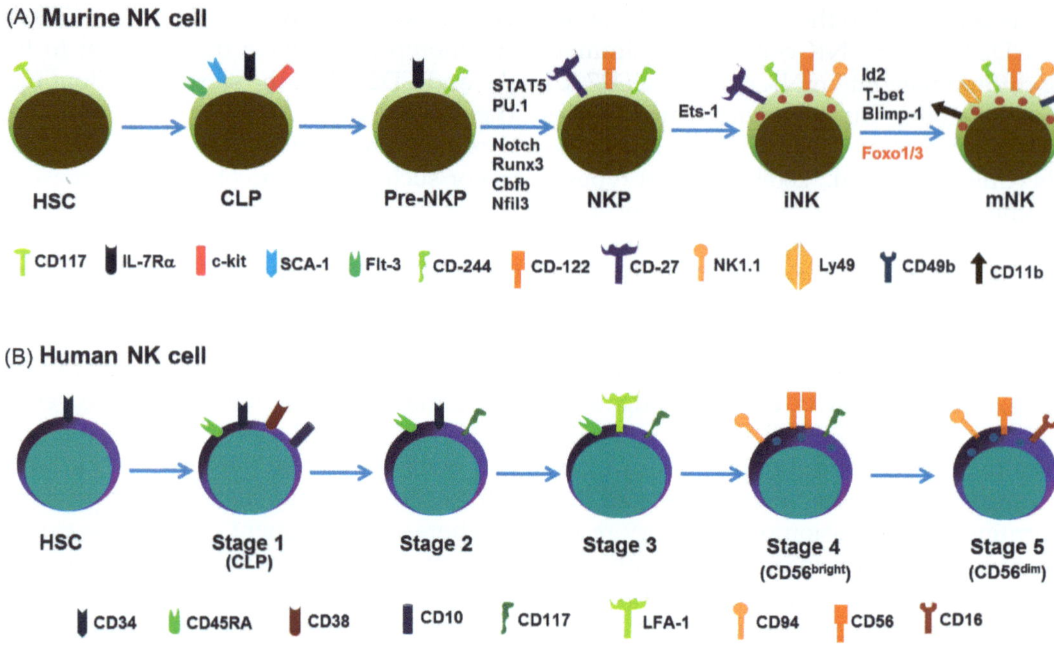

FIGURE 9.1 **Developmental stages of the NK cell.** (A) Mouse and (B) human NK cells develop in the bone marrow from hematopoietic stem cells to mature NK cells through a series of developmental intermediates. Each stage is characterized by the presence of a specific surface antigen, which is used to distinguish one intermediate from the others.

Human NK cell development follows a similar pattern as that of mice, and arises sequentially through four discrete stages (Fig. 9.1). NK cell development originates in the bone marrow from hematopoietic stem cells (HSCs), which are $Lin^-CD34^+CD38^{lo}CD45RA^-CD10^-$ and give rise to CLPs or stage 1 cells ($CD34^+CD38^+CD45RA^+CD10^+CD117^-$) (Freud et al., 2006; Galy et al., 1995). These cells have the potential to differentiate into NK cells, T cells, and DCs. Stage 1 cells further sequentially give rise to stage 2 cells ($CD34^+CD45RA^+CD10^-CD117^+CD94^-$) and intermediate stage 3 cells ($CD34^-CD45RA^+CD117^+CD94^-LFA\text{-}1^+$) (Freud et al., 2005, 2006; Freud and Caligiuri, 2006). Stage 2 intermediates can give rise to DCs, T cells and stage 3 cells but stage 3 cells lose the potential to give rise to DCs and T cells. Both stage 2 and 3 cells are IL-15 responsive and can give rise to mNK cells (Freud et al., 2006). Human stage 3 cells are attributed to iNK cells and give rise to stage 4 cells ($CD34^-CD117^{+/-}CD94^+CD56^{bright}$ $CD16^-$. $CD56^{bright}$). The stage 4 cells finally give rise to stage 5 cells ($CD34^-CD117^-CD94^{+/-}CD56^{dim}CD16^+$) (Chan et al., 2007; Freud et al., 2006; Grzywacz et al., 2006; Loza and Perussia, 2001). Both $CD56^{bright}$ and $CD56^{dim}$ subsets represent the mNK cell subsets in humans. $CD56^{bright}$ NK cells express CD62L and CCR7 and they are predominantly present in the secondary lymphoid tissues, neonatal tissues, and umbilical cord blood while $CD56^{dim}$ subset is prevalent in the peripheral blood and inflamed tissues. $CD56^{bright}$ NK cells display higher capacity for cytokine production, a higher proliferation rate, and less cytotoxic activity as compared to the $CD56^{dim}$

subset (Caligiuri, 2008). CD56^{dim} NK cells represent better terminally differentiated cells and are similar to mouse CD27^{lo} NK cells while CD56^{bright} subset is equivalent to murine CD27^{high} NK cells.

NK cell development from HSC requires integration of signaling from different transcription factors and cytokines. Notch signaling plays an important role in both mouse and human NK cell development. Notch-Jagged 1 and Notch-Jagged 2 signaling promotes NK cell development from CLP in mice (DeHart et al., 2005) while in humans Jagged 2, Delta 1, and Delta 4 signaling induce NK cell development from cord blood HSC (Beck et al., 2009). IL-3, IL-7, c-kit ligand, and Flt3 ligand also promote the generation of IL-15 responsive NK cell developmental intermediates from CD34^{+} bone marrow-derived hematopoietic progenitor cells as well as human stage 1 cells (Carson et al., 1994b; Freud et al., 2006; Yu et al., 1998). In humans, adhesion related kinase (Ark)/growth arrest-specific gene 6 (Gxl) pathway promotes IL-15, c-kit ligand, and Flt3 ligand-induced NK cell differentiation (Park et al., 2009, 2013). In contrast, transforming growth factor (TGF)-β suppresses NK cell development from CD34^{+} progenitors (Allan et al., 2010). NK cell lineage commitment from CLP commences with the gain of CD122 expression and loss of IL-7Rα expression. Murine NKP as well as human stage 2 and stage 3 cells express CD122 and also respond to IL-15 for development into the mNK cells (Di Santo, 2006; Freud et al., 2006; Yokoyama et al., 2004). In agreement with this, IL-15$^{-/-}$, JAK3$^{-/-}$, and STAT5b$^{-/-}$ mice or mice with deletion of STAT5b in NKP are shown to be deficient of NK cells (Eckelhart et al., 2011; Imada et al., 1998; Kennedy et al., 2000) while overexpression of IL-15 increases the frequency of NK cells (Fehniger et al., 2001). The other transcription factors such as Runx3 and, its binding partner core-binding factor beta subunit (Cbfb) induces CD122 expression, and mice devoid of these transcription factors are deficient in NK cells (Ebihara et al., 2015; Guo et al., 2008). Nfil3 also acts as a crucial factor in NK cell development as Nfil3 deficient mice have a reduced number of NK cells but have no defect on the number of B cells, CD4^{+}, and CD8^{+} T cells (Gascoyne et al., 2009; Kamizono et al., 2009). Nfil3 is required for the development of NKP cells from CLP while they are redundant during iNK stage (Firth et al., 2013). The Nfil3 expression is dependent on IL-15 signaling through PDK1 and mTOR. Mice with NK cells with specific deficiency of mTOR show a defective NK cell maturation in bone marrow and loss of peripheral NK cells (Firth et al., 2013; Gascoyne et al., 2009; Marcais et al., 2014). PU.1, an Ets family of transcription factor, regulates the homeostasis of NK cells, and mice deficient of PU.1 have a reduced number of NKP and iNK cells (Colucci et al., 2001). Ets-1, an Ets family of transcription factor, positively regulates the NK cell development from NKP to iNK stage. Mice deficient in Ets-1 show impaired frequency of peripheral NK cells (Barton et al., 1998; Ramirez et al., 2012). Maturation of NK cells is governed by a set of other transcription factors and one such factor is Id2. The Id family of transcription factors is required for progression from iNK to mNK development stage. Id2 deficient mice show reduced peripheral NK cells (Boos et al., 2007). Id2 and Id3 are expressed by NKP and promote NK cell development (Heemskerk et al., 1997; Schotte et al., 2010). The T-box family of transcription factors also plays an important role in regulating NK cell maturation. T-bet, a T-box family of transcription factor, maintains NK cells to an immature TNF-related apoptosis-inducing ligand (TRAIL)$^{+}$DX5^{-} state during fetal hematopoiesis in the liver. In contrast, NK cells maturation in bone marrow requires the loss of TRAIL expression and acquisition of Ly49 receptors and DX5, which is strictly dependent on the expression of Eomes, a

T-box family of transcription factor (Gordon et al., 2012). T-bet expression is also induced by GATA-3 as a result of which GATA-3 deficient NK cells also display an immature phenotype similar to T-bet deficient NK cells (Samson et al., 2003). The transcription factors Foxo1 and Foxo3 suppress terminal NK cell maturation by inhibiting T-bet expression as mice deficient of these transcription factors have a higher number of mNK cells (Deng et al., 2015; Huntington et al., 2007). In contrast, mice deficient in Tox (a HMG box transcription factor) lack mNK cells (Aliahmad et al., 2010). Aiolos, an Ikaros family of transcription factor, and interferon regulatory factor 2 (IRF-2), are also required for generation of terminally mature $CD27^-CD11b^+$ NK cells, and mice deficient of these factors have iNK cells but lack mNK cells (Lohoff et al., 2000; Taki et al., 2005). In contrast to the transcription factors, the role of activating receptors in NK cell development is not well studied. Although NK cell development is normal in $CD27^{-/-}$, $CD94^{-/-}$, and $Klrk1^{-/-}$ mice, they show reduced frequency of mNK cells (De Colvenaer et al., 2011; Orr et al., 2010; Zafirova et al., 2009). These findings suggest that development and maturation of NK cells require a complex molecular interaction and integration of signaling at different stages. How all these factors integrate into a complex cytoplasmic and nuclear environment requires studies involving a systems biology approach for a better understanding of NK cell development.

Although bone marrow is the primary site for the development of conventional NK cells in humans and mice, they can also develop in the thymus and liver. Thymic NK cells have been shown to develop from $CD4^-CD8^-$ DN thymocytes (Vargas et al., 2011). Vosshenrich et al. (2006) showed that thymic NK cells have $Ly49^{low}CD11b^{low}CD69^{high}$ phenotypes and resemble iNK cells in the bone marrow, and are mostly dependent on GATA-3 and IL-7 for development, and these cells also display reduced cytotoxicity but produce high amounts of cytokines. The murine liver also harbors a population of iNK cells, marked as $CD11b^{lo}CD49b(DX5)^-TRAIL^{hi}$ (Kim et al., 2002). A study by Peng et al. (2013) showed that liver NK cells are composed of two distinct populations, $CD49a^+DX5^-$ and $CD49a^-DX5^+$. $CD49a^+DX5^-$ NK cells are distinct from splenic NK cells, display memory potential, and represent noncirculating liver-resident NK cells while $CD49a^-DX5^+$ NK cells exhibit phenotypes similar to splenic NK cells and serve as circulating NK cells (Peng et al., 2013). The murine peripheral and splenic NK cells are mostly $CD27^-CD11b^+$ phenotypes but liver NK cells contain a significant population of immature $CD27^+CD11b^-$ cells and show high expression of inhibitory receptor NKG2A (Lassen et al., 2010). The human liver has been shown to harbor a population of stage 3 like $Lin^-CD34^-CD117^+CD94^-$ NK cells which show expression of CD122 and Id2, and can give rise to mNK cells (Moroso et al., 2011). Stage 1–5 NK cells and early NK cell developmental intermediates have been observed in human secondary lymphoid tissues and mice lymph nodes, suggesting that NK cell maturation takes place in these tissues (Freud and Caligiuri, 2006; Luther et al., 2011). In addition to the liver and thymus, NK cells are also found in the uterus. Uterine NK cells are phenotypically distinct from peripheral NK cells and express Ly49 receptors, a low level of CD11b, a high level of CD69 and c-kit, but lack DX5 expression (Yadi et al., 2008). Human uterine NK cells show predominant $CD56^{bright}$ phenotype with high expression of KIRs and CD9, and distinct from peripheral $CD56^{bright}$ NK cells (Koopman et al., 2003). Skin NK cells are mostly localized in the dermis and express multiple chemokine receptors for their migration into the skin (Ottaviani et al., 2006). Intestinal NK cells are predominantly found in lamina propria and to some extent in Payer's patches and mesenteric lymph nodes, and show

NKp46$^+$RORγt$^-$ phenotype (Reynders et al., 2011). The phenotypic and functional differences of NK cells in different anatomical sites might be due to change in activation or maturation signal in local tissue microenvironments. Interaction of NK cells with stromal cells into the local tissue might also shape the repertoire of tissue resident NK cells.

MECHANISM OF NK CELL FUNCTION

The effector function of the NK cell is dictated by the balance of signaling from two distinct types of receptors known as activating receptors and inhibitory receptors. The inhibitory receptors recognize MHC class I molecules expressed by normal host cells and deliver inhibitory signals which render NK cells tolerant to host cells. Most inhibitory receptors express immunoreceptor tyrosine-based inhibitory motifs (ITIMs) in their cytoplasmic tail that undergo phosphorylation upon ligand binding. These phosphorylated ITIMs serve as the docking site for phosphatases such as Src homology (SH)-containing tyrosine phosphatase 1 (SHP-1), SHP-2, and lipid phosphatase SH2 domain-containing inositol-5-phosphatase (SHIP) which negate downstream activating signaling molecules such as Vav-1 (Stebbins et al., 2003; Veillette et al., 2002). NK cells express a wide array of inhibitory receptors and are listed in Table 9.1. The most studied family of inhibitory receptors is the Ly49 receptors, which belong to the C-type lectin superfamily and include Ly49C, Ly49A, Ly49I, and Ly49P molecules. Ly49A binds to the MHC class I molecule H-2D^d and H-2D^k while Ly49C binds to H-2K^b and H-2D^b molecules (Dam et al., 2003). The expression of the Ly49

TABLE 9.1 Activating and Inhibitory Receptors of Natural Killer (NK) Cells

Activating Receptors	Inhibitory Receptors
NKG2D (human/mouse)	CD94-NKG2A (human/mouse)
Ly49D (mouse)	Ly49A (mouse)
Ly49H (mouse)	Ly49C (mouse)
CD94-NKG2C (human/mouse)	Ly49P (mouse)
NKp30 (human)	Ly49I (mouse)
NKp44 (human)	CD244 (human/mouse)
NKp46 (human/mouse)	KIR2DL1 (human)
DNAM-1 (human/mouse)	KIR2DL2 (human)
KIR2DS1 (human)	KIR2DL3 (human)
KIR2DS2 (human)	KIR3DL1 (human)
KIR2DS3 (human)	KIR2DL2 (human)
KIR2DS4 (human)	
KIR2DS5 (human)	
KIR3DS1 (human)	

inhibitory receptor is strictly restricted to murine NK cells. Humans express a homolog of Ly49 family, known as the killer cell immunoglobulin-like receptor (KIR) family of proteins. KIRs include KIR2DL1-3, KIR3DL1-3, and KIR2DL5, and recognize HLA-A, -B, and -C molecules (Rajalingam, 2012). Recognition of MHC class I molecules by Ly49 and KIR inhibitory receptors during NK cell development is essential for the complete functional response of activating receptors and rejection of cells undergoing stress. This process is known as NK cell education. NK cells from MHC class I-deficient mice lack the ability to kill MHC class I-deficient tumor cell lines or MHC class I-deficient allogeneic bone marrow cells due to the absence of NK cell education (Bix et al., 1991; Liao et al., 1991). NK cell education has been explained by several different models. The licensing or arming model states that NK cells are originally hyporesponsive and interaction of their inhibitory receptors with MHC class I molecules during development renders them licensed or armed effector cells (Kim et al., 2005). The second model is known as the disarming model which suggests that NK cells are originally responsive and interaction of activating receptors with their cognate ligand allows these cells to kill target cells. However, when these NK cells start acquiring self-MHC-specific inhibitory receptors during development, inhibitory signaling pathways make them tolerant to normal healthy cells. NK cells which fail to acquire inhibitory receptors are rendered hyporesponsive by the chronic stimulatory signal from normal cells (Gasser and Raulet, 2006). According to a third model, known as the tuning or rheostat model, response of NK cells is dependent on the number and strength of signaling from inhibitory and activating receptors. In a situation when inhibitory signaling dominates chronic activating stimuli, NK cells are highly responsive. The strong activating signal which overrides inhibitory signals results in hyporesponsive NK cells while intermediate stimulation maintains NK cells in the intermediate responsive state (Brodin et al., 2009).

Most tumor cells downregulate expression of surface MHC class I molecules. This results in reduced inhibitory signaling in the NK cells. However, lower inhibitory signals are not sufficient enough to trigger the cytotoxic activity of NK cells against tumor cells. Along with downregulation of MHC class I molecules, tumor cells often upregulate stress-associated molecules that are recognized by activating receptors of NK cells. Consequently, activating signals from NK cells now override inhibitory signals which direct NK cells to kill the tumor cells (Lanier, 2008). Activating receptors contain immunoreceptor tyrosine-based activating motifs (ITAMs) in their cytoplasmic tail. Engagement of activating receptor on the NK cell with its cognate ligand on target cells leads to the phosphorylation of ITAM and activation of Syk and Zap70 molecules. Downstream of this signaling molecules, mitogen-activated protein kinase and extracellular signal-regulated kinase are activated, which leads to the release of cytotoxic molecules and transcription of cytokine genes. One of the major activating receptors expressed by the majority of NK cells is natural-killer group 2, member D (NKG2D), which is a C-type lectin-like protein expressed as a homodimer (Jamieson et al., 2002). NKG2D exists in two isoforms, a short isoform (NKG2D-S) and a long isoform (NKG2D-L) in mice, while in humans only NKG2D-L is present. NKG2D-S can associate with both DAP-10 and DAP-12 adapter proteins while NKG2D-L associates only with DAP-10 (Gilfillan et al., 2002). NKG2D binds to retinoic acid early inducible-1 family of proteins (Rae-1α-ε), murine UL16-binding protein-like transcript 1 (MULT1), and H60 group of proteins (H60a, H60b, H60c) in mice while in humans it recognizes MHC class I chain-related protein A (MICA) and B (MICB) and UL16-binding proteins (ULBP1-6) (Carayannopoulos et al., 2002; Cerwenka et al., 2000; Cosman et al., 2001; Diefenbach et al., 2003). Most of these NKG2D ligands are a structural homolog of MHC class I molecules and are expressed at low levels on normal healthy cells but are upregulated in various tumor tissues and cell lines (Diefenbach et al., 2000; Groh et al., 1999). Signaling pathways such as DNA damage repair pathway and TLR signaling have been shown to regulate NKG2D-ligand expression (Eissmann et al., 2010). Besides regulation of ligand expression, the expression of the NKG2D receptor can also be regulated by cytokines such as IFN-γ, IL-15, IL-12, and TGF-β (Zhang et al., 2008). In addition to NKG2D, NKG2C and NKG2E also act as an activating receptor. However, unlike NKG2D, they are expressed as a heterodimer with CD94. Some of the Ly49 family of proteins also act as an activating receptor in NK cells. These receptors are Ly49D and Ly49H that associate with the DAP-12 adaptor protein (Smith et al., 1998). Ly49H recognizes glycoprotein m157 protein of murine cytomegalovirus (MCMV) while Ly49D binds H-2D^d molecules (Nakamura et al., 1999; Smith et al., 2002). Another group of widely studied activating receptors are natural cytotoxicity receptors (NCRs), which include NKp46 (NCR1), NKp44 (NCR2), and NKp30 (NCR3). All three NCRs are expressed by human NK cells while murine NK cell only express NKp46. The ligands for NCRs include viral and bacterial protein as well tumor cell proteins. NCRs signal through the ITAM present in the adapter protein FcεRI-γ and CD3-ζ associated with them.

Elimination of tumor cells by NK cells can be through direct NK cell-mediated cytotoxicity or indirectly through secretion of cytokines (Fig. 9.2). The direct cytotoxic response of NK cells involves the release of cytotoxic molecules such as perforin and granzymes into the immunological synapse between NK cells and a tumor cell. In normal resting NK cells, these molecules are stored in secretory lysosomes which express lysosomal-associated membrane protein-1 (LAMP-1 or CD107a) and LAMP-2 (CD107b). In response to the activating signal, these secretory lysosomes fuse with the plasma membrane of tumor cells and release cytotoxic molecules, and this process is known as degranulation (Topham and Hewitt, 2009). This leads to the transient expression of CD107a on the surface of degranulating NK cells before being internalized back into the cell and is often used to indirectly measure NK cell cytotoxic activity (Alter et al., 2004; Paul et al., 2016). Perforin undergoes polymerization and develops a pore in the tumor cell which promotes entry of granzyme into the target tumor cell. Granzymes activate caspase-dependent apoptosis of tumor cells. Perforin-mediated cytotoxicity is a central component of NK cell-dependent control of various tumors (Street et al., 2001). In addition to perforin and granzyme mediated cytotoxicity, the NK cell also utilizes death receptor ligands such as tumor necrosis factor (TNF), FasL, and TRAIL, to induce killing of tumor cells. These ligands bind to their receptors on tumor cells and recruit adaptor proteins which induce tumor cell apoptosis (Screpanti et al., 2001). TRAIL has been shown to play an important role in NK cell-mediated control of cancer metastasis (Takeda et al., 2002). The indirect activity of NK cells relies on the secretion of cytokines such as IFN-γ, IL-10, TNF-α, GM-CSF, and IL-13, and chemokines such as MIP-1α, MIP-1β, IL-8, and RANTES (Bluman et al., 1996; Fauriat et al., 2010; Roda et al., 2006). Cytokine such as IFN-γ exerts antiangiogenic, antiproliferative and proapoptotic effect against tumor cells (Beatty and Paterson, 2001; Chin et al., 1996).

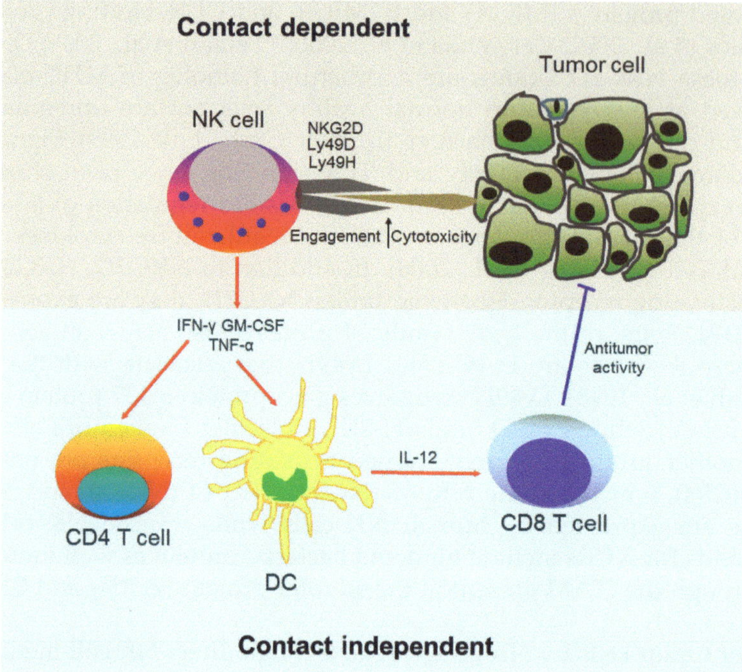

FIGURE 9.2 **Antitumor response of the tumor cell.** NK cells can interact with tumor cells in a contact-dependent and independent manner. In the contact-dependent pathway, the interaction of the NK cell activating receptor with its cognate ligand expressed by tumor cells activates signaling which leads to the killing of tumor cells. The NK cell can also exert its antitumor function in a contact-independent manner. Activated NK cells secrete cytokines such as IFN-γ, TNF-α, and GM-CSF that in turn activate DCs and prime cytotoxic $CD8^+$ T cells for their antitumor activity.

INTERACTION OF NK CELLS WITH THE ADAPTIVE IMMUNE CELLS

NK cells can interact with various components of the immune system and act as a bridge between the innate and adaptive immune systems. NK cell activation occurs very early during infection or malignant transformation and coincides with T cell activation in the secondary lymphoid organs. NK cells can regulate T cell response in a direct or indirect manner and this regulation can be positive or negative. NK cells can regulate T cell response by modulating antigen presenting cells (APCs) number and function, and indirectly control the immune response. Direct contact between NK cell and DC during an in vitro NK–DC coculture system leads to maturation of DCs. In this in vitro system, NK cell derived TNF-α and IFN-γ promotes upregulation of costimulatory molecules CD86, CD83, CD80, human leukocyte antigen-DR (HLA-DR), and CCR7, and also enhances IL-12 production by DCs (Piccioli et al., 2002). In murine tumor transplantation models, NK cells-derived IFN-γ activates DCs and induces IL-12 production, leading to enhanced $CD8^+$ T cell response and tumor rejection (Adam et al., 2005; Mocikat et al., 2003). Cytolysis of target cells by NK cells can also lead to uptake and presentation of target antigen by DC and enhanced $CD8^+$ T cell activation. In support of this concept, it has been shown that administration of ovalbumin (OVA)-expressing target cells induced antigen-specific $CD4^+$ T cell response to NK cell-mediated killing of OVA-expressing target cells (Krebs et al., 2009). In addition to the positive outcome of NK–DC interaction, occasionally these interactions can also lead to reducing T cell response. NK cells may dampen T cell response by restricting the availability of APCs. Depletion of NK cells during the initiation phase of lymphocytic choriomeningitis virus (LCMV) infection enhanced the ability of APC to stimulate $CD8^+$ T cells due to lack of killing of LCMV infected DCs by NK cells (Cook and Whitmire, 2013). During MCMV infection, NK cell-mediated elimination of infected DCs diminishes the T cell response (Andrews et al., 2010). Interestingly, the ratio of NK cell to DC determines the outcome of NK–DC interaction. The lower ratio favors DC activation, maturation, and enhanced T cell response whereas higher ratios lead to the elimination of DCs and curtailed the T cell response (Mandaric et al., 2012). The ability of NK cells to promote or inhibit the T cell response is also affected by cytokines present in the microenvironment. Human NK cells exposed to IL-2, IL-12, or IL-18 promote Th1 priming while IL-4 conditioning inhibits this process (Agaugue et al., 2008).

Besides indirect regulation, NK cells can also directly regulate $CD4^+$ T cell response through secretion of cytokines or direct cytotoxic activity. NK cells can promote Th1 differentiation in lymph nodes in response to OVA administration (Martin-Fontecha et al., 2004). Inhibition of TGF-β signaling in NK cells during *Leishmania major* infection promoted NK cell-mediated IFN-γ production and Th1 differentiation (Laouar et al., 2005). However, NK cell interaction with $CD4^+$ T cells during MCMV, *Toxoplasma gondii*, and *Listeria monocytogenes* infection have impaired T cell response through secretion of IL-10 (Perona-Wright et al., 2009). Direct killing of activated T cells is another mechanism by which NK cells can restrain T cell response. NK cell recognition of in vitro activated $CD4^+$ and $CD8^+$ T cells through NKG2D result in the killing of these cells in a perforin-dependent manner (Rabinovich et al., 2003). In vivo, NK cells have also been shown to eliminate antigen-specific $CD8^+$ T cells in NKG2D- and perforin-dependent manner. Transient NK cell depletion increases the antigen-specific memory $CD8^+$ T cells and provide better protection against tumor growth (Soderquest et al., 2011). Studies have also demonstrated that NK cells can kill activated T cells during viral infection such as LCMV in a perforin-dependent manner. Thus, transient depletion of NK cells during the initial phase of LCMV infection promotes activation of $CD4^+$ T cells, leading to subsequent $CD8^+$ T cell activation and virus elimination (Waggoner et al., 2011). Regulatory CD4 T cells have been shown to be targeted by the activated NK cells and killed in the NKG2D or NCR-dependent manner (Roy et al., 2008). Activated T cells upregulate NKG2D-ligand expression in an ATM-NF-κB-dependent manner as compared to naïve T cells, and this makes them susceptible to NK cell-mediated killing (Cerboni et al., 2007). Additionally, T cells can become susceptible to NK cell-mediated killing through the TRAIL-TRAIL-R pathway. During chronic hepatitis B virus (HBV) infection, HBV-specific $CD8^+$ T cells express enhanced level of TRAIL-R2 and prone to be killed by NK cell in a TRAIL-dependent manner (Peppa et al., 2013). T cell protection from killing by NK cells is conferred by type I IFNs through regulation of inhibitory receptor ligands on T cells. Using IFN-$\alpha R^{-/-}$ mice, Crouse et al. showed that activated IFN-$\alpha R^{-/-}$ T cells display a higher expression of ligands for activating receptor, NCR, and a lower expression of inhibitory receptor ligand such as MHC class I and Qa-1 as compared to IFN-αR sufficient T cells, leading to their NK cell-mediated elimination in a perforin-dependent manner (Crouse et al., 2014; Xu et al., 2014). Thus, NK cells can modulate T cell response in both a positive and negative manner depending on the microenvironment, and detailed knowledge about the molecular mechanism of this cross-talk will open new strategies to modulate this interaction for therapeutic purposes.

INTERACTION OF NK CELLS WITH OTHER INNATE IMMUNE CELLS

NK cells also collaborate with other innate immune cells such as macrophages to fight infection. Coculture of NK cells with macrophages activates NK cells in a contact-dependent manner and induces secretion of IFN-γ by NK cells. The interaction between macrophages and NK cells can be mediated by various receptors or soluble factors. CD40-CD40L engagement between NK cells and LPS-activated macrophages leads to activation of NK cells (Atochina and Harn, 2005). However, a high dose of LPS also induces expression of NKG2D ligands such as ULBP and MICA on human macrophages and results in lysis of the macrophage by NK cells in an NKG2D-dependent manner (Nedvetzki et al., 2007). In contrast, a low dose of LPS triggers 2B4 pathway leading to macrophage-mediated NK cell activation. Ligation of 2B4 with CD48 expressed on macrophage induces proliferation and cytokine secretion by NK cells (Nedvetzki et al., 2007). Other activating receptors such as NKp46 and DNAM-1 also participate in lysis of LPS stimulated, CMV-infected or *Mycobacterium tuberculosis*-infected macrophage through induction of cytotoxicity and cytokine secretion by NK cells (Bellora et al., 2010; Romo et al., 2011; Vankayalapati et al., 2005). In contrast, Qa-1 upregulation on macrophage by poly I:C treatment allows protection of macrophage from lysis by NK cells through inhibitory receptor NKG2A (Zhou et al., 2012). Human NK-macrophage coculture also enhanced the expression of several activating receptors such as NKG2D, 2B4, NKp30, NKp44, and NKp46 (Nedvetzki et al., 2007). However, expression of murine NKp46 and DNAM-1 is not affected in the presence of poly I:C activated peritoneal macrophage (Zhou et al., 2012).

In addition to receptor-mediated modulation, cytokines secreted by macrophage also affect NK cell function. IL-12 and IL-18 secreted by *Bacillus anthracis*-infected macrophages or liver-resident macrophages induce IFN-γ production by NK cells (Klezovich-Benard et al., 2012; Zhou et al., 2012). IL-12 and IL-18 secreted by Salmonella-infected macrophages are crucial for promoting NK cell cytotoxicity, resulting in better bacterial clearance (Lapaque et al., 2009). Poly I:C activated macrophages also release IL-15 that induce CD69 and NKG2D expression and IFN-γ secretion by NK cells (Zhou et al., 2012). Recently, Takayma et al. (2010) showed that $NKp46^+$ NK cells from Crohn's disease patients produce IFN-γ in response to IL-23 secreted by intestinal inflammatory macrophages. NK cell–macrophage interaction can also lead to inhibition of IFN-γ secretion and downregulation of costimulatory and adhesion molecules by NK cells through IL-10 released by *M. tuberculosis*-infected monocytes (Schierloh et al., 2005).

The outcome of macrophage–NK cell interaction also depends on the tissue microenvironment. Tissue-specific macrophages have different phenotypes and differ in their ability to modulate NK cell function. Mouse spleen macrophages exhibit a greater capacity to promote NK cell cytotoxicity compared to lung-resident macrophages (Michel et al., 2012).

Together, these studies suggest that NK cell interacts with several innate immune cells and modulates the effector immune response.

NK CELLS IN CANCER IMMUNOSURVEILLANCE

Cancer immunosurveillance is a complex process by which tumor cells are spontaneously recognized and continuously eliminated by the host immune system. This hypothesis was supported by various key experimental observations. Several studies showed that endogenous IFN-γ was protective against tumor growth, and lack of IFN-γ or its signaling components led to enhanced tumor growth (Dighe et al., 1994; Kaplan et al., 1998). Another key component of cancer immunosurveillance was found to be perforin, as mice lacking the perforin gene were observed to be prone to tumor formation as compared to wild-type mice (Street et al., 2001; van den Broek et al., 1996). Other molecules such as NKG2D and TRAIL were also observed to be important components of cancer immunosurveillance. Further investigations suggested that both innate and adaptive immune cells of the immune system play an important role in cancer immunosurveillance. Deficiency or depletion of various immune cells of the immune system such as αβ T cells, γδ T cells, NK cells, and NKT cells leads to a higher incidence of tumor growth (Girardi et al., 2001; Paul et al., 2016; Paul and Lal, 2016; Shankaran et al., 2001; Smyth et al., 2000a). Extensive work over the years has refined the concept of immunosurveillance and it is now more accurately described as "cancer immunoediting"—the process by which interaction between tumor cells and components of the immune system not only protect the host against tumor development but also modify the immunogenicity of tumor cells. This process is composed of three distinct steps—elimination, equilibrium, and escape (Fig. 9.3).

Elimination

In this phase, tumor cells are eliminated by innate and adaptive immune cells and prevent the development of the full-blown tumor. As tumor cells grow up to a certain size, they undergo angiogenesis and start invading neighboring tissues. This invasion induces inflammatory signals leading to recruitment of innate and adaptive immune cells. Immunodeficient mice develop spontaneous as well as carcinogen-induced tumors more frequently as compared to wild-type mice. Intact lymphocytic compartments in the secondary lymphoid tissues and effector molecules such as TRAIL, Fas/FasL, perforin, and NKG2D provide protection against tumor cells in the body (Vesely et al., 2011). IFN-α/β and IFN-γ both play an important role in the antitumor immune response. IFN-α/β has been shown to enhance antigen presentation to $CD8^+$ T cells by acting on $CD8\alpha^+$ DCs while IFN-γ targets tumor cells (Fuertes et al., 2011; Kaplan et al., 1998). Aneuploidy also enhances immunogenicity of tumor cells by increasing surface expression of calreticulin resulting in their elimination by $CD4^+$ and $CD8^+$ T cells (Senovilla et al., 2012). In addition to the role of adaptive immune cells in elimination of tumor cells, innate immune cells are also instrumental in tumor recognition and elimination (Paul and Lal, 2016). NK cells are pivotal among innate immune components during the elimination phase (Fig. 9.3). Removal of NK cells has been shown to increase the incidence of methylcholanthrene (MCA)-induced sarcoma, suggesting that NK cells are involved in tumor cell elimination (Smyth et al.,

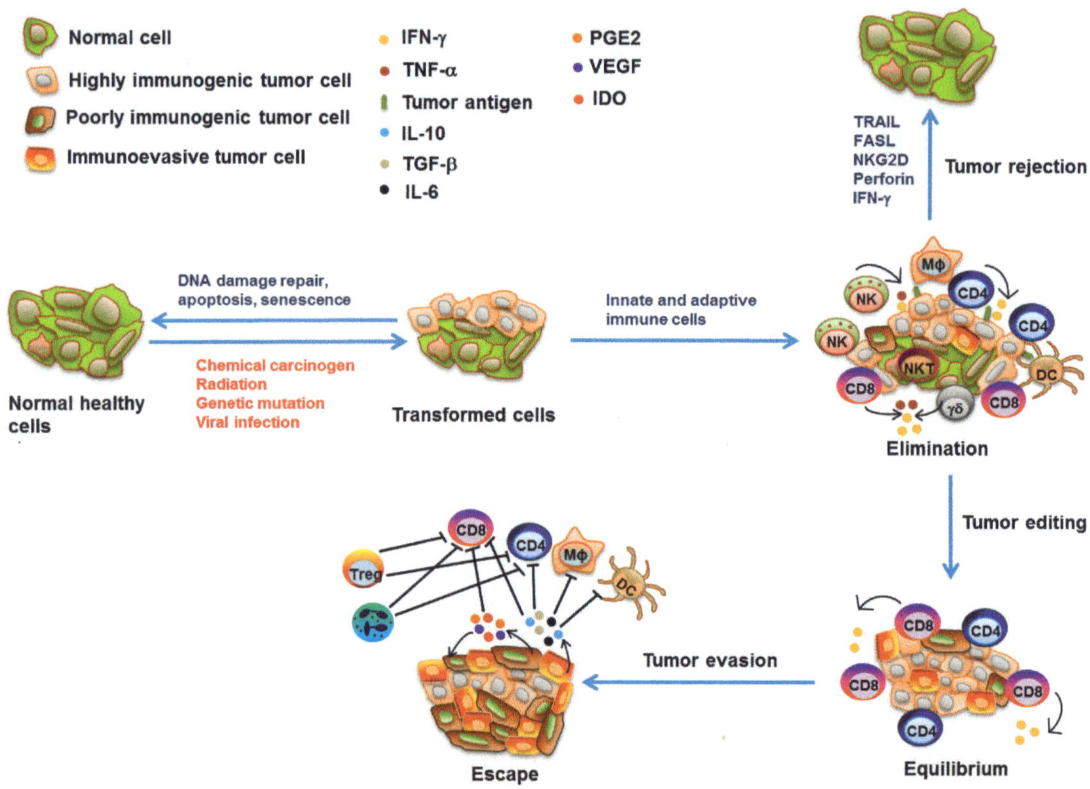

FIGURE 9.3 **The three phases of cancer immunoediting.** Normal healthy cells may give rise to transformed cells as a result of mutation due to exposure to a chemical carcinogen, radiation, or viral infection. Various cell-intrinsic tumor suppressor mechanisms such as DNA damage repair or apoptosis pathways detect the abnormal cells and will either try to revert the damage or eliminate the mutant cells. However, when these tumor suppressor mechanisms fail, cancer immunoediting is initiated. In the first phase, called elimination, tumor cells are detected and eliminated by innate and adaptive immune cells. Cells that escape the elimination phase are restrained in a dormant equilibrium state by immune cells, and this phase is called equilibrium phase. During equilibrium phase, tumor cells undergo immunoediting, resulting in the loss of tumor antigen and/or MHC class I expression or lack of antiapoptotic machinery. All these events give rise to poorly immunogenic transformed cells, which establish an immunosuppressive microenvironment and escape immune-mediated control, leading to the development of the clinically apparent tumor. These tumor cells further recruit immunosuppressive cells as well as secreting immunosuppressive factors that inhibit antitumor immune response in the tumor microenvironment.

2001). Further evidence of the role of NK cell in immunosurveillance was provided by the study by O'Sullivan et al. (2012) where the incidence of MCA-induced sarcoma was greater in RAG2$^{-/-}$γc$^{-/-}$ mice (lacking both adaptive immunity and NK cells) as compared to RAG2$^{-/-}$ mice (lacking only adaptive immunity). The mechanism of NK cell elimination of MCA-induced sarcomas involves NKG2D, IFN-γ, and perforin, and deficiency of these molecules lead to greater incidence of sarcomas (Chinkin, 1977; Kaplan et al., 1998; Smyth et al., 2005) while the absence of NKp46 did not affect incidence of sarcoma (Elboim et al., 2010). TRAMP prostate cancer and Eu-Myc-induced lymphoma were also amplified in NKG2D-deficient mice compared to controls (Guerra et al., 2008). NK cell activity is also required for control of B cell lymphomas and mammary carcinoma, and this protection is perforin-dependent (Smyth et al., 2000b). In a mouse model of liver carcinoma, restoration of endogenous p53 in tumor cells promotes NK cell-mediated elimination of senescent tumor cells (Iannello et al., 2013).

Equilibrium

It is the second stage where tumor cells and immune cells are in dynamic equilibrium. When the tumor cells are not completely destroyed in the elimination phase, they proceed into the equilibrium phase where the immune system is able to restrain the tumor growth but cannot completely eliminate it. Evidence of the existence of a discrete and distinguishable immunologically mediated equilibrium stage came from a study by Koebel et al. (2007). They showed that $CD4^+$ and $CD8^+$ T cells play an important role in controlling late-stage MCA-induced sarcoma growth, and depletion of NK cells using the anti-NK1.1 antibody or blocking NK cell function through inhibition of NKG2D and TRAIL did not affect the growth of late-stage MCA-induced sarcoma. This study also demonstrated that tumor cells, which are restrained by adaptive immunity in an equilibrium stage, are transformed but show poor proliferation and are also highly immunogenic compared to tumor cells that spontaneously escape equilibrium (Koebel et al., 2007). Another study by Teng et al. reiterated the concept that the equilibrium phase of cancer immunoediting is temporally and mechanistically distinct from the elimination stage. The same study also showed that IL-23p19 promotes persistence of tumor cells during the equilibrium stage while IL-12/23 p40 maintains cancer cells in a dormant state during the equilibrium stage. Other cytokines such as IL-17, IL-4, TNF, and IFNα/β, which showed critical role in the elimination stage, were dispensable for maintaining the equilibrium phase (Teng et al., 2012). A comparative study of the microenvironment profile of the equilibrium and escape phases of the MCA-induced sarcoma showed that antitumor effector immune cells such as $CD8^+$ T cells, NK cells and γδ T cells are required for maintaining the tumor cells in a dormant state, whereas regulatory $CD4^+$ T cells (Treg) and polymorphonuclear myeloid-derived suppressor cells promote tumor cell escape from the dormant equilibrium phase (Wu et al., 2013). Further study in T antigen (Tag)-induced pancreatic cancer demonstrated that Tag-specific $CD4^+$ T cells promote angiogenesis and multistage carcinogenesis in the absence of TNFR or IFN-γ signaling (Muller-Hermelink et al., 2008). In contrast, Tag-specific Th1 cells drive Tag-expressing tumor cells into senescence by inducing G0/G1 growth arrest, and this process was TNFR1 and IFN-γ-dependent (Braumuller et al., 2013). Although immune cells maintain tumor cells in a dormant phase, some tumor cells escape selection pressure due to the acquisition of immunoevasive mutations resulting from genomic instability, and these edited tumor cells progress into escape phase.

Escape

This is the third phase, where edited tumor cells display low immunogenicity and grow progressively, and are hard to restrain by the immune system. The progression from equilibrium to escape phase can be due to a change in tumor cell immunogenicity due to host immune editing functions, or tumor-induced suppression of the immune system. Loss of tumor antigen expression, lack of MHC class I protein, loss of antigen processing machinery in tumor cells, and induction of antiapoptotic mechanisms contribute to the selection of poorly immunogenic tumor cells that evade the immune system and grow progressively (Dunn et al., 2002; Khong and Restifo, 2002). Alternatively, tumor cells secrete immunosuppressive factors such as TGF-β, vascular endothelial growth factor, indoleamine 2,3-dioxygenase (IDO), prostaglandin E2 (PGE2), and adenosine, which inhibit antitumor immune functions (Baginska et al., 2013). One of the early infiltrating antitumor immune cells is the NK cell. Pietra et al. (2012) have shown that the cytolytic activity of NK cells is suppressed by melanoma cell-derived IDO and PGE2. We also demonstrated that intratumoral NK cells in melanoma tumor display a lower expression of activating receptors and a higher expression of inhibitory receptors as compared to their counterparts in secondary lymphoid organs which resulted in the poor antitumor response (Paul et al., 2016).

NK CELL-BASED CELLULAR THERAPY AND FUTURE PERSPECTIVE

The application of NK cells for cancer immunotherapy has attracted a lot of attention in recent years (Table 9.2). NK cells can be activated by various cytokines such as IL-2, IL-15, IL-12, IL-18, and IL-21. IL-12 and IL-18 have been shown to suppress the tumor metastasis in NKG2D and FasL-dependent manners, respectively (Smyth et al., 2004). IL-15 also activates

TABLE 9.2 Clinical Trials Using Natural Killer (NK) Cells as Adoptive Cellular Therapy

Sr No.	Clinical Trial Registry No.	NK Cells and Strategies	Tumor Type
1.	NCT02809092	Infusion of IL-21 expanded NK cells	Acute myeloid leukemia
2.	NCT02573896	Immunotherapy using expanded NK cells along with drug dinutuximab (ch.14.18; anti-IL-2 mAb) and lenalidomide (a derivative of thalidomide)	Neuroblastoma
3.	NCT00846833	Haploidentical NK cells infusion	Malignant melanoma
4.	NCT02030561	NK cell infusion together with trastuzumab (mAb binds to HER2 receptor)	$Her2^+$ breast and gastric cancer
5.	NCT03056339	Umbilical and cord blood-derived CAR-engineered NK cells	B cell malignancies
6.	NCT02507154	NK cells along with cetuximab (anti-EGFR mAb)	Head and neck cancer
7.	NCT03068819	Cytokine-induced memory-like NK cell adoptive therapy	Acute myeloid leukemia
8.	NCT00995137	Infusions of genetically modified haploidentical NK cells	B-lineage acute lymphoblastic leukemia
9.	NCT02734524	NK cell infusion combined with chemotherapy	Nonsmall cell lung cancer
10.	NCT02280278	Cytokine-induced NK cell immunotherapy	Surgical resected stage III colorectal cancer
11.	NCT02845999	Adoptive transfer of NK cells after cetuximab treatment	Metastatic gastrointestinal carcinoma
12.	NCT00383994	Immunotherapy with NK cell, Rituximab (anti-CD20 mAb), and rhu-GM-CSF	Nonmyeloablative allogeneic stem cell transplantation

the cytotoxic activity of NK cells and contributes to tumor immunosurveillance (Carson et al., 1994a; Kobayashi et al., 2005). IL-21 is known to promote NK cell activation and also results in the suppression of various murine tumors (Frederiksen et al., 2008; Paul et al., 2016; Takaki et al., 2005). Adoptive cellular therapy of NK cells activated and expanded by these cytokines in vitro can be a good choice for solid tumor patients. Although IL-2 supports the proliferation of NK cells, adoptive transfer of NK cells along with IL-2 did not lead to a favorable clinical outcome that may be due to increased Treg generation which shows protumorogenic function (Burns et al., 2003). A high concentration of serum IL-15 in non-Hodgkin lymphoma patients has been associated with improved survival post hematopoietic stem cell transplantation (Porrata et al., 2010). Administration of recombinant human IL-15 (rhIL-15) to metastatic melanoma patients resulted in increased proliferation of NK cells and also achieved clearance of lung lesions in two patients (Conlon et al., 2015). Recently, phase I clinical trials with autologous NK cells along with rhIL-15 are being conducted to treat both acute myelogenous leukemia patients (clinical trial registry no. NCT01385423) and children with refractory solid tumors (NCT01875601). Another IL-15-based superagonist known as ALT-803 which is IL-15 and IL-15RaSu/Fc-fusion protein has shown impressive antitumor activity against several murine solid and hematological tumors, and recently entered into a clinical trial for the treatment of multiple myeloma patients (NCT02099539). Another cytokine such as IL-21 when infused with cetuximab (a monoclonal antibody against epidermal growth factor receptor, EGFR) results in activation of NK cells (Steele et al., 2012). Infusion of IL-21 expanded NK cells in acute myeloid leukemia patients is currently in phase I/II clinical trial (NCT01787474). Recent literature also suggested evidence of NK cell phenotype and function being modulated by tumor-derived suppressor cell types and/or factors. The effect of these cytokines on suppressive factors in the tumor microenvironment needs to be studied in detail so as to design a better protocol in future therapies.

NKG2D is known to act as a primary activation signal and have the potential to override inhibitory signals received by other NK cell receptors (Gilfillan et al., 2002; Spear et al., 2013). It has been shown that suppressor cells such as regulatory CD4 T cells (Tregs) and myeloid-derived suppressor cells can inhibit the expression of NKG2D and the production of IFN-γ in the tumor microenvironment (Li et al., 2009). We showed that NK cells in the melanoma microenvironment represent activated cell phenotypes but have reduced NKG2D expression and also reduced cytolytic properties as compared to the splenic NK cells (Paul et al., 2016). These intratumoral NK cells also have very low activating receptors but higher inhibitory receptors. Activating the NK cells with NKG2D receptors increased the IFN-γ secretion and also enhanced the Th1 cell differentiation (Paul et al., 2016). This suggests that combining the adoptive NK cell-based cellular therapy with strategies to stimulate NK cell activating receptors could help in better antitumor response. Likewise, a better understanding of the migration potential of adoptively transferred NK cells into the tumor microenvironment and their potential to activate the other arms of the effector immune cells such as cytotoxic CD8 T cells and Th1 cells will be helpful.

Acknowledgments

GL laboratory is funded by the Department of Biotechnology (DBT), Government of India (Grants BT/RLF/Re-entry/41/2010; BT/PR15533/MED/30/1616/2015; BT/PR14156/BRB/10/1515/2016; BT/03/IYBA/2010), and NCCS intramural fund. SP received Senior Research Fellowship (SRF) from the Council of Scientific and Industrial Research (CSIR), Government of India.

References

Adam, C., King, S., Allgeier, T., et al., 2005. DC-NK cell cross talk as a novel CD4+ T-cell-independent pathway for antitumor CTL induction. Blood 106, 338–344.

Agaugue, S., Marcenaro, E., Ferranti, B., et al., 2008. Human natural killer cells exposed to IL-2, IL-12, IL-18, or IL-4 differently modulate priming of naive T cells by monocyte-derived dendritic cells. Blood 112, 1776–1783.

Aliahmad, P., de la Torre, B., and Kaye, J., 2010. Shared dependence on the DNA-binding factor TOX for the development of lymphoid tissue-inducer cell and NK cell lineages. Nat. Immunol. 11, 945–952.

Allan, D.S., Rybalov, B., Awong, G., et al., 2010. TGF-beta affects development and differentiation of human natural killer cell subsets. Eur. J. Immunol. 40, 2289–2295.

Alter, G., Malenfant, J.M., and Altfeld, M., 2004. CD107a as a functional marker for the identification of natural killer cell activity. J. Immunol. Methods 294, 15–22.

Andrews, D.M., Estcourt, M.J., Andoniou, C.E., et al., 2010. Innate immunity defines the capacity of antiviral T cells to limit persistent infection. J. Exp. Med. 207, 1333–1343.

Atochina, O., and Harn, D., 2005. LNFPIII/LeX-stimulated macrophages activate natural killer cells via CD40-CD40L interaction. Clin. Diagn. Lab. Immunol. 12, 1041–1049.

Baginska, J., Viry, E., Paggetti, J., et al., 2013. The critical role of the tumor microenvironment in shaping natural killer cell-mediated anti-tumor immunity. Front. Immunol. 4, 490.

Barton, K., Muthusamy, N., Fischer, C., et al., 1998. The Ets-1 transcription factor is required for the development of natural killer cells in mice. Immunity 9, 555–563.

Beatty, G., and Paterson, Y., 2001. IFN-gamma-dependent inhibition of tumor angiogenesis by tumor-infiltrating CD4+ T cells requires tumor responsiveness to IFN-gamma. J. Immunol. 166, 2276–2282.

Beck, R.C., Padival, M., Yeh, D., et al., 2009. The Notch ligands Jagged2, Delta1, and Delta4 induce differentiation and expansion of functional human NK cells from CD34+ cord blood hematopoietic progenitor cells. Biol. Blood Marrow Transplant. 15, 1026–1037.

Bellora, F., Castriconi, R., Dondero, A., et al., 2010. The interaction of human natural killer cells with either unpolarized or polarized macrophages results in different functional outcomes. Proc. Natl. Acad. Sci. U.S.A. 107, 21659–21664.

Bix, M., Liao, N.S., Zijlstra, M., et al., 1991. Rejection of class I MHC-deficient haemopoietic cells by irradiated MHC-matched mice. Nature 349, 329–331.

Bluman, E.M., Bartynski, K.J., Avalos, B.R., et al., 1996. Human natural killer cells produce abundant macrophage inflammatory protein-1 alpha in response to monocyte-derived cytokines. J. Clin. Invest. 97, 2722–2727.

Boos, M.D., Yokota, Y., Eberl, G., et al., 2007. Mature natural killer cell and lymphoid tissue-inducing cell development requires Id2-mediated suppression of E protein activity. J. Exp. Med. 204, 1119–1130.

Braumuller, H., Wieder, T., Brenner, E., et al., 2013. T-helper-1-cell cytokines drive cancer into senescence. Nature 494, 361–365.

Brodin, P., Karre, K., and Hoglund, P., 2009. NK cell education: not an on-off switch but a tunable rheostat. Trends Immunol. 30, 143–149.

Burns, L.J., Weisdorf, D.J., DeFor, T.E., et al., 2003. IL-2-based immunotherapy after autologous transplantation for lymphoma and breast cancer induces immune activation and cytokine release: a phase I/II trial. Bone Marrow Transplant. 32, 177–186.

Caligiuri, M.A., 2008. Human natural killer cells. Blood 112, 461–469.

Carayannopoulos, L.N., Naidenko, O.V., Fremont, D.H., et al., 2002. Cutting edge: murine UL16-binding protein-like transcript 1: a newly described transcript encoding a high-affinity ligand for murine NKG2D. J. Immunol. 169, 4079–4083.

Carotta, S., Pang, S.H., Nutt, S.L., et al., 2011. Identification of the earliest NK-cell precursor in the mouse BM. Blood 117, 5449–5452.

Carson, W.E., Giri, J.G., Lindemann, M.J., et al., 1994a. Interleukin (IL) 15 is a novel cytokine that activates human natural killer cells via components of the IL-2 receptor. J. Exp. Med. 180, 1395–1403.

Carson, W.E., Haldar, S., Baiocchi, R.A., et al., 1994b. The c-kit ligand suppresses apoptosis of human natural killer cells through the upregulation of bcl-2. Proc. Natl. Acad. Sci. U.S.A. 91, 7553–7557.

Cerboni, C., Zingoni, A., Cippitelli, M., et al., 2007. Antigen-activated human T lymphocytes express cell-surface NKG2D ligands via an ATM/ATR-dependent mechanism and become susceptible to autologous NK- cell lysis. Blood 110, 606–615.

Cerwenka, A., Bakker, A.B., McClanahan, T., et al., 2000. Retinoic acid early inducible genes define a ligand family for the activating NKG2D receptor in mice. Immunity 12, 721–727.

Chan, A., Hong, D.L., Atzberger, A., et al., 2007. CD56bright human NK cells differentiate into CD56dim cells: role of contact with peripheral fibroblasts. J. Immunol. 179, 89–94.

Chin, Y.E., Kitagawa, M., Su, W.C., et al., 1996. Cell growth arrest and induction of cyclin-dependent kinase inhibitor p21 WAF1/CIP1 mediated by STAT1. Science 272, 719–722.

Chinkin, A.S., 1977. [Influence of different regimes of muscular activity on catecholamine concentration in the myocardium]. Fiziol. Zh. SSSR Im I M Sechenova 63, 1016–1020.

Chiossone, L., Chaix, J., Fuseri, N., et al., 2009. Maturation of mouse NK cells is a 4-stage developmental program. Blood 113, 5488–5496.

Colucci, F., Samson, S.I., DeKoter, R.P., et al., 2001. Differential requirement for the transcription factor PU.1 in the generation of natural killer cells versus B and T cells. Blood 97, 2625–2632.

Conlon, K.C., Lugli, E., Welles, H.C., et al., 2015. Redistribution, hyperproliferation, activation of natural killer cells and CD8 T cells, and cytokine production during first-in-human clinical trial of recombinant human interleukin-15 in patients with cancer. J. Clin. Oncol. 33, 74–82.

Cook, K.D., and Whitmire, J.K., 2013. The depletion of NK cells prevents T cell exhaustion to efficiently control disseminating virus infection. J. Immunol. 190, 641–649.

Cosman, D., Mullberg, J., Sutherland, C.L., et al., 2001. ULBPs, novel MHC class I-related molecules, bind to CMV glycoprotein UL16 and stimulate NK cytotoxicity through the NKG2D receptor. Immunity 14, 123–133.

Crouse, J., Bedenikovic, G., Wiesel, M., et al., 2014. Type I interferons protect T cells against NK cell attack mediated by the activating receptor NCR1. Immunity 40, 961–973.

Dam, J., Guan, R., Natarajan, K., et al., 2003. Variable MHC class I engagement by Ly49 natural killer cell receptors demonstrated by the crystal structure of Ly49C bound to H-2K(b). Nat. Immunol. 4, 1213–1222.

De Colvenaer, V., Taveirne, S., Delforche, M., et al., 2011. CD27-deficient mice show normal NK-cell differentiation but impaired function upon stimulation. Immunol. Cell Biol. 89, 803–811.

DeHart, S.L., Heikens, M.J., and Tsai, S., 2005. Jagged2 promotes the development of natural killer cells and the establishment of functional natural killer cell lines. Blood 105, 3521–3527.

Deng, Y., Kerdiles, Y., Chu, J., et al., 2015. Transcription factor foxo1 is a negative regulator of natural killer cell maturation and function. Immunity 42, 457–470.

Diefenbach, A., Jamieson, A.M., Liu, S.D., et al., 2000. Ligands for the murine NKG2D receptor: expression by tumor cells and activation of NK cells and macrophages. Nat. Immunol. 1, 119–126.

Diefenbach, A., Hsia, J.K., Hsiung, M.Y., et al., 2003. A novel ligand for the NKG2D receptor activates NK cells and macrophages and induces tumor immunity. Eur. J. Immunol. 33, 381–391.

Dighe, A.S., Richards, E., Old, L.J., et al., 1994. Enhanced in vivo growth and resistance to rejection of tumor cells expressing dominant negative IFN gamma receptors. Immunity 1, 447–456.

Di Santo, J.P., 2006. Natural killer cell developmental pathways: a question of balance. Annu. Rev. Immunol. 24, 257–286.

Dunn, G.P., Bruce, A.T., Ikeda, H., et al., 2002. Cancer immunoediting: from immunosurveillance to tumor escape. Nat. Immunol. 3, 991–998.

Ebert, L.M., Meuter, S., and Moser, B., 2006. Homing and function of human skin gammadelta T cells and NK cells: relevance for tumor surveillance. J. Immunol. 176, 4331–4336.

Ebihara, T., Song, C., Ryu, S.H., et al., 2015. Runx3 specifies lineage commitment of innate lymphoid cells. Nat. Immunol. 16, 1124–1133.

Eckelhart, E., Warsch, W., Zebedin, E., et al., 2011. A novel Ncr1-Cre mouse reveals the essential role of STAT5 for NK-cell survival and development. Blood 117, 1565–1573.

Eissmann, P., Evans, J.H., Mehrabi, M., et al., 2010. Multiple mechanisms downstream of TLR-4 stimulation allow expression of NKG2D ligands to facilitate macrophage/NK cell crosstalk. J. Immunol. 184, 6901–6909.

Elboim, M., Gazit, R., Gur, C., et al., 2010. Tumor immunoediting by NKp46. J Immunol. 184, 5637–5644.

Fathman, J.W., Bhattacharya, D., Inlay, M.A., et al., 2011. Identification of the earliest natural killer cell-committed progenitor in murine bone marrow. Blood 118, 5439–5447.

Fauriat, C., Long, E.O., Ljunggren, H.G., et al., 2010. Regulation of human NK-cell cytokine and chemokine production by target cell recognition. Blood 115, 2167–2176.

Fehniger, T.A., Suzuki, K., VanDeusen, J.B., et al., 2001. Fatal leukemia in interleukin-15 transgenic mice. Blood Cells Mol. Dis. 27, 223–230.

Firth, M.A., Madera, S., Beaulieu, A.M., et al., 2013. Nfil3-independent lineage maintenance and antiviral response of natural killer cells. J. Exp. Med. 210, 2981–2990.

Frederiksen, K.S., Lundsgaard, D., Freeman, J.A., et al., 2008. IL-21 induces in vivo immune activation of NK cells and CD8(+) T cells in patients with metastatic melanoma and renal cell carcinoma. Cancer Immunol. Immunother. 57, 1439–1449.

Freud, A.G., and Caligiuri, M.A., 2006. Human natural killer cell development. Immunol. Rev. 214, 56–72.

Freud, A.G., Becknell, B., Roychowdhury, S., et al., 2005. A human CD34(+) subset resides in lymph nodes and differentiates into CD56bright natural killer cells. Immunity 22, 295–304.

Freud, A.G., Yokohama, A., Becknell, B., et al., 2006. Evidence for discrete stages of human natural killer cell differentiation in vivo. J. Exp. Med. 203, 1033–1043.

Fuertes, M.B., Kacha, A.K., Kline, J., et al., 2011. Host type I IFN signals are required for antitumor $CD8^+$ T cell responses through CD8{alpha}+ dendritic cells. J. Exp. Med. 208, 2005–2016.

Galy, A., Travis, M., Cen, D., et al., 1995. Human T, B, natural killer, and dendritic cells arise from a common bone marrow progenitor cell subset. Immunity 3, 459–473.

Gascoyne, D.M., Long, E., Veiga-Fernandes, H., et al., 2009. The basic leucine zipper transcription factor E4BP4 is essential for natural killer cell development. Nat. Immunol. 10, 1118–1124.

Gasser, S., and Raulet, D.H., 2006. Activation and self-tolerance of natural killer cells. Immunol. Rev. 214, 130–142.

Gilfillan, S., Ho, E.L., Cella, M., et al., 2002. NKG2D recruits two distinct adapters to trigger NK cell activation and costimulation. Nat. Immunol. 3, 1150–1155.

Girardi, M., Oppenheim, D.E., Steele, C.R., et al., 2001. Regulation of cutaneous malignancy by gammadelta T cells. Science 294, 605–609.

Gordon, S.M., Chaix, J., Rupp, L.J., et al., 2012. The transcription factors T-bet and Eomes control key checkpoints of natural killer cell maturation. Immunity 36, 55–67.

Gregoire, C., Chasson, L., Luci, C., et al., 2007. The trafficking of natural killer cells. Immunol. Rev. 220, 169–182.

Groh, V., Rhinehart, R., Secrist, H., et al., 1999. Broad tumor-associated expression and recognition by tumor-derived gamma delta T cells of MICA and MICB. Proc. Natl. Acad. Sci. U.S.A. 96, 6879–6884.

Grzywacz, B., Kataria, N., Sikora, M., et al., 2006. Coordinated acquisition of inhibitory and activating receptors and functional properties by developing human natural killer cells. Blood 108, 3824–3833.

Guerra, N., Tan, Y.X., Joncker, N.T., et al., 2008. NKG2D-deficient mice are defective in tumor surveillance in models of spontaneous malignancy. Immunity 28, 571–580.

Guo, Y., Maillard, I., Chakraborti, S., et al., 2008. Core binding factors are necessary for natural killer cell development and cooperate with Notch signaling during T-cell specification. Blood 112, 480–492.

Hayakawa, Y., and Smyth, M.J., 2006. CD27 dissects mature NK cells into two subsets with distinct responsiveness and migratory capacity. J. Immunol. 176, 1517–1524.

Hayakawa, Y., Sato-Matsushita, M., Takeda, K., et al., 2011. Early activation and interferon-gamma production of tumor-infiltrating mature CD27 high natural killer cells. Cancer Sci. 102, 1967–1971.

Heemskerk, M.H., Blom, B., Nolan, G., et al., 1997. Inhibition of T cell and promotion of natural killer cell development by the dominant negative helix loop helix factor Id3. J. Exp. Med. 186, 1597–1602.

Horowitz, A., Strauss-Albee, D.M., Leipold, M., et al., 2013. Genetic and environmental determinants of human NK cell diversity revealed by mass cytometry. Sci. Transl. Med. 5 208ra145.

Huntington, N.D., Puthalakath, H., Gunn, P., et al., 2007. Interleukin 15-mediated survival of natural killer cells is determined by interactions among Bim, Noxa and Mcl-1. Nat. Immunol. 8, 856–863.

Iannello, A., Thompson, T.W., Ardolino, M., et al., 2013. p53-dependent chemokine production by senescent tumor cells supports NKG2D-dependent tumor elimination by natural killer cells. J. Exp. Med. 210, 2057–2069.

Imada, K., Bloom, E.T., Nakajima, H., et al., 1998. Stat5b is essential for natural killer cell-mediated proliferation and cytolytic activity. J. Exp. Med. 188, 2067–2074.

Ishigami, S., Natsugoe, S., Tokuda, K., et al., 2000. Clinical impact of intratumoral natural killer cell and dendritic cell infiltration in gastric cancer. Cancer Lett. 159, 103–108.

Jamieson, A.M., Diefenbach, A., McMahon, C.W., et al., 2002. The role of the NKG2D immunoreceptor in immune cell activation and natural killing. Immunity 17, 19–29.

Jin, J., Fu, B., Mei, X., et al., 2013. CD11b(−)CD27(−) NK cells are associated with the progression of lung carcinoma. PLoS One 8, e61024.

Jinushi, M., Takehara, T., Tatsumi, T., et al., 2007. Natural killer cell and hepatic cell interaction via NKG2A leads to dendritic cell-mediated induction of CD4 CD25 T cells with PD-1-dependent regulatory activities. Immunology 120, 73–82.

Kamizono, S., Duncan, G.S., Seidel, M.G., et al., 2009. Nfil3/E4bp4 is required for the development and maturation of NK cells in vivo. J. Exp. Med. 206, 2977–2986.

Kaplan, D.H., Shankaran, V., Dighe, A.S., et al., 1998. Demonstration of an interferon gamma-dependent tumor surveillance system in immunocompetent mice. Proc. Natl. Acad. Sci. U.S.A. 95, 7556–7561.

Kennedy, M.K., Glaccum, M., Brown, S.N., et al., 2000. Reversible defects in natural killer and memory CD8 T cell lineages in interleukin 15-deficient mice. J. Exp. Med. 191, 771–780.

Khong, H.T., and Restifo, N.P., 2002. Natural selection of tumor variants in the generation of "tumor escape" phenotypes. Nat. Immunol. 3, 999–1005.

Kim, S., Iizuka, K., Kang, H.S., et al., 2002. In vivo developmental stages in murine natural killer cell maturation. Nat. Immunol. 3, 523–528.

Kim, S., Poursine-Laurent, J., Truscott, S.M., et al., 2005. Licensing of natural killer cells by host major histocompatibility complex class I molecules. Nature 436, 709–713.

King, A., Balendran, N., Wooding, P., et al., 1991. CD3-leukocytes present in the human uterus during early placentation: phenotypic and morphologic characterization of the $CD56^{++}$ population. Dev. Immunol. 1, 169–190.

Klezovich-Benard, M., Corre, J.P., Jusforgues-Saklani, H., et al., 2012. Mechanisms of NK cell-macrophage *Bacillus anthracis* crosstalk: a balance between stimulation by spores and differential disruption by toxins. PLoS Pathog. 8, e1002481.

Kobayashi, H., Dubois, S., Sato, N., et al., 2005. Role of trans-cellular IL-15 presentation in the activation of NK cell-mediated killing, which leads to enhanced tumor immunosurveillance. Blood 105, 721–727.

Koebel, C.M., Vermi, W., Swann, J.B., et al., 2007. Adaptive immunity maintains occult cancer in an equilibrium state. Nature 450, 903–907.

Kondo, M., Weissman, I.L., and Akashi, K., 1997. Identification of clonogenic common lymphoid progenitors in mouse bone marrow. Cell 91, 661–672.

Koopman, L.A., Kopcow, H.D., Rybalov, B., et al., 2003. Human decidual natural killer cells are a unique NK cell subset with immunomodulatory potential. J. Exp. Med. 198, 1201–1212.

Krebs, P., Barnes, M.J., Lampe, K., et al., 2009. NK-cell-mediated killing of target cells triggers robust antigen-specific T-cell-mediated and humoral responses. Blood 113, 6593–6602.

Lanier, L.L., 2008. Up on the tightrope: natural killer cell activation and inhibition. Nat. Immunol. 9, 495–502.

Laouar, Y., Sutterwala, F.S., Gorelik, L., et al., 2005. Transforming growth factor-beta controls T helper type 1 cell development through regulation of natural killer cell interferon-gamma. Nat. Immunol. 6, 600–607.

Lapaque, N., Walzer, T., Meresse, S., et al., 2009. Interactions between human NK cells and macrophages in response to Salmonella infection. J. Immunol. 182, 4339–4348.

Lassen, M.G., Lukens, J.R., Dolina, J.S., et al., 2010. Intrahepatic IL-10 maintains NKG2A + Ly49- liver NK cells in a functionally hyporesponsive state. J. Immunol. 184, 2693–2701.

Li, H., Han, Y., Guo, Q., et al., 2009. Cancer-expanded myeloid-derived suppressor cells induce anergy of NK cells through membrane-bound TGF-beta 1. J. Immunol. 182, 240–249.

Liao, N.S., Bix, M., Zijlstra, M., et al., 1991. MHC class I deficiency: susceptibility to natural killer (NK) cells and impaired NK activity. Science 253, 199–202.

Lohoff, M., Duncan, G.S., Ferrick, D., et al., 2000. Deficiency in the transcription factor interferon regulatory factor (IRF)-2 leads to severely compromised development of natural killer and T helper type 1 cells. J. Exp. Med. 192, 325–336.

Loza, M.J., and Perussia, B., 2001. Final steps of natural killer cell maturation: a model for type 1-type 2 differentiation? Nat. Immunol. 2, 917–924.

Luther, C., Warner, K., and Takei, F., 2011. Unique progenitors in mouse lymph node develop into CD127+ NK cells: thymus-dependent and thymus-independent pathways. Blood 117, 4012–4021.

Mandaric, S., Walton, S.M., Rulicke, T., et al., 2012. IL-10 suppression of NK/DC crosstalk leads to poor priming of MCMV-specific CD4 T cells and prolonged MCMV persistence. PLoS Pathog. 8, e1002846.

Marcais, A., Cherfils-Vicini, J., Viant, C., et al., 2014. The metabolic checkpoint kinase mTOR is essential for IL-15 signaling during the development and activation of NK cells. Nat. Immunol. 15, 749–757.

Martin-Fontecha, A., Thomsen, L.L., Brett, S., et al., 2004. Induced recruitment of NK cells to lymph nodes provides IFN-gamma for T(H)1 priming. Nat. Immunol. 5, 1260–1265.

Michel, T., Poli, A., Domingues, O., et al., 2012. Mouse lung and spleen natural killer cells have phenotypic and functional differences, in part influenced by macrophages. PLoS One 7, e51230.

Mocikat, R., Braumuller, H., Gumy, A., et al., 2003. Natural killer cells activated by MHC class I(low) targets prime dendritic cells to induce protective $CD8^+$ T cell responses. Immunity 19, 561–569.

Moroso, V., Famili, F., Papazian, N., et al., 2011. NK cells can generate from precursors in the adult human liver. Eur. J. Immunol. 41, 3340–3350.

Muller-Hermelink, N., Braumuller, H., Pichler, B., et al., 2008. TNFR1 signaling and IFN-gamma signaling determine whether T cells induce tumor dormancy or promote multistage carcinogenesis. Cancer Cell 13, 507–518.

Nakamura, M.C., Linnemeyer, P.A., Niemi, E.C., et al., 1999. Mouse Ly-49D recognizes H-2Dd and activates natural killer cell cytotoxicity. J. Exp. Med. 189, 493–500.

Nedvetzki, S., Sowinski, S., Eagle, R.A., et al., 2007. Reciprocal regulation of human natural killer cells and macrophages associated with distinct immune synapses. Blood 109, 3776–3785.

Newell, E.W., and Cheng, Y., 2016. Mass cytometry: blessed with the curse of dimensionality. Nat. Immunol. 17, 890–895.

O'Sullivan, T., Saddawi-Konefka, R., Vermi, W., et al., 2012. Cancer immunoediting by the innate immune system in the absence of adaptive immunity. J. Exp. Med. 209, 1869–1882.

Orr, M.T., and Lanier, L.L., 2010. Natural killer cell education and tolerance. Cell 142, 847–856.

Orr, M.T., Wu, J., Fang, M., et al., 2010. Development and function of CD94-deficient natural killer cells. PLoS One 5, e15184.

Ottaviani, C., Nasorri, F., Bedini, C., et al., 2006. CD56brightCD16(−) NK cells accumulate in psoriatic skin in response to CXCL10 and CCL5 and exacerbate skin inflammation. Eur. J. Immunol. 36, 118–128.

Park, I.K., Giovenzana, C., Hughes, T.L., et al., 2009. The Axl/Gas6 pathway is required for optimal cytokine signaling during human natural killer cell development. Blood 113, 2470–2477.

Park, I.K., Trotta, R., Yu, J., et al., 2013. Axl/Gas6 pathway positively regulates FLT3 activation in human natural killer cell development. Eur. J. Immunol. 43, 2750–2755.

Paul, S., Kulkarni, N., Shilpi, 2016. Intratumoral natural killer cells show reduced effector and cytolytic properties and control the differentiation of effector Th1 cells. Oncoimmunology 5, e1235106.

Paul, S., and Lal, G., 2016. Regulatory and effector functions of gamma-delta (gammadelta) T cells and their therapeutic potential in adoptive cellular therapy for cancer. Int. J. Cancer 139, 976–985.

Peng, H., Jiang, X., Chen, Y., et al., 2013. Liver-resident NK cells confer adaptive immunity in skin-contact inflammation. J. Clin. Invest. 123, 1444–1456.

Peppa, D., Gill, U.S., Reynolds, G., et al., 2013. Up-regulation of a death receptor renders antiviral T cells susceptible to NK cell-mediated deletion. J. Exp. Med. 210, 99–114.

Perona-Wright, G., Mohrs, K., Szaba, F.M., et al., 2009. Systemic but not local infections elicit immunosuppressive IL-10 production by natural killer cells. Cell Host Microbe 6, 503–512.

Piccioli, D., Sbrana, S., Melandri, E., et al., 2002. Contact-dependent stimulation and inhibition of dendritic cells by natural killer cells. J. Exp. Med. 195, 335–341.

Pietra, G., Manzini, C., Rivara, S., et al., 2012. Melanoma cells inhibit natural killer cell function by modulating the expression of activating receptors and cytolytic activity. Cancer Res. 72, 1407–1415.

Porrata, L.F., Inwards, D.J., Micallef, I.N., et al., 2010. Interleukin-15 affects patient survival through natural killer cell recovery after autologous hematopoietic stem cell transplantation for non-Hodgkin lymphomas. Clin. Dev. Immunol. 2010, 914945.

Rabinovich, B.A., Li, J., Shannon, J., et al., 2003. Activated, but not resting, T cells can be recognized and killed by syngeneic NK cells. J. Immunol. 170, 3572–3576.

Rajalingam, R., 2012. Overview of the killer cell immunoglobulin-like receptor system. Methods Mol. Biol 882, 391–414.

Ramirez, K., Chandler, K.J., Spaulding, C., et al., 2012. Gene deregulation and chronic activation in natural killer cells deficient in the transcription factor ETS1. Immunity 36, 921–932.

Reynders, A., Yessaad, N., Vu Manh, T.P., et al., 2011. Identity, regulation and in vivo function of gut NKp46 + RORgammat+ and NKp46 + RORgammat- lymphoid cells. EMBO J. 30, 2934–2947.

Roda, J.M., Parihar, R., Magro, C., et al., 2006. Natural killer cells produce T cell-recruiting chemokines in response to antibody-coated tumor cells. Cancer Res. 66, 517–526.

Romo, N., Magri, G., Muntasell, A., et al., 2011. Natural killer cell-mediated response to human cytomegalovirus-infected macrophages is modulated by their functional polarization. J. Leukoc. Biol. 90, 717–726.
Rosmaraki, E.E., Douagi, I., Roth, C., et al., 2001. Identification of committed NK cell progenitors in adult murine bone marrow. Eur. J. Immunol. 31, 1900–1909.
Roy, S., Barnes, P.F., Garg, A., et al., 2008. NK cells lyse T regulatory cells that expand in response to an intracellular pathogen. J. Immunol. 180, 1729–1736.
Samson, S.I., Richard, O., Tavian, M., et al., 2003. GATA-3 promotes maturation, IFN-gamma production, and liver-specific homing of NK cells. Immunity 19, 701–711.
Schierloh, P., Aleman, M., Yokobori, N., et al., 2005. NK cell activity in tuberculosis is associated with impaired CD11a and ICAM-1 expression: a regulatory role of monocytes in NK activation. Immunology 116, 541–552.
Schotte, R., Dontje, W., Nagasawa, M., et al., 2010. Synergy between IL-15 and Id2 promotes the expansion of human NK progenitor cells, which can be counteracted by the E protein HEB required to drive T cell development. J. Immunol. 184, 6670–6679.
Screpanti, V., Wallin, R.P., Ljunggren, H.G., et al., 2001. A central role for death receptor-mediated apoptosis in the rejection of tumors by NK cells. J. Immunol. 167, 2068–2073.
Senovilla, L., Vitale, I., Martins, I., et al., 2012. An immunosurveillance mechanism controls cancer cell ploidy. Science 337, 1678–1684.
Shankaran, V., Ikeda, H., Bruce, A.T., et al., 2001. IFNgamma and lymphocytes prevent primary tumour development and shape tumour immunogenicity. Nature 410, 1107–1111.
Small, C.L., McCormick, S., Gill, N., et al., 2008. NK cells play a critical protective role in host defense against acute extracellular *Staphylococcus aureus* bacterial infection in the lung. J. Immunol. 180, 5558–5568.
Smith, H.R., Heusel, J.W., Mehta, I.K., et al., 2002. Recognition of a virus-encoded ligand by a natural killer cell activation receptor. Proc. Natl. Acad. Sci. U.S.A. 99, 8826–8831.
Smith, K.M., Wu, J., Bakker, A.B., et al., 1998. Ly-49D and Ly-49H associate with mouse DAP12 and form activating receptors. J. Immunol. 161, 7–10.
Smyth, M.J., Thia, K.Y., Street, S.E., et al., 2000a. Differential tumor surveillance by natural killer (NK) and NKT cells. J. Exp. Med. 191, 661–668.
Smyth, M.J., Thia, K.Y., Street, S.E., et al., 2000b. Perforin-mediated cytotoxicity is critical for surveillance of spontaneous lymphoma. J. Exp. Med. 192, 755–760.
Smyth, M.J., Crowe, N.Y., and Godfrey, D.I., 2001. NK cells and NKT cells collaborate in host protection from methylcholanthrene-induced fibrosarcoma. Int. Immunol. 13, 459–463.
Smyth, M.J., Swann, J., Kelly, J.M., et al., 2004. NKG2D recognition and perforin effector function mediate effective cytokine immunotherapy of cancer. J. Exp. Med. 200, 1325–1335.
Smyth, M.J., Swann, J., Cretney, E., et al., 2005. NKG2D function protects the host from tumor initiation. J. Exp. Med. 202, 583–588.
Soderquest, K., Walzer, T., Zafirova, B., et al., 2011. Cutting edge: $CD8^+$ T cell priming in the absence of NK cells leads to enhanced memory responses. J. Immunol. 186, 3304–3308.
Spear, P., Wu, M.R., Sentman, M.L., et al., 2013. NKG2D ligands as therapeutic targets. Cancer Immunity 13, 8.
Stebbins, C.C., Watzl, C., Billadeau, D.D., et al., 2003. Vav1 dephosphorylation by the tyrosine phosphatase SHP-1 as a mechanism for inhibition of cellular cytotoxicity. Mol. Cell. Biol. 23, 6291–6299.
Steele, N., Anthony, A., Saunders, M., et al., 2012. A phase 1 trial of recombinant human IL-21 in combination with cetuximab in patients with metastatic colorectal cancer. Br. J. Cancer 106, 793–798.
Street, S.E., Cretney, E., and Smyth, M.J., 2001. Perforin and interferon-gamma activities independently control tumor initiation, growth, and metastasis. Blood 97, 192–197.
Takaki, R., Hayakawa, Y., Nelson, A., et al., 2005. IL-21 enhances tumor rejection through a NKG2D-dependent mechanism. J. Immunol. 175, 2167–2173.
Takayama, T., Kamada, N., Chinen, H., et al., 2010. Imbalance of NKp44(+)NKp46(−) and NKp44(−)NKp46(+) natural killer cells in the intestinal mucosa of patients with Crohn's disease. Gastroenterology 139 882–892, 892 e881–883.
Takeda, K., Smyth, M.J., Cretney, E., et al., 2002. Critical role for tumor necrosis factor-related apoptosis-inducing ligand in immune surveillance against tumor development. J. Exp. Med. 195, 161–169.
Taki, S., Nakajima, S., Ichikawa, E., et al., 2005. IFN regulatory factor-2 deficiency revealed a novel checkpoint critical for the generation of peripheral NK cells. J. Immunol. 174, 6005–6012.
Teng, M.W., Vesely, M.D., Duret, H., et al., 2012. Opposing roles for IL-23 and IL-12 in maintaining occult cancer in an equilibrium state. Cancer Res. 72, 3987–3996.

Topham, N.J., and Hewitt, E.W., 2009. Natural killer cell cytotoxicity: how do they pull the trigger? Immunology 128, 7–15.

van den Broek, M.E., Kagi, D., Ossendorp, F., et al., 1996. Decreased tumor surveillance in perforin-deficient mice. J. Exp. Med. 184, 1781–1790.

Vankayalapati, R., Garg, A., Porgador, A., et al., 2005. Role of NK cell-activating receptors and their ligands in the lysis of mononuclear phagocytes infected with an intracellular bacterium. J. Immunol. 175, 4611–4617.

Vargas, C.L., Poursine-Laurent, J., Yang, L., et al., 2011. Development of thymic NK cells from double negative 1 thymocyte precursors. Blood 118, 3570–3578.

Veillette, A., Latour, S., and Davidson, D., 2002. Negative regulation of immunoreceptor signaling. Annu. Rev. Immunol. 20, 669–707.

Vesely, M.D., Kershaw, M.H., Schreiber, R.D., et al., 2011. Natural innate and adaptive immunity to cancer. Annu. Rev. Immunol. 29, 235–271.

Vosshenrich, C.A., Garcia-Ojeda, M.E., Samson-Villeger, S.I., et al., 2006. A thymic pathway of mouse natural killer cell development characterized by expression of GATA-3 and CD127. Nat. Immunol. 7, 1217–1224.

Waggoner, S.N., Cornberg, M., Selin, L.K., et al., 2011. Natural killer cells act as rheostats modulating antiviral T cells. Nature 481, 394–398.

Wu, X., Peng, M., Huang, B., et al., 2013. Immune microenvironment profiles of tumor immune equilibrium and immune escape states of mouse sarcoma. Cancer Lett. 340, 124–133.

Xu, H.C., Grusdat, M., Pandyra, A.A., et al., 2014. Type I interferon protects antiviral $CD8^+$ T cells from NK cell cytotoxicity. Immunity 40, 949–960.

Yadi, H., Burke, S., Madeja, Z., et al., 2008. Unique receptor repertoire in mouse uterine NK cells. J. Immunol. 181, 6140–6147.

Yokoyama, W.M., Kim, S., and French, A.R., 2004. The dynamic life of natural killer cells. Annu. Rev. Immunol. 22, 405–429.

Yu, H., Fehniger, T.A., Fuchshuber, P., et al., 1998. Flt3 ligand promotes the generation of a distinct CD34(+) human natural killer cell progenitor that responds to interleukin-15. Blood 92, 3647–3657.

Zafirova, B., Mandaric, S., Antulov, R., et al., 2009. Altered NK cell development and enhanced NK cell-mediated resistance to mouse cytomegalovirus in NKG2D-deficient mice. Immunity 31, 270–282.

Zhang, C., Zhang, J., Niu, J., et al., 2008. Interleukin-12 improves cytotoxicity of natural killer cells via upregulated expression of NKG2D. Hum. Immunol. 69, 490–500.

Zhou, Z., Zhang, C., Zhang, J., et al., 2012. Macrophages help NK cells to attack tumor cells by stimulatory NKG2D ligand but protect themselves from NK killing by inhibitory ligand Qa-1. PLoS One 7, e36928.

PART II

ROLE IN DISEASE AND TREATMENT

CHAPTER

10

Varicella Zoster Vaccine Experience in HIV-Infected Children and Adolescents

William Borkowsky and Anne A. Gershon

OUTLINE

Abstract

Live-attenuated Varicella Zoster Virus (VZV) Vaccine was first developed in Japan and tested for safety in children who were steroid treated. It proved safe and immunogenic. This led to its testing in the United States in a population of children with acute lymphoblastic leukemia who were in remission. This study too showed that the vaccine was safe and immunogenic in this population. After the vaccine was approved for healthy children, it was tested in a population of HIV-infected children and adolescents with minimal to moderate immunosuppression (i.e., CD4 percentage of >25 and a count of >350). It also proved safe and relatively immunogenic in this population. Finally, it has been evaluated in a more immunocompromised population of HIV-infected children and adolescents (i.e., CD4 percentage >15 and a count >200). VZV vaccine also proved safe and relatively immunogenic in this population. It is suggested that two immunizations, 12 weeks apart, be the vaccination strategy for these HIV populations.

M.A. Hayat (ed): *Immunology, Volume 1.*
DOI: http://dx.doi.org/10.1016/B978-0-12-809819-6.00010-1

© 2018 Elsevier Inc. All rights reserved.

INTRODUCTION

Varicella Zoster virus (VZV) was and, in many parts of the world, still is a major cause of common childhood exanthems (Hussey et al., 2016). In situations of acute or sustained immunodeficiency, infection was also a cause of significant morbidity and death. The effect of common immunosuppressive intervention such as corticotrophin-like drugs in asthma was sufficient to result in progressive disease if given during the first week following exposure to VZV. In the case of leukemia in children, as many as 7% died of VZV infection (Feldman and Lott, 1987). This changed dramatically with the introduction of effective antivirals and the development of a live-attenuated VZV vaccine.

VARICELLA ZOSTER VACCINE

VZV Vaccine in Leukemic Children

The VZV vaccine was first tested in the United States in children with acute lymphoblastic leukemia in remission, defying the usual testing procedure which starts with adults and then children, all healthy. The initial trial demonstrated significant safety and efficacy, as well as protection from varicella zoster (Gershon et al., 1984; Lawrence et al., 1988; Hardy et al., 1991).

VZV Vaccine in HIV-Infected VZV-Naïve Children and Adolescents

The early 1980s saw the emergence of Acquired Immunodeficiency Syndrome (AIDS) caused by HIV-1. Although the initial victims were adults with risk factors such as intravenous drug use, sex (particularly men who have sex with men) and exposure to blood products via transfusion or clotting factors, it soon became clear that pediatric populations were also included. The majority of these were infected by maternal-to-child transmission either in utero or peripartum. Others were infected by transfusion of blood or blood products. The majority of children in these groups were too young to have been potential recipients of the VZV vaccine as it wasn't licensed for "healthy" children aged 1–12 years until 1995. While some of these individuals experienced uncomplicated infections (Srugo et al., 1993), others noted chronic and disseminated infections (Jura et al., 1989; Leibovitz et al., 1992, 1993). In addition, VZV infection in HIV-infected children was associated with a high rate of zoster (70% in those with low CD4 T cell levels) (Gershon et al., 1997). With the approval of a VZV vaccine for "healthy" children, it now appeared possible that vaccinating HIV-infected children with this vaccine at a time before they experienced a significant loss of CD4 T cells (i.e., when still >25%), would be likely safe and immunogenic, resulting in protection from infection.

In order to test this premise a team of investigators within the Pediatric AIDS Clinical Trial Group developed a protocol (P265). It had already been determined that normal adults may require two vaccinations to develop protective levels of antibodies to VZV (Gershon and Steinberg, 1990; Hardy and Gershon, 1990). Consequently, the protocol was designed to deliver 2 doses of vaccine 12 weeks apart. The findings of the study were that vaccination was safe and relatively immunogenic. Two months after the second vaccination, 60%

developed antivaricella antibodies measured by the sensitive fluorescent antibody to membrane antigen (FAMA) method and 83% demonstrated lymphoproliferative responses to varicella antigens (Levin et al., 2001). This team subsequently evaluated the safety and efficacy of VZV vaccine in a population of children with less well preserved CD4 T cells (i.e., >15% and a count >200). These were divided into those with no clinical symptomatology (CDC clinical category status of A) and those with minor symptoms, having a CDC clinical category B status. Also studied were those individuals with a more severe clinical category C status 3 months prior to vaccination who, after antiretroviral therapy, were now classified as CDC class A. The controls in the study were those individuals first reported in P265 and those with wild-type VZV infection. They found the vaccine safe and immunogenic. In excess of 79% developed specific antibody and/or cell-mediated immunity (CMI) 2 months after the last vaccination. One year after vaccination, antibody levels had declined in all groups so that only 43–73% still maintained a positive FAMA titer. The lymphocyte proliferation assay (LPA) responses remained stable. Some of the individuals that were no longer FAMA positive received a booster vaccine at that time without evidence of immunologic boosting. Further assessment of immunogenicity at 2 and 3 years postvaccination showed no additional fall off in responses and comparable response rates in both vaccine and wild-type exposure groups.

Because of the issue of high spontaneous in vitro proliferation noted in a subset of patients, a novel CMI assay, measuring responder cell frequency (RCF) was piloted in this second study. This assay employs limiting dilution techniques to detect antigen-specific replication capacity. Using RCF, in excess of 80% of vaccines demonstrated CMI to VZV antigens. This response declined somewhat when measured at later time points. In addition, although the second and third vaccinations did not seem to provide additional immunity, the authors suggested that the second vaccination be maintained in the future.

A second study looked at a comparable immunologic grouping of children (i.e., >15% and 200 CD4 T cells) receiving 2 vaccines 12 weeks apart. These children had been receiving antiretroviral therapy for a median of 27 months. Using an ELISA assay they found that 34 initially seronegative children seroconverted (12% after the first vaccine and 79% after the second dose) while 26 initially seropositive showed boosting from the vaccines as the geometric mean titer increased from 57 at baseline to 108 after the first and 135 after the second vaccination. There was a trend toward better responses among the younger, with higher CD4 and better viral suppression (Taweesith et al., 2011).

Why do antibody titers wane with time after vaccination? A Swiss study compared initial median anti-VZV titers after vaccination of 97 HIV-infected children and 78 HIV-infected adults. They reported that the children developed lower initial titers than the adults (264 vs 1535 IU/L, $p < .001$). The median titer of the infected children was also significantly lower ($p < .001$) than that achieved by uninfected children (1151 IU/L). Over the course of up to 10 years, a regression slope of the children's antibodies increased but not to the level of antibody shown by the infected adults, whose titer slope was unchanged. A univariate analysis showed that a higher HIV RNA and lower CD4 percent were significantly associated with loss of antibody. In the multivariate analysis, only the HIV RNA level remained significant ($p = .01$). Not only were the initial titers lower in the HIV-infected children, but antibody avidity, a measurement of B cell maturation dependent on T cell help, was also significantly lower ($p < .001$) in the infected versus the uninfected children. Lower avidity translates into poorer neutralization of the virus (L'Huillier et al., 2012).

There were insufficient numbers of vaccines to determine whether the vaccinees were "protected" from disease. A retrospective analysis of vaccine effectiveness was undertaken by investigators from Columbia University and Saint Jude Children's Research Hospital. Effectiveness was measured by looking at 64 cases of "natural" varicella in the prevaccine era (1989–1998) versus the vaccine era (1999–2007) with 19 cases of chicken pox. In addition there were 72 vaccinated children. The incident cases of VZV infection in the three groups showed a reduction in cases in the unvaccinated vaccine era group (36.8 incident cases per 1000 patient years) compared to the prevaccine era (103.3 incident cases per patient years) and yet a further reduction to 6.8 among the vaccinated group. Thus the availability of highly active antiretroviral therapy (HAART) post-1996 probably helped to reduce the incidence of varicella by itself (Son et al., 2010). The addition of vaccination resulted in a more significant reduction in cases both by reducing community exposure to other children and by inducing immunity in the HIV-infected.

A recent review evaluated immunity among HIV-infected adolescents who were recipients of varicella vaccine compared to an HIV-uninfected cohort. VZV positivity was seen in 82% of the 432 HIV-infected population while virtually all of the uninfected cohort was seropositive. The infected group serostatus was positive in 100% within 3 years of vaccination. This decreased to 73% and 77% seropositivity when assayed 3–6 years and >6 years postvaccination. The greater contribution toward seropositivity was undoubtedly due to natural VZV infection as 71% of the unvaccinated demonstrated seropositivity. The presence of antibody was associated with higher CD4, higher nadir CD4 and lower levels of viremia over time. Lower seropositivity correlated with more significant HIV associated clinical illness, as measured by the presence of a CDC clinical class C status (Purswani et al., 2016).

The availability of HAART by 1996 allowed for significant immune reconstitution in HIV-infected children, previously failing earlier antiretroviral regimens (Essajee et al., 1999; Nikolic-Djokic et al., 2002). Paradoxically, this increase in CD4 T cells was associated with the development of zoster in the initial months of reconstitution (Tangsinmankong et al., 2004). Thus VZV appears to behave like other latent infections such as mycobacteria tuberculosis, hepatitis B, and CMV, which have been shown to reactivate as a consequence of immune reconstitution in those with very low CD4 T cells who respond dramatically after HAART with undetectable plasma HIV RNA and normalizing CD4 cells. Affected individuals appear more likely to have had nonprotective levels of anti-VZV antibody prior to the onset of zoster.

Why is the VZV seropositivity rate so high in the HIV-infected in the absence of vaccination? VZV infection is latent with intermittent reactivation boosting immune responses, even among immunologically normal individuals. In the case of individuals with immune suppression, reactivation occurs more commonly. One study looked for factors that correlated with the development of zoster. Subjects studied were 47 individuals with zoster and 141 matched controls without zoster. Using an in vitro VZV stimulation assay, those with zoster had higher levels of Th1-like response, both stimulated and unstimulated. However, VZV-stimulated $CD8^+CD107a^+$ (a marker of cytotoxic effector cells) were significantly lower ($p = .02$) in cases versus controls. Regulatory T cells (Treg), often described as $CD4^+IL10^+$ or $CD4^+TGF\text{-}beta^+$ and $CD4^+PD1^+$ were associated with an increased risk of zoster. Thus Treg cells may be restraining the effector T cell responses that are necessary to control VZV reactivation (Weinberg et al., 2012).

VZV Vaccine in HIV-Infected Children and Adolescents With Previous VZV Infection or Vaccination

The role VZV-specific T cells play in recovery from primary varicella and the protection from zoster is well established (Arvin, 1996; Arvin et al., 1996). To assess the comparison of CMI in HIV-infected individuals and uninfected individuals with a history of VZV infection, Weinberg et al. again employed the RCF assay in 200 HIV-positive and 752 HIV-negative persons. The RCF frequency was seen to peak in healthy individuals at 34 years of age with natural infection and increased until 43 years of age in those with varicella vaccine immunization. This increase may represent the boosting of responses with subclinical reactivation of VZV. HIV-infected children and adults had significantly lower frequencies at various time points than healthy individuals, although by 59 years of age, this was no longer significant in the adults due to decreasing CMI in healthy aging individuals. Recipients of the VZV zoster vaccine showed significant boosting of their CMI, particularly in the individuals aged 65–75 years compared to those without vaccination. HIV-infected children immunized with 2 doses of VZV vaccine had lower RCF frequencies than healthy children with wild-type infection, although the differences were not significant (Weinberg et al., 2010).

CONCLUSION

In conclusion, it appears that vaccination with a live-attenuated varicella vaccine in HIV-infected individuals with mild to moderate immunosuppression is both safe and effective in inducing immunity and reducing cases of primary varicella and herpes zoster.

References

Arvin, A.M., 1996. Immune responses to varicella-zoster virus. Inf. Dis. Clin. North Am. 10, 529–570.

Arvin, A.M., Moffat, J.F., and Redman, R., 1996. Varicella-zoster virus: aspects of pathogenesis and host response to natural infection and varicella vaccine. Adv. Virus Res. 46, 263–309.

Essajee, S.M., Kim, M., Gonzalez, C., et al., 1999. Immunologic and virologic responses to HAART in severely immunocompromised HIV-1-infected children. AIDS 13, 2523–2532.

Feldman, S., and Lott, L., 1987. Varicella in children with cancer: impact of antiviral therapy and prophylaxis. Pediatrics 80, 465–472.

Gershon, A.A., Steinberg, S., Galasso, G., et al., 1984. Live attenuated varicella vaccine in children with leukemia in remission. Biken J. 27, 77–81.

Gershon, A.A., and Steinberg, S.P., 1990. Live attenuated varicella vaccine: protection in healthy adults compared with leukemic children. National Institute of Allergy and Infectious Diseases Varicella Vaccine Collaborative Study Group. J. Infect. Dis. 161, 661–666.

Gershon, A.A., Mervish, N., LaRussa, P., et al., 1997. Varicella-zoster virus infection in children with underlying human immunodeficiency virus infection. J. Infect. Dis. 176, 1496–1500.

Hardy, I.R., and Gershon, A.A., 1990. Prospects for use of a varicella vaccine in adults. Inf. Dis. Clin. North Am. 4, 159–173.

Hardy, I.R., Gershon, A.A., Steinberg, S.P., et al., 1991. The incidence of zoster after immunization with live attenuated varicella vaccine. A study in children with leukemia. Varicella Vaccine Collaborative Study Group. N. Engl. J. Med. 325, 1545–1550.

Hussey, H.S., Abdullahi, L.H., Collins, J.E., et al., 2016. Varicella zoster virus-associated morbidity and mortality in Africa: a systematic review protocol. BMJ Open 6, e010213.

Jura, E., Chadwick, E.G., Josephs, S.H., et al., 1989. Varicella-zoster virus infections in children infected with human immunodeficiency virus. Ped. Infect. Dis. J. 8, 586–590.

Lawrence, R., Gershon, A.A., Holzman, R., et al., 1988. The risk of zoster after varicella vaccination in children with leukemia. N. Engl. J. Med. 318, 543–548.

Leibovitz, E., Kaul, A., Rigaud, M., et al., 1992. Chronic varicella zoster in a child infected with human immunodeficiency virus: case report and review of the literature. Cutis 49, 27–31.

Leibovitz, E., Cooper, D., Giurgiutiu, D., et al., 1993. Varicella-zoster virus infection in Romanian children infected with the human immunodeficiency virus. Pediatrics 92, 838–842.

Levin, M.J., Gershon, A.A., Weinberg, A., et al., 2001. Immunization of HIV-infected children with varicella vaccine. J. Ped. 139, 305–310.

L'Huillier, A.G., Ferry, T., Courvoisier, D.S., et al., 2012. Impaired antibody memory to varicella zoster virus in HIV-infected children: low antibody levels and avidity*. HIV Med. 13, 54–61.

Nikolic-Djokic, D., Essajee, S., Rigaud, M., et al., 2002. Immunoreconstitution in children receiving highly active antiretroviral therapy depends on the CD4 cell percentage at baseline. J. Infect. Dis. 185, 290–298.

Purswani, M.U., Karalius, B., Yao, T.J., et al., 2016. Prevalence and persistence of varicella antibodies in previously immunized children and youth with perinatal HIV-1 infection. Clin. Infect. Dis. 62, 106–114.

Son, M., Shapiro, E.D., LaRussa, P., et al., 2010. Effectiveness of varicella vaccine in children infected with HIV. J. Infect. Dis. 201, 1806–1810.

Srugo, I., Israele, V., Wittek, A.E., et al., 1993. Clinical manifestations of varicella-zoster virus infections in human immunodeficiency virus-infected children. Am. J. Dis. Child 147, 742–745.

Tangsinmankong, N., Kamchaisatian, W., Lujan-Zilbermann, J., et al., 2004. Varicella zoster as a manifestation of immune restoration disease in HIV-infected children. J. Allergy Clin. Immun. 113, 742–746.

Taweesith, W., Puthanakit, T., Kowitdamrong, E., et al., 2011. The immunogenicity and safety of live attenuated varicella-zoster virus vaccine in human immunodeficiency virus-infected children. Ped. Infect. Dis. J. 30, 320–324.

Weinberg, A., Lazar, A.A., Zerbe, G.O., et al., 2010. Influence of age and nature of primary infection on varicella-zoster virus-specific cell-mediated immune responses. J. Infect. Dis. 201, 1024–1030.

Weinberg, A., Huang, S., Song, L.Y., et al., 2012. Immune correlates of herpes zoster in HIV-infected children and youth. J. Virol. 86, 2878–2881.

CHAPTER

11

Development of an Effective Alzheimer's Vaccine

Dante J. Marciani

OUTLINE

Abstract

The presence of a protective immunity against Alzheimer's disease (AD), which decreases with age, supports the concept of preventive immune therapy or vaccination against this disease. Yet, all of the attempts to develop a vaccine using amyloid-β (Aβ) have ended in disappointment; setbacks due to the vaccines having the wrong immunogen, i.e., the B-cell epitope of Aβ, which stimulates an antibody response that releases the neurotoxic Aβ oligomers trapped in plaque, and proinflammatory adjuvants that induce a damaging inflammatory autoimmune response, or adjuvants that are ineffective in the elderly. Hence, an effective Alzheimer's vaccine should have an immunogen that has the conformational epitope characteristic of toxic oligomers, which has B and T-cell epitopes and is the effective therapeutic target; also, to avert an inflammatory immune response, the vaccine should have a sole Th2 adjuvant that while inhibits without

M.A. Hayat (ed): Immunology, Volume 1.
DOI: http://dx.doi.org/10.1016/B978-0-12-809819-6.00011-3

© 2018 Elsevier Inc. All rights reserved.

abrogating Th1 immunity, it may also ameliorate the immune decline associated with aging. Because of the cooperativity between Aβ and tau protein, an effective preventive vaccine should aim to have both immunogens. To prevent the problems encountered with protein carriers, AD vaccines should use the whole protein, or a large segment of it, structurally stabilized to preserve the conformational epitope required to induce a protective immune response. Finally, AD vaccines should be used in a preventive mode, because due to immunosenescence and the degenerative nature of AD, a therapeutic vaccine may be unable to reverse the damage.

INTRODUCTION

Alzheimer's disease (AD) is a neurodegenerative proteopathy that afflicts largely those over 65 years old; i.e., it is estimated that currently there are worldwide around 46 million people suffering of AD and related disorders (ADRD), a number that will almost double every 20 years (www.alz.co.uk/worldreport2015). Proteopathies are diseases characterized by the production of aberrant conformers of certain proteins that are misfolded and aggregate in a crystallization-like seeding mechanism—changes that lead to a disturbance of their cellular functions and disease. There is significant evidence that a competitive immune system produces protective natural antibodies (nAbs) that recognize and remove the aberrant forms of the normally present proteins that cause AD; i.e., the immune system starts at an early age a surveillance to neutralize and eliminate those aberrant proteins responsible for AD, a process that has been called the body's housekeeping. This immune surveillance decreases with age due to immunosenescence, a decline that coincides with the onset of AD (reviewed in Marciani, 2015a). Convincing support for the protective role of nAbs against AD is provided by the clinical studies with aducanumab (Biogen), a human monoclonal antibody (mAb) directed against the neurotoxic Ab oligomers, which in prodromal AD slows down the decline in cognitive functions (reviewed in Marciani, 2016). While aducanumab has showed therapeutic benefits, other mAbs as well as vaccines targeting the amyloid-β (Aβ) amino-terminal region have failed to show any benefits in humans; yet, they were effective in improving the condition of AD transgenic mouse models. These findings stress the significant differences between the immune responses from humans and transgenic animal models, a situation that explains the constant clinical failures despite the promising preclinical results.

That most of the therapeutic approaches to prevent or treat AD have been focused on Aβ is a consequence of the "amyloid cascade hypothesis" or ACH, proposed in 1992 to explain the Aβ role in the pathological events leading to AD by Hardy and Higgins (Selkoe and Hardy, 2016). This hypothesis has been for years the subject of intense discussions, largely a consequence of the poor clinical outcome obtained with AD therapies targeting Aβ. Since its inception, the ACH has gone through a series of changes to incorporate new findings related to Aβ and its interactions with other proteins in causing AD; i.e., the evidence supports the notion that an imbalance between the production and clearance of Aβ and some of its derived peptides is often an early factor in the initiation of AD (Selkoe and Hardy, 2016). In effect Aβ, particularly Aβ42 that has 42 amino acid residues in its sequence, has a tendency to form soluble oligomers that are neurotoxic. Indeed, it is now being assumed that Aβ plaques have a protective role against AD by sequestering and immobilizing the neurotoxic soluble Aβ oligomers. In fact, the deposit and oligomerization of Aβ42 starts a sequence of events leading to extensive neuronal damage, as proposed by the cascade hypothesis (Selkoe and Hardy, 2016; Reitz, 2012). Also that new information shows that Aβ can interact with other proteins, e.g., tau protein and α-synuclein, indicates that Aβ functions and pathological activity can include other proteins. Yet, Aβ is a normal protein component present in the brains of healthy individuals as well as those affected by AD, which plays various roles such as in memory formation, neuronal survival, physiological control of neuronal, and several others (Fedele et al., 2016). Moreover, the precursor of Aβ or β-amyloid precursor protein (APP), apparently has a role in the growth and maturation of various cells, including a tissue-repair function, but the mechanism by which APP exerts these actions is unclear (Dawkins and Small, 2014). Hence, that Aβ and APP have normal functions in the body, suggests that their nonselective removal may interfere with some cellular functions—a situation that may limit the options available in the development of products to remove Aβ neurotoxic forms.

RELEVANT PROTEIN IMMUNOGENS IN ALZHEIMER'S DISEASE

Amyloid β

As previously discussed, Aβ seems to be the most important protein in AD pathology. It is a relatively small protein of molecular weight around 44,000, which is found mainly in two isoforms, one having 40 amino acid residues, Aβ40, and the other having two extra amino acid residues at the C-terminus, Aβ42—the latter being the main component of amyloid plaques. Because of the two additional amino acid residues, the Aβ42 isoform has higher associative properties than Aβ40, rapidly aggregating to form oligomers, protofibrils, and fibrils in its way to form the characteristic plaques associated with AD. An important characteristic of Aβ and some other proteins is their capacity to form—as a result of misfolding—structures called amyloids, which are aggregates characterized by being closely packed, highly ordered, and rich in β-sheet structures (Knowles et al., 2014) that tend to form fibrils. A characteristic of these aggregates is the presence of a conformational epitope, which is common to most amyloidogenic proteins and is amino acid sequence independent (O'Nuallain and Wetzel, 2002). Soluble Aβ is found in both monomeric as well as oligomeric forms, where the monomer is nontoxic, while the soluble dimers and higher oligomers are neurotoxic (Kayed et al., 2003); in effect, the current opinion is that insoluble plaques, by sequestering and immobilizing the soluble Aβ oligomers, may be acting as a protection system (Lee et al., 2004). While Aβ40 and Aβ42 are the most common forms, there is a newly reported Aβ43 specie, an intermediate in the processing of APP toward Aβ40; this isoform has higher aggregative properties, is more neurotoxic than the other isoforms, and triggers the aggregation of the typically nontoxic Aβ40, all indications that this form may be a factor in AD (Burnouf et al., 2015).

Structurally, the Aβ amino-terminal region where the B-cell epitope is located is exposed to the aqueous milieu, while the most hydrophobic central and C-terminus regions are buried inside the oligomers (Dodel et al., 2011). There is evidence that misfolding of amyloidogenic proteins can facilitate interactions with the lipid bilayer present in cell membranes, leading to membrane damage and disruption of the cellular homeostasis (Relini et al., 2014). Indeed, soluble Aβ oligomers can act as synaptotoxins, causing disruption of receptors and signaling pathways that result in synaptic injury and memory impairment (Larson and Lesné, 2012). Nonetheless, Aβ has the capacity of interacting with other proteins important for AD pathology, such as tau protein and α-synuclein (Miller et al., 2011); hence, it is possible that a close association and cooperativity between these proteins accelerates the development and progression of AD. Besides, it has been proposed that Aβ is a key damage-associated molecular pattern (DAMP), which is recognized by some toll-like receptor, inducing the production of inflammatory cytokines (Clark and Vissel, 2015). Hence, Aβ plays an active role in inducing a damaging proinflammatory immune response that promotes microglia activation and increases plaque burden (Browne et al., 2013), which supports the standpoint that AD vaccines inducing Th1 immunity will be damaging rather than beneficial.

Tau Protein

Tau, a protein that promotes assembly stabilizing microtubules in neurons, is a highly hydrophilic protein characterized for adopting a random coil-type conformation, with six isoforms having 350–440 amino acid residues. Tau has been considered an amyloidogenic protein where oligomeric intermediates are cytotoxic by mediating deleterious effects on cell membranes (Flach et al., 2012). Tau has various posttranslational modifications like phosphorylation, and it has been projected that the largest isoform has 80 serine and threonine residues that can be phosphorylated, i.e., the normal function of tau requires some phosphorylation, while hyperphosphorylation leads to a loss of its biological function. In fact, it has been shown that soluble Aβ dimers isolated from the brains of late-onset AD patients induce tau hyperphosphorylation and neuritic degeneration (reviewed in Marciani, 2016). Consequently, it has been assumed that vaccines having phosphorylated tau peptides would be effective in preventing AD. But, it has clearly been demonstrated that the pathological effects of tau are due to its oligomers, i.e., administration to animal models of antibodies against tau soluble oligomers reverses tauopathy phenotypes without affecting hyperphosphorylated tau (Castillo-Carranza et al., 2014). That vaccination with a tau peptide that is essential for pathological tau–tau interactions improves the AD phenotype and reduces hyperphosphorylation by 95% (Kontsekova et al., 2014a), supports the concept that like with Aβ, the pathological effects are due to soluble tau oligomers rather than its hyperphosphorylated isoforms.

α-Synuclein

A relatively small size protein that has 140 amino acid residues, is an intrinsically disordered protein lacking stable three-dimensional (3D) structures and that, similar to tau protein, has a random coil conformation. This protein, mainly found in the brain, is an amyloidogenic one that can interact with the lipids from the cell membrane, as well as with Aβ and tau protein. Although α-synuclein has a predominant role in Parkinson's disease, it seems to act in a synergistic manner with Aβ and tau, accelerating neuropathology in AD. Indeed, it has been proposed that tau and α-synuclein may interact with each other to promote the solubility and fibrillization of each other, contributing in this way to the development of neurodegeneration. Actually, a new report indicates that α-synuclein fibrils and protofibrils are the ones that interact with tau, inhibiting its function to stabilize microtubules and promoting tau aggregation—changes that lead to dysfunction of the neuronal cells (Oikawa et al., 2016).

IMMUNE RESPONSE AGAINST ALZHEIMER'S DISEASE PROTEIN IMMUNOGENS

Several studies have shown that there is a natural protective immunity that recognizes aberrant isoforms of proteins such as Aβ and tau, neutralizing and removing them from the system. This immune surveillance starts to decline with age, when AD and other diseases linked to aging like cancer begin to appear, a consequence of the immune deterioration associated with aging or immunosenescence (reviewed in Marciani, 2016). Hence, it is apparent that any effective active immune therapy approach to prevent AD would need also to ameliorate the effects of immunosenescence, to allow the immune system to mount an effective protective immune response.

Natural Immune Response Against Aβ

Initial evidence of a protective natural immunity against AD was provided in 2009 by a retrospective study by Fillit et al. of the effects of prolonged administration of intravenous gamma globulins (IVIG) pooled from healthy young adult donors to late middle age individuals free of AD and ADRD (reviewed in Marciani, 2016). In fact, those individuals treated with IVIG showed 5 years later a 42% lower risk of being diagnosed with ADRD than the untreated ones. Subsequent analysis of IVIG preparations have shown that they contain nAbs that recognize the oligomeric, but not the monomeric, forms of Aβ (Dodel et al., 2011; O'Nuallain et al., 2010); nAbs that decrease with age, as well as the anti-tau nAbs, due to immunosenescence (reviewed in Marciani, 2015a, 2016). Moreover, administration of anti-Aβ nAbs to an AD transgenic mouse model resulted in an improvement of their cognitive functions, supporting the proposed protective role of these antibodies (Dodel et al., 2011). Hence, these nAbs may explain the IVIG's efficacy in averting AD reported by Fillit et al.; but, IVIG treatment has failed clinically to improve the cognitive functions of patients with early AD, indicating that antibodies are more effective when used preventively rather than therapeutically.

Another study has shown that administration of anti-Aβ oligomers nAbs to a transgenic mouse model for AD resulted in a reversal of synaptic dysfunction, modulation of inflammatory cytokines, reduction in toxic Aβ oligomers, and improvement of cognitive functions (Mengel et al., 2013). It is of interest that IVIG preparations, while having nAbs that recognize toxic Aβ oligomers, usually do not recognize normal Aβ monomers; i.e., nAbs recognize the region encompassing the Aβ residues 28–40, but not the N-terminal region (Dodel et al., 2011); which means that the monomers have important physiological functions. The importance of the absence of nAbs against the Aβ monomers that recognize the N-terminal region is shown by the fact that antibodies against that region dissociate Aβ fibrils, releasing neurotoxic Aβ soluble oligomers that have been sequestered as plaque—a process that has been called the "dust-raising effect" (Liu et al., 2012, 2015a). Confirming this observation is the fact that treatment of an AD mouse transgenic model with a mouse mAb 3D6 targeting the N-terminal region of Aβ caused an increased neuronal hyperactivity as compared to wild-type mice expressing normal levels of Aβ (Busche et al., 2015). This damaging effect was observed in animals with plaque as well as with animals too young to have developed plaques, which implies that the mAb 3D6 favors the release of soluble neurotoxic Aβ oligomers rather than their sequestration as plaques, aggravating the disease. This fact was corroborated with a different mAb, β1 antibody, also directed against the N-terminus of Aβ and capable of preventing Aβ deposition in an AD transgenic mouse model (reviewed in Marciani, 2016).

Another piece of information that points at the potentially damaging effects of antibodies against the Aβ N-terminal region, $A\beta_{1-17}$, is that they promote the amyloidogenic processing of the parent APP. This processing may drive up the production of Aβ, contributing in this way to the development of sporadic AD (Deng et al., 2012). Hence, it is evident that a protective nAbs response must target the oligomeric rather than monomeric forms of Aβ, i.e., it should recognize this "generic" conformational epitope which is characteristic of the soluble neurotoxic oligomers and that is shared by other amyloids (O'Nuallain and Wetzel, 2002).

Natural Immune Response Against Tau Protein

Further examination of the nAbs present in IVIG has shown the presence of autoantibodies that recognize different regions of the tau protein, but the content of these antibodies varies with different preparations. Indeed, a study of tau nAbs in an animal model has shown that IVIG treatment while reducing tau aggregation did not change the preformed tau aggregates (Esteves-Villanueva et al., 2015). In a manner reminiscent of that of Aβ, administration of IVIG to animal models for AD has resulted in a depletion of the aberrant tau isoforms and an improvement of the pathology and/or behavioral parameters (Rothstein, 2016). Yet, in spite of the promising results in AD transgenic animal models, the clinical application of IVIG to AD patients has failed to deliver positive results—an indicator that antibodies in general are more effective when used in a preventive mode. Hence, it is evident that active immune therapy or vaccination would be more effective before the onset of disease and occurrence of immunosenescence.

Passive and Active AD Immune Therapy in Transgenic Mouse Models

While not a perfect replica of AD, the transgenic mouse models have delivered relevant information about the different processes leading to this disease, e.g., they have help to elucidate the nature of the pathogenic species and have shown that Aβ and tau have a significant cooperation in inducing their pathological effects. For instance, while tau immune therapy was initially focused on the hyperphosphorylated regions of that protein (after the finding that tau soluble oligomers are the neurotoxic species (Kayed, 2010; Kontsekova et al., 2014a, b), it has shifted its focus toward the oligomers. That a mAb that prevents tau's oligomerization in a transgenic mouse model precludes its hyperphosphorylation and reduces tau pathology (Kontsekova et al., 2014b), and that treatment with a mAb specific for tau oligomers reverses the tauopathy phenotypes reducing the oligomers, but not the hyperphosphorylated neurofibrillary tangles, confirms that the oligomers are the relevant therapeutic target, rather than the hyperphosphorylated tau regions.

Hence, it has been proposed that mAbs against tau oligomers, like those against Aβ oligomers, may be effective in passive immune therapy (Kayed, 2010; Kontsekova et al., 2014b). In fact, the efficacy of anti-tau immune therapy has been confirmed in an AD mouse model by active vaccination to induce an antibody response against those tau epitopes responsible for its oligomerization (Kontsekova et al., 2014a). That either active or passive immunization against tau oligomers reverses tauopathy phenotype, regardless of tau's phosphorylation state, implies that the actual therapeutic target is oligomeric tau rather than its phosphorylated isoforms. However, we should be cautious about the absolute validity of AD transgenic mouse models; i.e., it is important to remember that all of the anti-Aβ immune therapy methods that failed clinical studies did well in transgenic mouse models.

While the relative importance of Aβ versus tau as therapeutic targets in the treatment of AD has been the subject of debate (Reitz, 2012), there is evidence that both proteins may cooperate in causing and aggravating AD. In fact, intracerebral infusion of Aβ-containing brain extracts induced tau pathology in a transgenic AD mouse model—an effect that can be explained by the fact that soluble Aβ dimers isolated from AD cortex, at subnanomolar concentrations, induced hyperphosphorylation of tau, followed by disruption of the microtubule cytoskeleton and neuritic degeneration in the absence of Aβ fibrils (reviewed in Marciani, 2016). Furthermore, a comparison of the effects of Aβ40 and Aβ42 on a live cellular model showed that Aβ42 induced cleavage, phosphorylation, and aggregation of full length tau, but not Aβ40. While the evidence seems to indicate that Aβ, as proposed by the ACH, plays a pivotal role in AD development (Selkoe and Hardy, 2016), due to the previous Aβ vaccine clinical failures it has been proposed that tau may be a better therapeutic target than Aβ (Pedersen and Sigurdsson, 2015). But there are close interactions and synergism between Aβ and tau in their neurotoxic effects leading to AD; i.e., one conclusion is that Aβ is upstream of tau in AD pathogenesis, triggering tau's conversion from a normal to toxic state (Stancu et al., 2014), yet it is evident that toxic tau enhances Aβ toxicity via a feedback loop (Bloom, 2014). Clearly, there is what it has been called an "intense crosstalk" between these two molecules, as well as the fact that the Aβ toxicity is mediated at least in part by tau (Nisbet et al., 2015).

An indication of a close link between Aβ and tau is provided by the fact that Aβ vaccination of an AD mouse model resulted in a reduction of Aβ as well as tau pathology and its hyperphosphorylation; changes that resulted in protection from neuron loss (Wilcock et al., 2009). Yet, similar effects were obtained upon removal of pathological tau oligomers with a tau oligomer-specific mAb in an AD mouse model, i.e., the treatment reversed memory deficits while accelerating plaque formation in the brain (Castillo-Carranza et al., 2015). Of significance is that treatment with this mAb resulted in a decrease in the levels of neurotoxic Aβ oligomers; indeed, a reduction of tau oligomers shifted the Aβ aggregation pathway toward the formation of plaque, which is apparently protective (Lee et al., 2004; Castillo-Carranza et al., 2015). These findings reinforce the case for tau oligomers being effective therapeutic targets in AD treatment. However, since Aβ and tau have pathogenic roles in AD, either separately or in a cooperative manner, an effective strategy would be to target simultaneously both proteins' toxic forms, an approach that has been previously suggested, but never tested (Wisniewski and Boutajangout, 2010; Lambracht-Washington and Rosenberg, 2013; Nisbet et al., 2015).

Validation of the AD Immune Therapy Concept

While preclinical work with immune therapy in AD transgenic mouse models has delivered promising results, these methods have consistently failed when tested in humans (reviewed in Marciani, 2015a, 2016). This outcome, despite the significant evidence about a natural anti-AD protective immunity, has raised skepticism about the validity of such an approach. Nonetheless, a promising indication of the value of antibodies as active or passive immune therapy is found in the clinical results obtained with aducanumab, a high affinity human IgG1 mAb (http://www.alzforum.org/therapeutics/aducanumab) directed against a conformational epitope present in Aβ oligomers (Sevigny et al., 2015). This mAb was derived from autoimmune B lymphocytes from older individuals that were cognitively normal, under the assumption that this antibody should protect against the neurotoxic Aβ oligomers. Indeed, treatment of prodromal AD with aducanumab resulted in a dose-dependent reduction in deposition of Aβ in the brain, as well as a slowdown in the decline of cognitive functions (reviewed in Marciani, 2016; Sullivan, 2016). It is of relevance that aducanumab seems to be equally effective in removing Aβ regardless of the ApoE genotype. That AD onset coincides with the initiation of immunosenescence indicates that any approach using the intrinsic protective immunity would need to ameliorate the age-linked immune decline. A comparison of aducanumab with other mAbs shows that while the latter mAbs were largely directed against the Aβ N-terminus region and recognized only monomers (Moreth et al., 2013), aducanumab was directed against a conformational epitope found solely in the aberrant toxic Aβ oligomers. Hence, as indicated before, nAbs can distinguish between the Aβ normal and its aberrant forms, which explains their protective role in humans.

While the clinical setbacks have usually been accepted as evidence that the ACH is wrong and the immune therapy approaches unfeasible, an examination of those studies reveals that, with the exception of aducanumab, in all cases the basic assumptions were erroneous (Marciani, 2016); which would explain the constant clinical failures. Although the objective in AD vaccines is to induce a solely antiinflammatory Th2 immunity to neutralize and remove misfolded proteins that are self-antigens, many of the AD vaccines induced Th1 immunity; a problem magnified by use of the wrong Aβ-derived immunogens, which resulted in a likely damaging rather than protective immune response (Marciani, 2016). Nonetheless, new preclinical evidence strengthens the case for preventive AD vaccines, as the report that cellular implants containing cells genetically engineered to produce anti-Aβ antibodies effectively reduce the levels of Aβ40 and Aβ42 in the brain validates the effectiveness of endogenously produced antibodies to reduce Aβ levels and tau pathology in an AD mouse model (Lathuilière et al., 2016). Hence, development of safe and effective AD vaccines that induce the production of protective antibodies would require new strategies that should take into account the lessons learned from past vaccine failures.

DEVELOPMENT OF AN ALZHEIMER'S DISEASE PREVENTIVE VACCINE

Since in AD the brain damage is most likely irreversible and due to immunosenescence the patients' immune system is weakened, it is unlikely that therapeutic vaccines would be of benefit, which may explain in part the consistent clinical failures. Hence, AD vaccine development should focus on prevention rather than treatment (Cribbs, 2010); this way it would be possible to prolong the natural anti-AD immunity by boosting the immune system when it is still competent to mount an effective protective response. This vaccination strategy should prevent or delay the onset of AD, allowing for a better management of this epidemic at the medical and socioeconomic levels. Because of the AD vaccine requirements, apparently there are two promising approaches to its development: the subunit and DNA vaccines, which may be used separately or alternatively in conjunction (Wang et al., 2014; Lambracht-Washington and Rosenberg, 2015). Here, we would analyze both types of vaccines.

AD Subunit Vaccines

Similar to viral and cancer subunit vaccines, the AD vaccines have two crucial components: the immunogen and the immune modulator or adjuvant. The immunogen, known also as the antigen, is responsible for targeting the immune response, while the adjuvant decides the nature, i.e., Th1 or Th2, as well as the intensity of the response. However, there are significant differences between these vaccines. Because the immunity elicited by viral/cancer vaccines is a Th1/Th2 immunity designed to kill those cells carrying abnormal surface antigens via cytokines and cytotoxic T lymphocytes (CTLs), their immunogens are usually short linear T-cell epitopes that are presented on the surface of antigen-presenting cells (APCs), such as dendritic cells (DCs), complexed with major histocompatibility complex (MHC) molecules of Class I or II (Elsawa et al., 2004). Depending on the nature of the MHC molecules, Class I or II, there will be production of CTLs or Helper T lymphocytes, respectively, with both types of lymphocytes participating in the killing of cells carrying foreign T-cell epitopes. In contrast, in vaccines against proteopathies like AD, the immunogen is not foreign but a self-antigen that differs from its normal form in its misfolding, resulting in a deviant 3D structure or conformational epitope, having unique associative properties leading to the formation of cytotoxic oligomers (Hayden and Teplow, 2013). However, that these conformational epitopes include both T and B-cell epitopes, complicates their use as immunogens due to the possible induction of a damaging Th1 autoimmune response, such as that observed with the AN1792 AD vaccine (reviewed in Marciani, 2014). Indeed, as previously indicated, nAbs recognize these distinctive conformational epitope(s) found in the soluble toxic oligomers of Aβ, but apparently not the normal monomeric isoforms (Dodel et al., 2011).

The Immunogen

Attempts to use only the B-cell epitopes from Aβ, to prevent a Th1 immune response with either Th1 or Th2 adjuvants, have resulted in disappointments; the results of trying to treat patients with already diagnosed AD as well as using Aβ immunogens that instead of protecting could cause an immunopotentiation of the disease (reviewed in Marciani, 2016). Indeed, in the case of Aβ the conformational epitope has a new 3D structure formed by the contact between monomers forming oligomers, which requires T and B-cell epitopes, and is independent on the amino sequence of the protein, i.e., a common structural motif shared by different amyloidogenic proteins (O'Nuallain and Wetzel, 2002; Kayed et al., 2003; Glabe, 2006). Therefore, different strategies have been tried in order to prevent an undesirable Th1 immune, such as nonhuman random sequence amyloid oligomer mimics (Rasool et al., 2012), peptides selected by molecular mimicry also known as mimotopes (Mandler et al., 2015), and synthetic Aβ dimeric peptides called miniamyloids that are bound by human nAbs (Roeder et al., 2013). These relatively small peptides are conjugated to a carrier protein to increase their immunogenicity, an approach that has the intrinsic problem of carrier-induced epitope-specific suppression (CIESS), where the antipeptide antibody response is suppressed—an occurrence dependent of the recipient's genotype (Marciani, 2016). This CIESS can be aggravated by the presence of certain chemical groups in the spacers linking the peptide to the carrier protein, e.g., maleimide—all factors contributing to a weak immune response (Marciani, 2016).

As previously indicated, Aβ derived immunogens, having only the B-cell epitopes region, while apparently inducing a protective immunity in transgenic models in fact induce a damaging immune response that releases toxic soluble oligomers that have been sequestered as plaque (Liu et al., 2015a; Busche et al., 2015). Actually, patients suffering from early AD and immunized with CAD106—a vaccine composed of $A\beta_{1-6}$ peptides coupled to Qβ virus-like particles—showed an increase of Aβ in serum (Riviere et al., 2014). A similar increase of Aβ in plasma was observed in the patients immunized with Vanutide cridificar, another vaccine that like with CAD106 has as an immunogen the $A\beta_{1-7}$ N-terminal peptide (Pasquier et al., 2016). This increase in circulating Aβ following immunization with the Aβ N-terminal peptide raises questions about the possible cytotoxicity of that circulating form. Indeed, such Aβ was likely securely immobilized as plaque, until it was released by the antibodies against the Aβ N-terminal region. Also, central clearance of Aβ, i.e., at the brain, has other potential complications—particularly if the antibodies are against the N-terminal region or B-cell epitope—problems that can occur with passive or active immune therapy, i.e., mAbs or vaccination (Liu et al., 2014). These potential adverse effects include vessel-damage edema that can damage the blood–brain barrier (BBB), phagocytosis-related neuroinflammation caused by the release of proinflammatory cytokines and other factors by microglia upon phagocytosis of Aβ, and autoimmune-related concerns, found mainly with nAbs directed against the B-cell epitope. Therefore, all of these concerns would need to be addressed to develop effective preventive and/or therapeutic approaches for AD (Liu et al., 2012).

Thus, to prevent the complications encountered with Aβ derived immunogens, the best approach could be to use the whole Aβ, prepared in way to present that unique 3D motif characteristic of toxic soluble oligomers, while modifying the B-cell epitope to avert any immune responses against this epitope in the naturally occurring protein. This is a challenging endeavor, due to the metastability of the Aβ oligomers as well as their polydispersity (Rosensweig et al., 2012; Goure et al., 2014). Indeed, with time Aβ changes from monomeric to oligomeric and insoluble fibrillar forms. Therefore, the immune response induced by prolonged vaccination with whole Aβ would change over time, targeting initially monomers, followed by all forms of Aβ regardless of conformation, and finally a response that recognizes either monomers and fibrils or oligomeric forms in a conformation dependent manner (Vasilevko et al., 2010). This immune response must be a reflection of the Aβ transient aggregation changes, a situation that underscores the need for vaccines to have stable soluble toxic Aβ soluble oligomers, rather than monomers or fibrils, to selectively induce a protective immunity against those oligomers. An effective way to obtain stable soluble Aβ oligomers, e.g., dimers and tetramers, would be to use oligomers that have been stabilized by cross-linking. Indeed, there are Aβ oligomers, from dimers to pentamers, which have been stabilized by a photoinduced cross-linking of unmodified proteins and are neurotoxic (Bitan, 2006; Hayden and Teplow, 2013), that can be purified to homogeneity by gel permeation chromatography. Another approach is to form a globular Aβ42 oligomer or globulomer, composed of ~12 monomers, which is relatively stable and is toxic (Barghorn et al., 2005). However, this globulomer can be cross-linked by a mild treatment with a cross-linking agent like glutaraldehyde. This globulomer induces the generation of mAbs that are specific for oligomeric but not monomeric or fibrillar forms, and showed therapeutic value in an AD transgenic mouse model (Hillen et al., 2010) An interesting alternative is the stabilization of neurotoxic soluble Aβ oligomers by protein engineering, where the conformation of an Aβ double-cysteine mutant is stabilized by an intramolecular disulfide bond; this modified Aβ while able to form various oligomeric species is unable to convert into amyloid fibrils, however it is recognized by a conformation-specific mAb that also recognizes wild-type Aβ oligomers (Sandberg et al., 2010).

Hence, these and some other Aβ special constructs can be used in AD vaccines to display to the immune system the conformational epitope characteristic of the neurotoxic Aβ oligomers, while minimizing the presence of nontoxic monomers, in this way stimulating the production of antibodies against the therapeutically relevant conformational epitope. Due to their composition, indeed the whole Aβ peptide, in contrast to past AD vaccines these new immunogens will have both B and T-cell epitopes; hence, they would not need to be linked to a carrier protein. However, it would be advantageous to prevent a damaging immune response against the Aβ B-cell epitope. Moreover, because of the T-cell epitopes present in Aβ, these immunogens would require special adjuvants to elicit a safe and effective sole Th2 antiinflammatory immunity.

As there is significant evidence that Aβ and tau act in a cooperative manner aggravating AD pathology (Stancu et al., 2014), then a preventive vaccine should consider having both Aβ and tau immunogens, i.e., a bivalent vaccine. However, because of the large size of tau isoforms, about ten times the size of Aβ, this protein may be a good candidate for using peptides as immunogens. In fact, a conjugated vaccine with a tau peptide essential for tau–tau pathological interactions conjugated to keyhole limpet hemocyanin, and with alum as an adjuvant, has shown in a transgenic mouse model reduction of both tau toxic oligomers and hyperphosphorylation (Kontsekova et al., 2014a). An effective alternative would be to stabilize the tau toxic oligomers via chemical cross-linking and use them as immunogens (Patterson et al., 2011), alone or in combination with Aβ.

The Adjuvant

As previously indicated, the adjuvant is responsible for deciding the nature of the immune response induced by a vaccine, i.e., Th1 proinflammatory or Th2 antiinflammatory, as well as its strength (Marciani, 2015a). While an induced Th1 is always present with Th2 immunity, the latter can be found as a sole Th2 antiinflammatory immunity, perhaps as a result of Th2 immunity being a sort of repair mechanism to promote healing after a potentially damaging Th1 inflammatory response (Gause et al., 2013). In fact, the number of agents that elicit Th1/Th2 immunity is considerable and they are largely conserved compounds found in pathogens, which trigger an immediate immune response in the body known as innate immunity, which is followed by Th1/Th2 immunity. In contrast, the number of agents that induce a sole Th2 immunity is very small, currently there are only three groups of compounds that induce this antiinflammatory immunity: alum, sphingosine phosphate (S1P) derivatives, and fucosylated glycans (reviewed in Marciani, 2015a). It is perhaps due to the lack of effective Th2 adjuvants that many clinically tested AD vaccines used Th1 adjuvants combined with antigens having only Aβ B-cell epitopes, a strategy that assumes that the lack of T-cell epitopes would prevent an inflammatory immune response. But, regardless of the immunogen, Th1 adjuvants will induce a systemic inflammatory response that spreads across the body via cytokines, which by acting at the BBB can induce the production of inflammatory factors (Erickson et al., 2012); a situation that would aggravate neuropathies. Hence, Th1 adjuvants are unsafe immune modulators for any vaccine designed to prevent or treat proteopathies, as they will induce a damaging inflammatory autoimmunity, a situation that may be aggravated by microglia activation, which promotes production of inflammatory cytokines.

Alum, a Th2 adjuvant used in numerous infectious disease vaccines, performs poorly in the elderly population as it is unable to ameliorate immunosenescence. In fact, the effectiveness of the influenza vaccine, which is near 90% in young adults, can be 10% or less in those that are over 70 years old. Nonetheless, alum has been used in some of the AD vaccines clinically tested in elderly patients (Marciani, 2016). This situation may explain the modest immune response elicited by vaccines without or with weak adjuvants, the exception being the inflammatory Th1 adjuvant QS-21 (Arai et al., 2015). Hence, it is unlikely that alum would be a useful adjuvant for AD vaccines, which presumably would be administered to those near 60 years of age and older. The other Th2 adjuvants are S1P derivatives that favor Th17 immunity while dampening Th1 immune responses, but the type of immune response, i.e., absence or presence of inflammatory immunity, would depend on several factors that are difficult to control (reviewed in Marciani, 2015a). In fact, that S1P-based adjuvants can induce the powerful proinflammatory immunity Th17 may be problematic, as they could stimulate a damaging autoimmunity that would aggravate proteopathies like AD (reviewed in Marciani, 2015a). Nonetheless, the Th2 immunomodulatory properties of S1P have been used to develop an AD vaccine formulated with $A\beta_{1-42}$ in a liposomal form, a formulation that is being tested in various transgenic mouse models (reviewed in Marciani, 2015a).

A new family of Th2 adjuvants is that which is based on fucosylated glycans, which act at the DC-SIGN level by binding to this lectin found on the DC and microglia surfaces. DC-SIGN is a trans-membrane C-type lectin that can bind either mannose or fucose-containing glycans. While both sugars bind in the same pocket of the lectin they interact with different binding sites (reviewed in Marciani, 2015a). Actually, the types of immunities induced by these sugars are different. While binding of mannose oligomers promotes Th1/Th2 immunity, fucose glycans stimulate a sole Th2 antiinflammatory response inhibiting but not abrogating Th1 immunity, which is important as protection against pathogens (Marciani, 2014; reviewed in Marciani, 2015a). In effect, the Th2 immune response stimulated by fucosylated glycans is fairly similar to the immune tolerance induced during pregnancy and by some tumors having fucosylated glycans on their cell surfaces, frequently as part of the Lewis^x antigen motif (reviewed in Marciani, 2014, 2015a). But the fucosylated glycans have not been linked to Th17 immunity or other inflammatory responses, making them safe Th2 adjuvants.

Serendipitously, a likely Th2 adjuvant for AD vaccines is a deacylated derivative of QS-21, the adjuvant responsible for the damaging meningoencephalitis observed during the clinical studies of the AN1792 vaccine. This fucosylated triterpene glycoside, named QT-0101, apparently binds to DC-SIGN, inducing a sole Th2 immunity (Marciani, 2014, 2015a,b). Different from QS-21, the glycoside QT-0101 has a low acute toxicity with a good margin of safety, which is a requirement for a vaccine adjuvant. Also, as QT-0101 has an aldehyde group, it can ameliorate the immunosenescence effects by delivering a costimulatory signal to T-cells, which replaces the one missing due to a loss of T-cells' CD28 receptor and DCs' ligands CD80 and CD86 caused by aging, this way preventing T-cell anergy (Marciani, 2015a). Hence, QT-0101's immune modulatory effect is a result of the concerted actions of the fucosyl residue that biases the immune system toward a sole Th2 immunity, and the aldehyde group that prevents the failure by the immune system to mount an effective immunity, i.e., all characteristics required in an effective adjuvant for AD vaccines targeting the elderly population.

AD DNA Vaccines

A promising approach to AD vaccines is genetic immunization, where the immunogen is expressed by the host cells that have been transfected with a gene encoding for the whole or sectors of a protein, such as Aβ or tau (Kudrna and Ugen, 2015). Of relevance is that DNA vaccines for Aβ induce a Th2 immune response against the 3D motif characteristic of toxic soluble oligomers, confirming that the transiently expressed protein, albeit at very low levels, behaves like the one produced normally by the body (Matsumoto et al., 2013; Lambracht-Washington and Rosenberg, 2013). In contrast, DNA vaccines encoding for proteins from foreign agents induce Th1 immunity, however, the Th2 immunity induced by DNA vaccines can be changed to Th1 immunity by subsequent immunizations with Aβ plus a Th1 adjuvant like Quil A (Lambracht-Washington and Rosenberg, 2015). A consequence of the little protein expression by DNA vaccination is the low production of high affinity antibodies, which can be somewhat increased by subsequent DNA immunizations or by using methods such as electroporation (Kudrna and Ugen, 2015). While the antibody levels elicited by DNA vaccines would not approach those attained with subunit vaccines, the selectivity for the conformational epitope is high, which is not the case with subunit vaccines. Yet, like all vaccines, the immunity induced by DNA vaccines is affected by immunosenescence. Hence, while DNA vaccination induces the production of high affinity antibodies—a desirable property considering the low level of antibodies that pass across the BBB—the low antibody production may affect their protective functions. Therefore, in order to increase the production of high affinity antibodies, an effective approach has been to combine both the subunit and the DNA vaccines.

As indicated, the low production of antigen induced by DNA vaccination, e.g., Aβ, while favoring the clonal selection of B-cells producing high affinity antibodies, results in their limited expansion, which is reflected in the low antibody production (Marciani, 2016). Actually, it has been shown that coimmunization with DNA and an Aβ derived antigen resulted in higher antibody production (Liu et al., 2015b). Hence, subsequent immunizations with Aβ subunit vaccines will result in the expansion of those B-cell clones producing high affinity antibodies, and hence a higher antibody production. Yet, because of immunosenescence, AD vaccines would need an adjuvant to maintain the Th2 immunity while ameliorating the immune decline caused by aging. Thus, an effective strategy would be to follow the initial genetic immunization with subunit vaccines that have the oligomeric Aβ immunogen stabilized to present the characteristic 3D conformational epitope to those B-cells having on their surfaces high affinity antibodies specific for that 3D epitope, as well as an effective sole Th2 adjuvant like QT-0101 to ameliorate the effects of immunosenescence (Fig. 11.1). Although there is no direct evidence that genetic immunization with the tau gene stimulates antibodies against the oligomers, it is likely that it would result in the expression of multiple isoforms with the production of oligomers (Kanai et al., 1989). Thus, like with Aβ DNA vaccination, the tau oligomers would induce production of antibodies against the 3D conformational epitope. To enhance the production of antibodies against the tau oligomers, there would be a need for immunizations with subunit vaccines, with tau present as stable oligomers. Like with Aβ, the transient oligomers may be stabilized by their chemical cross-linking (Patterson et al., 2011), this way delivering immunogens that can present the conformational epitope to those B-cells having high affinity antibodies against that epitope as receptors, and this way stimulating their clonal expansion and higher antibody production.

DISCUSSION

It is evident that with vaccines against proteopathies like AD, where the aberrant antigen is a normal but misfolded self-protein, the immune response should be a sole Th2 anti-inflammatory one that produces antibodies which recognize, neutralize, and facilitate the removal of aberrant protein isoforms. In fact, AD vaccination should not induce the production of Th1 immunity mediators, which would aggravate the disease by initiating damaging inflammatory autoimmune responses, particularly because Aβ, which is a DAMP, and possibly tau, induce Th1 inflammatory immune responses (reviewed in Marciani, 2015a). Consequently, there would be a need for new strategies that will require novel immunogens and adjuvants.

As mentioned before, AD vaccines require immunogens with complex 3D conformational epitopes, which are the result of interactions between Aβ monomers, to yield soluble oligomers—interactions that require contacts between surfaces on the monomers that include B and T-cell epitopes. This situation partially explains the failures of AD vaccines

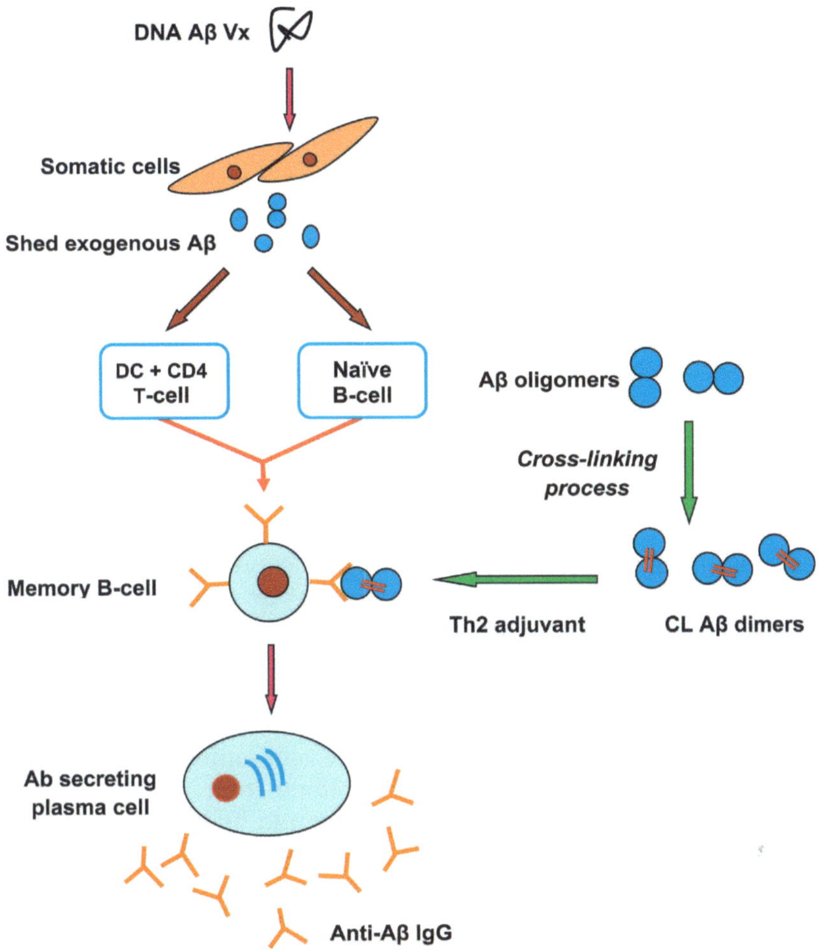

FIGURE 11.1 Abbreviated illustration of the proposed immunization protocol merging Aβ DNA with subunit AD vaccines. Transfection of somatic cells with the Aβ gene results in shedding of the expressed Aβ, and subsequent formation of neurotoxic oligomers having the conformational epitope. The shed Aβ is then processed by DCs and presented to a helper CD4 T-cell, which is activated. The Aβ is also bound by naïve B-cells carrying receptors for that antigen, which upon interaction with activated CD4 helper T-cells become activated B-cells that, after some changes, and producing antibodies (Ab), become memory B-cells, i.e., the primary immune response. Some memory B-cells, upon exposure to the conformational epitope found in the stabilized cross-linked Aβ oligomers (CL Aβ dimers), undergo clonal expansion to produce antibody secreting plasma cells. The CL Aβ dimers are delivered as subunit vaccine boosters to elicit a secondary immune response. To avoid the risk of inflammatory immunity induced by Aβ and its interactions with various cells from the immune system, all of the vaccines are formulated with a sole Th2 adjuvant, which is also capable of ameliorating the effects of immunosenescence.

having as immunogens linear Aβ B-cell epitopes. In fact, the information available shows that soluble oligomers of Aβ, tau, and perhaps α-synuclein, initiate the events leading to neurodegeneration. Thus, posttranslational modifications like phosphorylation are an effect rather than the cause of AD (Huang and Jiang, 2009). Consequently, the ideal AD vaccine immunogens would be the neurotoxic oligomers which, due to the metastable nature of proteins like Aβ and tau, are only transiently present unless the oligomers are stabilized by cross-linking (Sandberg et al., 2010; Patterson et al., 2011) or other alternative structures (Barghorn et al., 2005). Yet, in the case of Aβ, the N-terminal region should be unable to induce an antibody response, like that elicited by all of the clinically tested AD vaccines, which leads to release of neurotoxic oligomers from plaque.

Since in the DNA/subunit vaccination approach the proteins expressed after genetic immunization refold to adopt various conformations, including the aberrant 3D one, the immune system would be exposed to the different immunogen isoforms to yield memory B-cells specific to each epitope. Exposure of these memory B-cells to their corresponding epitopes will result in their clonal expansion, the reason why cross-linked immunogens should be advantageous in stabilizing the 3D conformational epitope needed to stimulate the expansion of those specific B-cells. Although cross-linking of the immunogen should have a relatively small effect on the induction of a secondary immune response, it may affect the primary response more. The reason being that to induce a primary immune response, the cross-linked proteins would be processed by the APCs, a processing that could be affected by the presence of cross-linked amino acid residues. Yet, that some of these cross-linked proteins have been used to prepare mAbs against the oligomeric forms indicates that such interference must not be significant and/or that would depend on the nature of the cross-linking; a situation to be evaluated at the preclinical phase. Nevertheless, priming with DNA vaccines results in a low production of high affinity antibodies, which can be expanded significantly by boosting with a stabilized subunit vaccine, all in the presence of a sole Th2 adjuvant, which may be an effective way to induce the production of high levels of protective antibodies having high affinity and specificity for the aberrant 3D conformational epitope from Aβ and quite possible tau (Fig. 11.1). Following the same train of thought, it would be beneficial if the immunogen does not induce an immune response against the Aβ B-cell epitope. Another requirement essential for an effective AD vaccine is the stimulation of a sole Th2 immunity, a requisite since Aβ and maybe tau can induce a proinflammatory Th1 immunity that aggravates AD, and microglia upon phagocytosis of Aβ produces inflammatory cytokines and other mediators of inflammation (Liu et al., 2014). An additional complication with AD vaccination is immunosenescence, which is responsible for the immune decline associated with aging. Thus, an effective adjuvant should, besides inducing a sole Th2 and inhibiting Th1 immunity, ameliorate immunosenescence. The stimulation of antiinflammatory immunity can be achieved by using fucosylated glycans as adjuvants, which by acting at the DC level bias the immune system to Th2 immunity, but these glycans cannot ameliorate the immunosenescence effects. However, presence of an aldehyde group would deliver an alternative T-cell costimulatory signal, which replaces the one lost due to aging, this way preventing T-cell anergy (Marciani, 2015a). This immune modulation, where the concerted actions of the fucose on DCs and the aldehyde on T-cells would allow the stimulation of an effective Th2 immune response despite aging, can be achieved with the glycoside QT-0101, which paradoxically is closely related to QS-21 (Marciani, 2014, 2015a).

Hence, an effective preventive AD vaccine should have immunogens, like Aβ and tau, which present to the immune system the 3D conformational epitopes characteristic of the toxic oligomeric isoforms. In order to minimize the effects of the metastability of these proteins, they should have some discrete cross-linking to stabilize the conformation of the conformational epitope. The use of a monovalent with a single immunogen, or a bivalent vaccine with two immunogens, should be decided after preclinical testing that should include an animal model(s) naturally susceptible to AD, like the canine or nonhuman primate models (Laurijssens et al., 2013), to complement the studies with transgenic mouse models. While a vaccination protocol using only a subunit vaccine would be easier to implement with relatively large populations, due to the apparent advantages of DNA vaccines as a primary immunization to induce high affinity antibodies, such an approach should be evaluated in parallel to the subunit vaccine-only method. For practical reasons, intramuscular delivery would be the preferential method.

CONCLUSIONS

Evidently, to avoid misleading information, all of the preclinical and clinical development should be done using similar formulation(s), i.e., while the immunogens structures would change due to variations in cross-linking and other variables, the adjuvant should be always the same. Indeed, many AD vaccine studies had assumed wrongly that information obtained with complete Freund's adjuvant (CFA) could be used with other Th1 adjuvants, which is not the case (Marciani, 2016). Moreover, because CFA cannot be used clinically, any information obtained with this adjuvant may be of little help in developing a product for human use. Taking into account the needs for a sole Th2 immunity, amelioration of immunosenescence, and safety, it is evident that a modulator like QT-0101 would be beneficial. As mentioned before, this immune modulator while inducing a sole antiinflammatory immunity would not abrogate Th1 immunity, critical for defense against pathogens and incipient cancers. In fact, the effects of QT-0101 would be transient, lasting long enough to induce the production of B-cells secreting antibodies against the selected target and memory B-cells, but no T-cells that can mediate a Th1 immune response. However, upon subsequent immunizations with the same vaccine formulation, the available memory B-cells will be induced to undergo clonal expansion once more, i.e., the reason for the additional immunizations would be to maintain the population of B-cell secreting antibodies, which may decrease due to immunosenescence. Indeed, it is possible that the effects of immunosenescence on the protective antibody response against AD will parallel the decrease in immunity observed with infectious disease vaccines. Thus, like with those vaccines, a strategy would be to maintain an acceptable level of protective anti-AD antibodies by increasing the frequency of booster shots. But the end-results would depend on the immune competence of each individual. Yet, a delay in the onset of AD of a few years would be helpful in dealing with this epidemic. Hence, the impact of an effective AD vaccine, because of its easy use and cost-effectiveness, should be worldwide.

Acknowledgments

I would like to thank Dr. Yan-Jiang Wang from the Third Military Medical University, Chongqing, China, for bringing to my attention the role and implications of peripheral clearance of Aβ in AD immune therapy.

References

Arai, H., Suziki, H., and Yoshiyama, T., 2015. Vanutide cridificar and the QS-21 adjuvant in Japanese subjects with mild to moderate Alzheimer's disease: results from two phase 2 studies. Curr. Alzheimer Res. 12, 242–254.

Barghorn, S., Nimmrich, V., Striebinger, A., et al., 2005. Globular amyloid β-peptide$_{1-42}$ oligomer – a homogeneous and stable neuropathological protein in Alzheimer's disease. J. Neurochem. 95, 834–847.

Bitan, G., 2006. Structural study of metastable amyloidogenic protein oligomers by photo-induced cross-linking of unmodified proteins. Methods Enzymol. 413, 217–236.

Bloom, G.S., 2014. Amyloid-β and tau: the trigger and bullet in Alzheimer disease pathogenesis. JAMA Neurol. 71, 505–508. http://dx.doi.org/10.1001/jamaneurol.2013.5847.

Browne, T.C., McQuillan, K., McManus, R.M., et al., 2013. IFN-γ production by amyloid β-specific Th1 cells promotes microglial activation and increases plaque burden in a mouse model of Alzheimer's disease. J. Immunol. 190, 2241–2251.

Burnouf, S., Gorsky, M.K., Dols, J., et al., 2015. $A\beta_{43}$ is neurotoxic and primes aggregation of $A\beta_{40}$ in vivo. Acta Neuropathol. 130, 35–47.

Busche, M.A., Grienberger, C., Keskin, A.D., et al., 2015. Decreased amyloid-β and increased neuronal hyperactivity in Alzheimer's models. Nat. Neurosci. 18, 1725–1727.

Castillo-Carranza, D.L., Sengupta, U., Guerrero-Muñoz, M.J., et al., 2014. Passive immunization with tau oligomer monoclonal antibody reverses tauopathy phenotypes without affecting hyperphosphorylated neurofibrillary tangles. J. Neurosci. 34, 4260–4272.

Castillo-Carranza, D.L., Guerrero-Muñoz, M.J., Sengupta, U., et al., 2015. Tau immunotherapy modulates both pathological tau and upstream amyloid pathology in an Alzheimer's disease mouse model. J. Neurosci. 35, 4857–4868.

Clark, A., and Vissel, B., 2015. Amyloid β: one of the three danger-associated molecules that are secondary inducers of the proinflammatory cytokines that mediate Alzheimer's disease. Br. J. Pharmacol. 172, 3714–3727.

Cribbs, D.H., 2010. Abeta DNA vaccination for Alzheimer's disease: focus on disease prevention. CNS Neurol. Disord. Drug Targets 9, 207–216.

Dawkins, E., and Small, D.H., 2014. Insights into the physiological function of the β-amyloid precursor protein: beyond Alzheimer's disease. J. Neurochem. 129, 756–769.

Deng, J., Hou, H., Giunta, B., et al., 2012. Autoreactive Aβ-antibodies promote APP β-secretase processing. J. Neurochem. 120, 732–740.

Dodel, R., Balakrishnan, K., Keyvani, K., et al., 2011. Naturally occurring autoantibodies against β-amyloid: investigating their role in transgenic animal and *in vitro* models of Alzheimer's disease. J. Neurosci. 31, 5847–5854.

Elsawa, S.F., Rodeberg, D.A., and Celis, E., 2004. T-cell epitope peptide vaccines. Expert Rev. Vaccines 3, 563–575.

Erickson, M.A., Dohi, K., and Banks, W.A., 2012. Neuroinflammation: a common pathway in CNS diseases as mediated at the blood-brain barrier. Neuroimmunomodulation 19, 121–130.

Esteves-Villanueva, J.O., Trzeciakiewicz, H., Loeffler, D.A., et al., 2015. Effects of tau domain-specific antibodies and intravenous immunoglobulin on tau aggregation and aggregate degradation. Biochemistry 54, 293–302.

Fedele, E., Rebosi, O.C., Pronzato, M.A., et al., 2016. Amyloid beta and the brain: where are we now? J. Biomol. Res. Ther. 5, 2.

Flach, K., Hilbrich, I., Schiffmann, A., et al., 2012. Tau oligomers impair artificial membrane integrity and cellular viability. J. Biol. Chem. 287, 43223–43233.

Gause, W.C., Wynn, T.A., and Allen, J.E., 2013. Type 2 immunity and wound healing: evolutionary refinement of adaptive immunity by helminths. Nat. Rev. Immunol. 13, 607–614. http://dx.doi.org/10.1038/nri3476.

Glabe, C.G., 2006. Common mechanisms of amyloid oligomer pathogenesis in degenerative disease. Neurobiol. Aging 27, 570–575.

Goure, W.F., Krafft, G.A., Jerecic, J., et al., 2014. Targeting the proper amyloid-beta neuronal toxins: a path forward for Alzheimer's disease immunotherapeutics. Alzheimers Res. Ther. 6, 42. http://alzres.com/content/6/4/42.

Hayden, E.Y., and Teplow, D.B., 2013. Amyloid β-protein oligomers and Alzheimer' disease. Alzheimers Res. Ther. 5, 60. http://dx.doi.org/10.1186/alzrt226. eCollection 2013.

Huang, H.-C., and Jiang, Z.-F., 2009. Accumulated amyloid-β peptide and hyperphosphorylated Tau protein: relationship and links in Alzheimer's disease. J. Alzheimers Dis. 16, 15–27.

Hillen, H., Barghorn, S., Striebinger, A., et al., 2010. Generation and therapeutic efficacy of highly oligomer-specific β-amyloid antibodies. J. Neurosci. 30, 10369–10379.
Kanai, Y., Takemura, R., Oshima, T., et al., 1989. Expression of multiple tau isoforms and microtubule bundle formation in fibroblasts transfected with a single tau cDNA. J. Cell Biol. 109, 1173–1184.
Kayed, R., 2010. Anti-tau oligomers passive vaccination for the treatment of Alzheimer disease. Hum. Vaccin. 6 (11), 931–935.
Kayed, R., Head, E., Thompson, J.L., et al., 2003. Common structure of soluble amyloid oligomers implies common mechanism of pathogenesis. Science 300, 486–489.
Kontsekova, E., Zilka, N., Kovacech, B., et al., 2014a. First-in-man tau vaccine targeting structural determinants essential for pathological tau-tau interaction reduces tau oligomerization and neurofibrillary degeneration in an Alzheimer's disease model. Alzheimers Res. Ther. 6, 44. http://dx.doi.org/10.1186/alzrt278. eCollection 2014.
Kontsekova, E., Zilka, N., Kovacech, B., et al., 2014b. Identification of structural determinants on tau protein essential for its pathological function: novel therapeutic target for tau immunotherapy in Alzheimer's disease. Alzheimers Res. Ther. 6, 45. http://dx.doi.org/10.1186/alzrt277.
Knowles, T.P.J., Vendruscolo, M., and Dobson, C.M., 2014. The amyloid state and its association with protein misfolding diseases. Nat. Rev. Mol. Cell Biol. 15, 384–396.
Kudrna, J.J., and Ugen, K.E., 2015. Gene-based vaccines and immunotherapeutic strategies against neurodegenerative diseases: potential utility and limitations. Hum. Vaccin. Immunother. 11, 1921–1926. http://dx.doi.org/10.1080/21645515.2015.1065364.
Lambracht-Washington, D., and Rosenberg, R.N., 2013. Anti-amyloid beta to tau-based immunization: developments in immunotherapy for Alzheimer disease. Immunotargets Ther. 2013 (2), 105–114.
Lambracht-Washington, D., and Rosenberg, R.N., 2015. A noninflammatory immune response in aged DNA Aβ42-immunized mice supports its safety for possible use as immunotherapy in AD patients. Neurobiol. Aging 36, 1274–1281. http://dx.doi.org/10.1016/j.neurobiolaging.2014.12.011.
Larson, M.E., and Lesné, S.E., 2012. Soluble Aβ oligomer production and toxicity. J. Neurochem. 120 (Suppl. 1), 125–139.
Lathuilière, A., Laversenne, V., Astolfo, A., et al., 2016. A subcutaneous cellular implant for passive immunization against amyloid-β reduces brain amyloid and tau pathologies. Brain 139 (Pt 5), 1587–1604. http://dx.doi.org/10.1093/brain/aww036.
Laurijssens, B., Aujard, F., and Rahman, A., 2013. Animal models of Alzheimer's disease and drug development. Drug Discov. Today Technol. 10 (3), e319–27.
Lee, H.G., Casadesus, G., Zhu, X., et al., 2004. Challenging the amyloid cascade hypothesis: senile plaques and amyloid-beta as protective adaptations to Alzheimer's disease. Ann. N.Y. Acad. Sci. 1019, 1–4.
Liu, Y.H., Giunta, B., Zhou, H.D., et al., 2012. Immunotherapy of Alzheimer's disease-the challenge of adverse effects. Nat. Rev. Neurol. 8, 465–469.
Liu, Y.H., Wang, R.Y., Xiang, Y., et al., 2014. Clearance of amyloid-beta in Alzheimer's disease: Shifting the action site from center to periphery. Mol. Neurobiol. 51, 1–7. http://dx.doi.org/10.1007/s12035-014-8694-9.
Liu, Y.H., Bu, X.L., Liang, C.R., et al., 2015a. An N-terminal antibody promotes the transformation of amyloid fibrils into oligomers and enhances the neurotoxicity of amyloid-beta: the dust-raising effect. J. Neuroinflammation 12, 153. http://dx.doi.org/10.1186/s12974-015-0379-4.
Liu, S., Shi, D.Y., Wang, H.C., et al., 2015b. Co-immunization with DNA and protein mixture: a safe and efficacious immunotherapeutic strategy for Alzheimer's disease in PDAPP mice. Sci. Rep. 5, 7771. http://dx.doi.org/10.1038/srep07771.
Mandler, M., Santic, R., Gruber, P., et al., 2015. Tailoring the antibody response to aggregated Aβ using novel Alzheimer-vaccines. PLoS One 10 (1), e0115237. http://dx.doi.org/10.1371/journal.pone.0115237.
Marciani, D.J., 2014. New Th2 adjuvants for preventive and active immunotherapy of neurodegenerative proteinopathies. Drug Discov. Today 19, 912–920.
Marciani, D.J., 2015a. Alzheimer's disease vaccine development: a new strategy focusing on immune modulation. J. Neuroimmunol. 287, 54–63.
Marciani, D.J., 2015b. Is fucose the answer to the immunomodulatory paradox of *Quillaja saponins*? Int. Immunopharmacol. 29, 908–913.
Marciani, D.J., 2016. A retrospective analysis of the Alzheimer's disease progress – the critical need for new development strategies. J. Neurochem. 137, 687–700.

Matsumoto, Y., Niimi, N., and Kohyama, K., 2013. Development of a new DNA vaccine for Alzheimer's disease targeting a wide range of Aβ species and amyloidogenic peptides. PLoS One 8 (9), e75203. http://dx.doi.org/10.1371/journal.pone.0075203.

Mengel, D., Röskam, S., Neff, F., et al., 2013. Naturally occurring autoantibodies interfere with β-amyloid metabolism and improve cognition in a transgenic mouse model of Alzheimer's disease 24 h after single treatment. Trans. Psychiatry 3, e236. http://dx.doi.org/10.1038/tp.2012.151.

Miller, Y., Ma, B., and Nussinov, R., 2011. Synergistic interactions between repeats in tau and Aβ amyloids may be responsible for accelerated aggregation via polymorphic states. Biochemistry 50, 5172–5181.

Moreth, J., Mavoungou, C., and Schindowski, K., 2013. Passive anti-amyloid immunotherapy in Alzheimer's disease: what are the most promising targets? Immun. Ageing 10, 18.

Nisbet, R.M., Polanco, J.C., Ittner, L.M., et al., 2015. Tau aggregation and its interplay with amyloid-β. Acta Neuropathol. 129, 207–220.

Oikawa, T., Nonanka, T., Terada, M., et al., 2016. A-Synuclein fibrils exhibit gain-of-toxic-function, promoting tau aggregation and inhibiting microtubule assembly. J. Biol. Chem. http://dx.doi.org/10.1074/jbc.M116.736355.

O'Nuallain, B., and Wetzel, R., 2002. Conformational Abs recognizing a generic amyloid fibril epitope. Proc. Natl. Acad. Sci. U.S.A. 99, 1485–1490.

O'Nuallain, B., Williams, A.D., McWilliams-Koeppen, H.P., et al., 2010. Anti-amyloidogenic activity in IgGs contained in normal plasma. J. Clin. Immunol. 30 (Suppl. 1), S37–S42.

Pasquier, F., Sadowski, C., Holstein, A., et al., 2016. Two phase 2 multiple ascending-dose studies of Vanutide cridificar (ACC-001) and QS-21 adjuvant in mild-to-moderate Alzheimer's disease. J. Alzheimers Dis. 51, 1131–1143.

Patterson, K., Remmers, C., Fu, Y., et al., 2011. Characterization of prefibrillar tau oligomers in vitro and in Alzheimer's disease. J. Biol. Chem. 286, 23063–23076.

Pedersen, J.T., and Sigurdsson, E.M., 2015. Tau immunotherapy for Alzheimer's disease. Trends Mol. Med. 21, 394–402.

Rasool, S., Albay III, R., Martinez-Coria, H., et al., 2012. Vaccination with a non-human random sequence amyloid oligomer mimic results in improved cognitive function and reduced plaque deposition and micro hemorrhage in Tg2576 mice. Mol. Neurodegener. 7, 37. http://dx.doi.org/10.1186/1750-1326-7-37.

Reitz, C., 2012. Alzheimer's disease and the amyloid cascade hypothesis: a critical review. Int. J. Alzheimers Dis. 2012, 369808. http://dx.doi.org/10.1155/2012/369808.

Relini, A., Marano, N., and Gliozzi, A., 2014. Misfolding of amyloidogenic proteins and their interactions with membranes. Biomolecules 4, 20–55. http://dx.doi.org/10.3390/biom4010020.

Riviere, M.-A., Caputo, A., Laurent, N., et al., 2014. Active Ab immunotherapy CAD106 phase II dose-adjuvant finding study: immune response. Alzheimers Dement. 10.4, P447–P448.

Roeder, A.M., Roettger, Y., Stünde, L.A., et al., 2013. Synthetic dimeric Aβ(28-40) mimics the complex epitope of the human anti-A β autoantibodies against toxic Aβ oligomers. J. Biol. Chem. 288, 27638–27645.

Rosensweig, C., Ono, K., Murakami, K., et al., 2012. Preparation of stable amyloid β-protein oligomers of defined assembly order. Methods Mol. Biol. 849, 23–31.

Rothstein, T.L., 2016. Natural antibodies as rheostats for susceptibility to chronic diseases in the aged. Front. Immunol. 7, 127.

Sandberg, A., Luheshi, L.M., Söllvander, S., et al., 2010. Stabilization of neurotoxic Alzheimer amyloid-β oligomers by protein engineering. Proc. Natl. Acad. Sci. U.S.A. 107, 15595–15600.

Selkoe, D.J., and Hardy, J., 2016. The amyloid hypothesis of Alzheimer's disease at 25 years. EMBO Mol. Med. 8, 595–608. http://dx.doi.org/10.15252/emmm.201606210.

Sevigny, J., Chiao, P., Williams, L., et al., 2015. Aducanumab (BIIB037), an anti-amyloid beta monoclonal antibody, in patients with prodromal or mild Alzheimer's disease: interim results of a randomized, double-blind, placebo-controlled, phase 1b study. Alzheimer's Dement. 11 (7), P277.

Stancu, I.C., Vasconcelos, B., Terwel, D., et al., 2014. Models of β-amyloid induced Tau-pathology: the long and "folded" road to understand the mechanism. Mol. Neurodegener. 9, 51. http://dx.doi.org/10.1186/1750-1326-9-51.

Sullivan, M.G., 2016. Aducanumab subanalysis bolsters phase III trials in very early Alzheimer's disease. Neurol. Rev. 24 (2), 8.

Vasilevko, V., Pop, V., Kim, H.J., et al., 2010. Linear and conformation specific antibodies in aged beagles after prolonged vaccination with aggregated Abeta. Neurobiol. Dis. 39, 301–310.

Wang, S., Yu, Y., Geng, S., et al., 2014. A coimmunization vaccine of Aβ42 ameliorates cognitive deficits without brain inflammation in an Alzheimer's disease model. Alzheimer's Res. Ther. 6 (3), 26. http://dx.doi.org/10.1186/alzrt256. eCollection 2014.
Wilcock, D.M., Gharkholonarehe, N., Van Nostrand, W.E., et al., 2009. Amyloid reduction by amyloid-beta vaccination also reduces mouse tau pathology and protects from neuron loss in two mouse models of Alzheimer's disease. J. Neurosci. 29, 7957–7965.
Wisniewski, T., and Boutajangout, A., 2010. Vaccination as a therapeutic approach to Alzheimer's disease. Mt. Sinai J. Med. 77, 17–31. http://dx.doi.org/10.1002/msj.20156.

CHAPTER

12

Obesity-Induced Defects in Dendritic Cell and T Cell Functions: Implications for Immunotherapeutic Efficacy in Cancer

Rachael Orlandella and Lyse Norian

OUTLINE

Abstract

Obesity induces chronic low-grade inflammation which fosters innate and adaptive immune dysfunctions. Additionally, obesity increases the risk of several types of cancer, including renal cell carcinoma, postmenopausal breast cancer, and endometrial cancer. The rising prevalence of obesity in US adults and adolescents will likely precede heightened percentages of obese patients with these cancers, which makes studying them in the context of obesity essential for developing effective therapies. Despite significant achievements of cancer immunotherapies in preclinical models, successful clinical translation of these therapies in patients has been underwhelming. Investigating the basis for immunotherapeutic failure in patients is crucial for improving the response rate. As obesity is a major comorbidity in many cancer patients, and a contributor to diminished immunity, understanding the extent to which obesity alters immune responses to tumors in the absence or presence of immune-based therapy is crucial. Here, we have summarized current knowledge regarding the multifaceted relationship between cancer, obesity, immunity, and immunotherapeutic efficacy.

M.A. *Hayat (ed): Immunology, Volume 1.*
DOI: http://dx.doi.org/10.1016/B978-0-12-809819-6.00012-5

© 2018 Elsevier Inc. All rights reserved.

INTRODUCTION

An epidemic in itself, obesity alters several metabolic parameters and increases the risk of developing cardiovascular disease and cancer, the two leading causes of death in the United States. As the second leading cause of death in the United States, cancer is undoubtedly one of the most significant health concerns of today. Numerous risk factors for cancer have been identified, including tobacco use, exposure to sunlight, and age. More recently, obesity has emerged as a risk factor for several types of cancer (Wolin et al., 2010), as the chronic low-grade inflammation associated with obesity has been shown to drive neoplasia (Deng et al., 2016). This association is strongest for renal, endometrial, esophageal, and postmenopausal breast cancers (Wolin et al., 2010). The impact of obesity on cancer-related mortality is not fully understood; several clinical studies across different cancer types reveal discrepancies. Some studies observed a significant correlation between obesity and diminished cancer survival rates (Secord et al., 2016; Reeves et al., 2007; Calle et al., 2003), whereas others found that higher body mass index (BMI) actually confers better prognosis (Mano et al., 2014; Donin et al., 2016). The finding that obesity increases the relative risk of developing certain types of cancers, but fosters an improved prognosis compared to nonobese patients, is known as the "obesity paradox" (Hakimi et al., 2013). More basic research is required to understand the complex relationship between obesity and cancer mortality.

Breakthroughs in immune checkpoint blockade therapies for cancer treatments have led to a renewed interest in cancer immunotherapy (Topalian et al., 2015). Immune-based therapies harness and stimulate the inherent properties of the immune system in order to overcome immunosuppressive tumor microenvironments and promote tumor rejection. Although highly efficacious in preclinical models, immunotherapies have failed to recapitulate high response rates in cancer patients (Klevorn and Teague, 2016). Preexisting patient comorbidities, such as obesity, may account for this inconsistency (Klevorn and Teague, 2016; James et al., 2012). Obesity also prompts baseline defects in several immune cell types, which may further compromise the efficacy of immunotherapeutic strategies for cancer treatment (Boi et al., 2016; Han and Levings, 2013; James et al., 2012). The extent to which obesity-induced immune dysfunction alters immune responses to cancer has not yet been fully elucidated. Here, we summarize current knowledge regarding the complex interplay between obesity and cancer growth, as related to immune dysfunction and net results on immunotherapeutic outcomes.

OBESITY AND CANCER

Cancer is a globally significant concern in the public health and biomedical research realms. As the second leading cause of death in the United States (Siegel et al., 2016), cancer has become a focal point of research and public awareness. Several risk factors for cancer have been identified, such as tobacco use, consumption of alcohol, exposure to sunlight, and obesity. It is widely accepted that certain risk factors give rise to specific cancers. For example, tobacco use is the major risk factor for developing lung cancer; alcohol consumption promotes liver cancer; and excess sunlight exposure induces skin cancer. In contrast, obesity increases the risk of multiple cancer types.

Obesity is a major risk factor for several morbidities. Currently, there are over 2.1 billion overweight or obese individuals worldwide. In the United States alone, nearly 35% of adults are considered to be obese (Smith and Smith, 2016) and projection studies predict this rate of occurrence will continue to rise. Given the strong association between obesity and cardiovascular disease, healthcare costs spent treating obesity-related comorbidities are significant. In addition to heart disease, it is estimated that weight gain and obesity account for roughly 20% of all cancer cases (Wolin et al., 2010; Eheman et al., 2012). The highest association between obesity and cancer development occurs in endometrial, esophageal, colon, and renal cancers (specifically, clear-cell renal cell carcinoma (RCC)) (Wolin et al., 2010; Lauby-Secretan et al., 2016; Chow et al., 2010). Obesity also increases the relative risk of developing liver, pancreatic, ovarian, and postmenopausal breast cancer as well (Wolin et al., 2010; Lauby-Secretan et al., 2016).

Although the association between obesity and cancer development is clear, the effects of obesity on mortality rates for specific cancer types are still widely debated. Some previous reports indicate that obesity correlates with poorer prognosis and higher cancer-related mortality rates (Secord et al., 2016; Reeves et al., 2007; Calle et al., 2003), whereas others show that increased BMI confers a survival advantage (Mano et al., 2014; Donin et al., 2016). Women with endometrial cancer experience higher mortality rates with increasing BMI (Secord et al., 2016). A recent review found that patients with a BMI >35 yielded higher cancer-related mortality rates when compared to their normal weight (NW) counterparts (Secord et al., 2016). Several sources support this trend with regard to endometrial cancer and esophageal cancer (Reeves et al., 2007; Calle et al., 2003). However, the role of obesity in renal cancer mortality is less straightforward. There is evidence that renal cancer-related mortality rises as BMI increases (Reeves et al., 2007). For example, it has been shown that obese female renal cancer patients (BMI $\geq$ 40) had a relative risk of death of 4.75 when compared to NW patients (BMI = 18.5–24.9) (Calle et al., 2003). In contrast, other sources found that increased BMI confers improved disease-free survival for nephrectomized high-risk patients with clear-cell RCC (Mano et al., 2014; Donin et al., 2016). Further research is required to understand the impact of obesity on survival in renal cancer patients.

OBESITY AND THE IMMUNE SYSTEM

In addition to a heightened risk of developing cardiovascular diseases, insulin resistance, and certain types of cancers, obesity also alters baseline immunity. It has been well documented that adipose tissue in obese individuals exhibits hallmarks of chronic inflammation when compared to adipose tissue from nonobese subjects (Han and Levings, 2013; Lumeng et al., 2007). Determining the mechanisms and cell types responsible for this inflammatory phenotype has become a focal point in obesity-related research. Although the origin and sequence of events preceding obesity-induced inflammation remain unclear, it appears that both innate and adaptive immune cells are involved (Han and Levings, 2013).

With regard to innate immunity, a vast amount of research has focused on macrophage infiltration of obesity-associated adipose tissue. Analysis of adipose tissue from obese and lean mice and humans has revealed a skewed macrophage population in obese individuals. Unlike the anti-inflammatory M2 polarized macrophages found in lean adipose tissue, classically activated proinflammatory M1 polarized macrophages predominate in obese adipose tissue (Lumeng et al., 2007; Ohashi et al., 2010). Adipose tissue from diet-induced obese (DIO) mice shows upregulation of characteristic M1 proteins (such as TNFα and iNOS), complemented with diminished expression of M2 associated proteins such as arginase-1 and IL-10 (Lumeng et al., 2007; Ohashi et al., 2010). The origin of adipose tissue infiltration by macrophages may be tied to a proliferation of myeloid progenitors in the bone marrow, as seen in obese mice (Nagareddy et al., 2014); it may also be linked to obesity-induced alterations in cytokine production from NK and NKT cells residing within adipose tissue (Lee et al., 2016; Lynch et al., 2012).

Along with a skewed proinflammatory M1 polarized macrophage profile, obesity also impacts dendritic cell (DC) function. We have shown that high-fat DIO BALB/c mice show higher percentages of splenic $CD11c^{high}/I\text{-}A^{d+}$ DCs in comparison to both their lean counterparts on a standard rodent chow (James et al., 2012), and an obese-resistant (OB-Res) cohort on the same high-fat diet (Boi et al., 2016). However, DCs isolated from DIO mice display a diminished ability to stimulate proliferation of naive $CD8^+$ T cells, whereas lean and OB-Res mice retain DC stimulatory capacity (James et al., 2012; Boi et al., 2016). Other sources report similar results in human subjects. O'Shea et al. (2013) found that in response to TLR ligation, DCs isolated from obese human subjects failed to upregulate CD83, a useful marker for DC maturity, as efficiently as their lean counterparts when stimulated in vitro. After stimulation with poly I:C, DCs from obese subjects produced less IL-12 and more IL-10, suggesting an impaired ability to optimally stimulate T cell immunity (O'Shea et al., 2013). High-fat DIO also impairs lymphatic transport and modifies lymph node structure. Wild type DIO mice fed a high-fat diet exhibit a reduction in DC trafficking to peripheral lymph nodes in comparison to NW controls. Additionally, lymph nodes from obese mice are smaller than those from lean mice (Weitman et al., 2013).

Obesity also alters several cell types involved in the adaptive immune response. Murine models of DIO correlate with a breakdown of characteristic T and B cell zones and CCL21 gradients in peripheral lymph nodes (Weitman et al., 2013). Furthermore, DIO mice show diminished CD4 helper T cells in the spleen (Boi et al., 2016) and the stromal vascular fraction (SVF) from epididymal fat pads (Nishimura et al., 2009). Foxp3+ Regulatory T cells (Tregs), which are important for downregulating inflammatory responses, are normally enriched in the abdominal fat and SVF of lean mice (Feuerer et al., 2009). Conversely, adipose tissue from obese mice fed a high-fat diet reveals reduced Treg percentages, which parallels obesity-induced inflammation and insulin resistance (Nishimura et al., 2009; Feuerer et al., 2009). Obese adipose tissue is also characterized by a rise in activated effector $CD8^+$ T cells in the SVF. Adipose tissue $CD8^+$ T cells are thought to promote subsequent macrophage accumulation, and ablation of this $CD8^+$ T cell population reverses adipose inflammation and obesity-induced insulin resistance (Nishimura et al., 2009).

Hormones produced by adipose tissue, called adipokines, are critical mediators of metabolism. Two such proteins are leptin and adiponectin, which have complementary functions in terms of regulating appetite and insulin sensitivity. Elevated levels of leptin and reduced levels of adiponectin are hallmarks of obesity. Several sources report that both leptin (Zarkesh-Esfahani et al., 2001; Mattioli et al., 2005; Naylor and Petri, 2016; De Rosa et al., 2007) and adiponectin (Ohashi et al., 2010; Jung et al., 2012) can act directly on immune cells via binding to their specific receptors. Leptin promotes IL-6 and TNF-α release in monocytes

(Zarkesh-Esfahani et al., 2001), activates DCs (Mattioli et al., 2005), polarizes naive $CD4^+$ T cells to the Th1 (Mattioli et al., 2005) and Th17 (Naylor and Petri, 2016) phenotypes, and suppresses Foxp3+ Treg proliferation (Naylor and Petri, 2016; De Rosa et al., 2007) to maintain an overall proinflammatory state. The impact of adiponectin on immune function is less clear; adiponectin has been shown to polarize macrophages to an M2 antiinflammatory phenotype (Ohashi et al., 2010). However, it has also been implicated in the promotion of proinflammatory Th1 and Th17 polarization (Jung et al., 2012).

OBESITY AND CANCER IMMUNOTHERAPY

The interaction between the immune system and cancer is a prominent area of current cancer-related research. It is well known that immune cells can recognize tumor-associated antigens and infiltrate tumors, and tumor antigen-specific T cells have been isolated from cancer patients (Schietinger et al., 2016). Despite this, T cells within tumors lack potent effector function and do not retain their cytotoxicity (Schietinger et al., 2016). Several mechanisms limit protective antitumor immunity: self-tolerance (Schietinger and Greenberg, 2014), persistent antigen exposure (Schietinger et al., 2016), tumor-induced local hypoxia (Wilson and Hay, 2011), and accumulation of immunosuppressive populations including Tregs, myeloid-derived suppressor cells (MDSCs), tumor-associated macrophages, and regulatory DCs (James et al., 2014; Ostrand-Rosenberg, 2010; Deleeuw, et al., 2012; Yeung et al., 2015; Norian et al., 2009) all mediate tumor-specific T cell exhaustion and dysfunction. Refer to Fig. 12.1A for a representation of key leukocyte populations within the tumor microenvironment.

Cancer immunotherapy aims to overcome the immunosuppressive tumor microenvironment and promote immunological tumor clearance. Immunotherapeutic approaches to cancer treatment exhibit several advantages over traditional chemotherapy and surgery. Chemotherapy nonspecifically eradicates all rapidly proliferating cells, and does not discriminate between cancer cells and healthy dividing self-cells, including immune cells (Zitvogel et al., 2008). The immune response demonstrates precision and ensures tumor-specific cytotoxicity, which can reduce the off-target toxicity associated with traditional therapies. Additionally, immune cells can traffic throughout the circulation to sites of distant metastases that frequently cannot be surgically resected. Finally, standard therapy too often results in cancer relapse. In contrast, lymphocytes form specific immunological memory upon antigenic clearance to prevent secondary infections, and these memory responses can also be used to control cancer recurrence. These qualities make immunotherapy a clinically relevant and favorable treatment option for cancer patients.

Several successful immune-based cancer therapies have progressed to the clinic, including checkpoint inhibitors (anti-PD-1, anti-PD-L1, anti-CTLA4), cytokine therapies (IL-2, IFNα), cell therapy (adoptive transfer, chimeric antigen receptor T cells), therapeutic peptide vaccines, and monoclonal antibody therapies (Farkona et al., 2016; Topalian et al., 2015, 2012; Maude et al., 2014; Mitchell et al., 2015). Despite its reputation as a breakthrough cancer treatment with promising results in preclinical mouse models, immunotherapy certainly has its drawbacks. Several immune therapies still exhibit a high degree of toxicity, and only a subset of treated patients achieve remission and improved cancer-related survival with

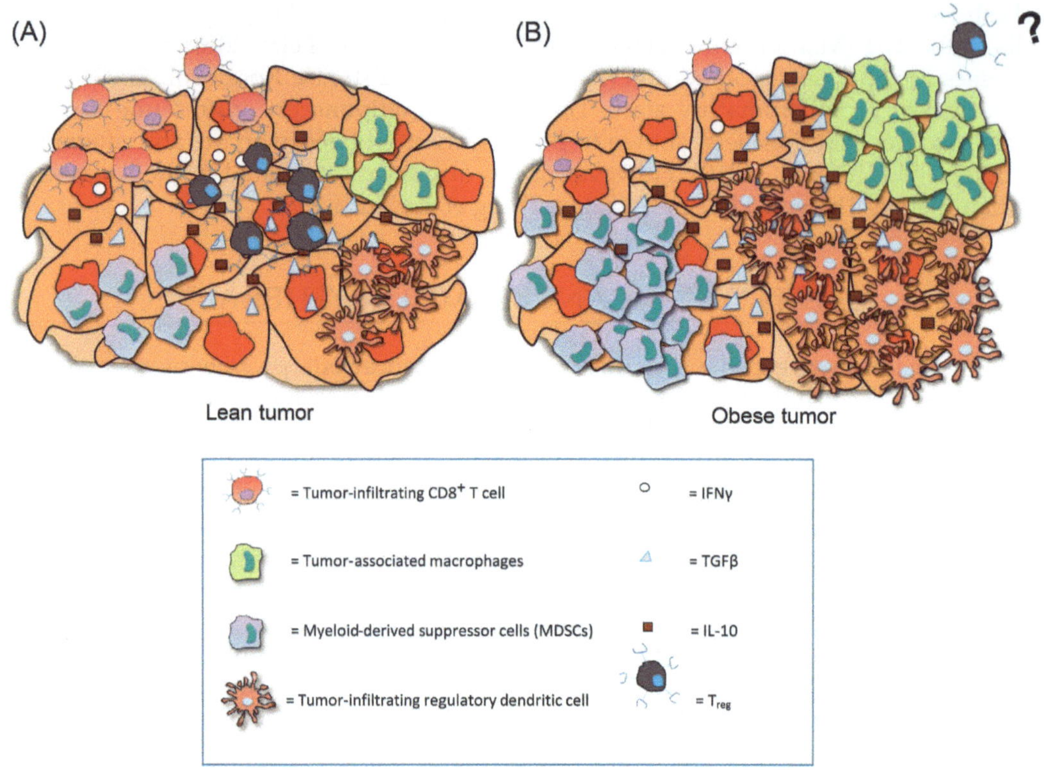

FIGURE 12.1 Obesity alters the tumor microenvironment. (A) Solid tumors from lean mice are frequently infiltrated by a variety of leukocytes including $CD8^+$ T cells, tumor-associated macrophages, myeloid-derived suppressor cells, and dendritic cells. Despite the presence of these cell types in tumors, their function is typically altered relative to their counterparts in other organs. (B) With the presence of host obesity, the relative percentages of leukocyte populations appear to shift for many tumor types, as do the phenotype and functions of these cells. For example, obesity has been linked to increased percentages of myeloid-derived suppressor cells in orthotopic renal tumors as well as increased percentages of immune-suppressive or regulatory dendritic cells. The effects of obesity on tumor-infiltrating Tregs is currently unclear. Simultaneous alterations in local cytokines such as IFNγ (decreased) and IL-10 (increased) may also contribute to the development of obesity-specific changes in the tumor microenvironment.

any immune-based therapy (Klevorn and Teague, 2016). Understanding which factors contribute to immunotherapeutic failure at the preclinical stage is critical. Given the known risks of obesity-induced cancer development, as well as the dysregulated immune parameters associated with the obesogenic environment, studying cancer immunotherapies in the context of obesity is essential for improving immunotherapeutic efficacy.

To date, the impact of obesity on immune responses to tumors and immunotherapeutic efficacy has been addressed in a limited number of studies. As previously discussed, obesity increases the risk for developing several cancer types, including RCC. Our laboratory has determined that obesity fundamentally alters antitumor immunity to RCC in mice with DIO. High-fat diet induces obesity in roughly 45–55% of BALB/c mice following

20 weeks of ad libitum feeding (Boi et al., 2016). In our model, NW mice on standard chow and DIO mice are intrarenally (IR) tumor challenged with a luciferase-expressing RCC cell line (Renca) and tumor outgrowth is monitored by bioluminescent imaging. Using this system, we determined that DIO mice show accelerated early tumor outgrowth, relative to NW controls, but equivalent tumor burdens at later time points (James et al., 2012). DIO mice also exhibit elevated percentages of tumor-infiltrating DCs (TIDCs) relative to NW controls at day 21 post-tumor challenge. However, rather than stimulating protective T cell immunity, TIDC from DIO mice act as regulatory cells. Specifically, ex vivo experiments using sort-purified, peptide-pulsed TIDCs revealed that TIDCs from DIO mice, but not NW mice, suppressed naive CD8 T cell expansion (James et al., 2012). Subsequent studies within this model have also demonstrated heightened MDSC accumulation in renal tumors of DIO mice in comparison to their NW counterparts (James et al., 2014; Hale et al., 2015). This rise in MDSCs is attributed to increased local secretion of the chemoattractant CCL2 by TIDCs, macrophages, B cells, and nonleukocytes (Hale et al., 2015). Fig. 12.1B demonstrates the augmented accumulation of immunosuppressive myeloid lineage populations in tumors in obesogenic environments. Clinically, higher percentages of MDSC tumor infiltration correlates with poor prognosis in cancer patients (Markowitz et al., 2013).

Similar findings linking obesity with increased tumor infiltration by myeloid lineage cells have been made in preclinical models of melanoma and breast cancer. Jung et al. (2014) found that C57BL/6N mice fed a high-fat diet exhibited enhanced tumor growth and an increased frequency of intratumoral M2 macrophages following a B16F10 melanoma challenge. Similarly, Arendt et al. (2013) demonstrated that the obesogenic environment activates macrophages, which go on to promote angiogenesis and accelerated tumorigenesis in a preclinical model of humanized breast cancer. Multiple other reviews have covered the role of altered cytokines and chemokines regulating myeloid lineage recruitment to tumors in obesogenic host environments (Gilbert and Slingerland, 2012; Gati et al., 2014; Okwan-Duodu et al., 2013), so these will not be discussed here. Collectively, there is now ample evidence to demonstrate that host obesity impairs protective antitumor immunity, and specifically favors the local accumulation of suppressive myeloid cell populations.

Multiple immunotherapies are now FDA-approved or in development for the treatment of metastatic RCC (mRCC), including checkpoint inhibitors, adoptive cell therapies, and therapeutic vaccines (Curtis et al., 2016). High-dose IL-2 has been used to treat mRCC clinically since its approval in 1992, but is highly toxic and only induces full responses in roughly 7% of patients (Curtis et al., 2016). The checkpoint blockade biologic Nivolumab (anti-PD-1, Optidovo) was FDA-approved in November of 2015, and thus far has proven clinically efficacious in 20–30% of treated RCC patients (McDermott et al., 2015; Motzer et al., 2015). Therefore, although clinical responses are occurring with immune therapies, the majority of patients still fail to respond, and it is possible that comorbidities such as obesity could provide an explanation for their low success rates. We have been addressing the impact of obesity on immunotherapeutic efficacy to renal cancer, and interest in this field is increasing rapidly.

Our model of murine RCC evaluates the DC and T cell-dependent immunotherapy AdTRAIL + CpG, which consists of an adenoviral vector encoding murine TRAIL (TNF-related apoptosis inducing ligand) given in combination with an immune-stimulatory unmethylated CpG-containing oligonucleotide and TLR9 agonist (VanOosten and Griffith,

2007; van Mierlo et al., 2004; Griffith et al., 2009). We previously demonstrated that AdTRAIL + CpG therapy, administered intrarenally 7 days post-tumor challenge, significantly reduces orthotopic renal tumors and lung metastases in NW mice (Norian et al., 2012). Strikingly, renal tumor-bearing DIO mice treated with the same immunotherapy show no reduction in tumor burden at day 28 when compared to DIO mice treated with PBS alone (James et al., 2012). Similarly, when a subcutaneous Renca challenge is treated with intraperitoneal AdTRAIL + CpG injection, this results in tumor clearance and long-term survival in 80% of NW mice, but DIO mice treated identically exhibit progressive tumor outgrowth and eventually succumb to their tumors (James et al., 2012).

As observed in the renal tumors of obese mice that did not receive therapy, tumors from AdTRAIL + CpG-treated DIO mice contained increased percentages of immunosuppressive TIDCs relative to their NW counterparts (James et al., 2012). When compared to TIDCs from therapy-treated NW mice, TIDCs from therapy-treated DIO tumor-bearing kidneys showed reduced expression of CD83 (a surface marker for DC maturity), and produced less of the DC effector cytokines IL-12p70 and TNFα. Additionally, sort-purified TIDCs from therapy-treated DIO mice acted as regulatory cells to inhibit T cell proliferation ex vivo. Finally, DIO mice had fewer renal tumor-infiltrating IFNγ+ $CD8^+$ T cells than NW mice on day 21 post-IR tumor challenge (James et al., 2012). Overall, these findings attribute immunotherapeutic failure in DIO mice to obesity-specific defects in DC and T cell function.

Research from another group indicates that the use of immunotherapies in cancer subjects with obesity may require additional caution. In this report, the authors determined that increased adiposity was associated with a systemic cytokine storm following administration of anti-CD40 + IL-2, leading to rapid lethality (Mirsoian et al., 2014). Death occurred in response to TNFa and M1 macrophages in adipose tissue, and could be prevented by either cytokine neutralization or macrophage depletion (Mirsoian et al., 2014). In this study, mice were both aged and obese, suggesting that the combination of both factors may increase toxicities associated with immunotherapy administration. Although this concept has not been thoroughly explored, at least one human subject study did find that obesity was associated with increased cardiotoxicity in response to anthracycline + trastuzumab treatment in breast cancer patients (Guenancia et al., 2016). These reports illustrate the possibility that in addition to decreased immunotherapeutic efficacy, obesity may prove to heighten immunotherapeutic toxicities in cancer patients, which could also serve to limit positive outcomes through early cessation of treatment.

SUMMARY

Cancer immunotherapy has shown great promise in treating patients with advanced disease. However, many patients still fail to respond, and this may be due to the detrimental effects of obesity on immune function. Given the prevalence of obesity in the United States today, understanding the effects of obesity and its associated comorbidities on cancer progression and immunotherapeutic efficacy will be crucial for improving treatment outcomes. Preclinical cancer models often develop immunotherapeutic regimens in young, healthy mice. Although this eliminates confounding variables, it fails to accurately represent cancer patient demographics. Obesity is a major risk factor for several types of cancer, including kidney cancer, and alters numerous metabolic parameters that often result in the onset of cardiovascular disease and/or type II diabetes. At this time, relatively little work has been done to identify the specific or cumulative effects of obesity and its associated comorbidities on immunotherapeutic efficacy. The few studies that have been published on this topic all suggest that obesity may negatively impact the ability of immune-stimulatory therapies to mediate tumor clearance. This may occur through obesity-derived defects in DC or T cell function, an altered host cytokine milieu, or enhanced treatment-associated toxicity. It is not yet known whether these findings in murine models will translate into the clinic. Because immunotherapy is poised to produce tremendous advances in clinical cancer management, a deeper understanding of the interactions between obesity, cancer growth, and immunotherapeutic efficacy is needed.

References

Arendt, L.M., McCready, J., Keller, P.J., et al., 2013. Obesity promotes breast cancer by CCL2-mediated macrophage recruitment and angiogenesis. Cancer Res. 73, 6080–6093.

Boi, S.K., Buchta, C.M., Pearson, N.A., et al., 2016. Obesity alters immune and metabolic profiles: New insight from obese-resistant mice on high-fat diet. Obesity (Silver Spring) 24, 2140–2149.

Calle, E.E., Rodriguez, C., Walker-Thurmond, K., et al., 2003. Overweight, obesity, and mortality from cancer in a prospectively studied cohort of U.S. adults. N. Engl. J. Med. 348, 1625–1638.

Curtis, S.A., Cohen, J.V., and Kluger, H.M., 2016. Evolving immunotherapy approaches for renal cell carcinoma. Curr. Oncol. Rep. 18, 57.

Chow, W.H., Dong, L.M., and Devesa, S.S., 2010. Epidemiology and risk factors for kidney cancer. Nat. Rev. Urol. 7, 245–257.

De Rosa, V., Procaccini, C., Calì, G., et al., 2007. A key role of leptin in the control of regulatory T cell proliferation. Immunity 26, 241–255.

Deleeuw, R.J., 2012. The prognostic value of FoxP3+ tumor infiltrating lymphocytes in cancer: a critical review of the literature. Clin. Cancer Res. 18, 3022–3029.

Deng, T., Lyon, C.J., Bergin, S., et al., 2016. Obesity, inflammation, and cancer. Ann. Rev. Pathol. 11, 421–449.

Donin, N.M., Pantuck, A., Klöpfer, P., et al., 2016. Body mass index and survival in a prospective randomized trial of localized high-risk renal cell carcinoma. Cancer Epidemiol. Biomarkers Prev. 25, 1326 1332.

Eheman, C., Henley, S.J., Ballard-Barbash, R., et al., 2012. Annual Report to the Nation on the status of cancer, 1975-2008, featuring cancers associated with excess weight and lack of sufficient physical activity. Cancer 118, 2338–2366.

Farkona, S., Diamandis, E.P., and Blasutig, I.M., 2016. Cancer immunotherapy: the beginning of the end of cancer? BMC Med. 14, 73.

Feuerer, M., Herrero, L., Cipolletta, D., et al., 2009. Lean, but not obese, fat is enriched for a unique population of regulatory T cells that affect metabolic parameters. Nat. Med. 15, 930–939.

Gati, A., Kouidhi, S., Marrakchi, R., et al., 2014. Obesity and renal cancer: role of adipokines in the tumor-immune system conflict. Oncoimmunology 3, e27810.

Gilbert, C.A., and Slingerland, J.M., 2012. Cytokines, obesity, and cancer: new insights on mechanisms linking obesity to cancer risk and progression. Annu. Rev. Med. 64, 22.1–22.13.

Griffith, T.S., Stokes, B., Kucaba, T.A., et al., 2009. TRAIL gene therapy: from preclinical development to clinical application. Curr. Gene Ther. 9, 9–19.

Guenancia, C., Lefebvre, A., Cardinale, D., et al., 2016. Obesity as a risk factor for anthracyclines and Trastuzumab cardiotoxicity in breast cancer: a systematic review and meta-analysis. J. Clin. Oncol. 34, 3157–3165.

Hakimi, A.A., Furberg, H., Zabor, E.C., et al., 2013. An epidemiologic and genomic investigation into the obesity paradox in renal cell carcinoma. J. Natl. Cancer Inst. 105, 1862–1870.

Hale, M., Itani, F., Buchta, C.M., et al., 2015. Obesity triggers enhanced MDSC accumulation in murine renal tumors via elevated local production of CCL2. PLoS One 10, e0118784.

Han, M.H., and Levings, M.K., 2013. Immune regulation in obesity-associated adipose inflammation. J. Immunol. 191, 527–532.

James, B.R., Tomanek-Chalkley, A., Askeland, E.J., et al., 2012. Diet-induced obesity alters dendritic cell function in the presence and absence of tumor growth. J. Immunol. 189, 1311–1321.
James, B.R., Anderson, K.G., Brincks, E.L., et al., 2014. CpG-mediated modulation of MDSC contributes to the efficacy of Ad5-TRAIL therapy against renal cell carcinoma. Cancer Immunol. Immunother. 63, 1213–1227.
Jung, J., Cho, H.J., Jung, Y.J., et al., 2014. High-fat diet-induced obesity increases lymphangiogenesis and lymph node metastasis in the B16F10 melanoma allograft model: roles of adipocytes and M2-macrophages. Int. J. Cancer 136, 258–270.
Jung, M.Y., Kim, H.S., Hong, H.J., et al., 2012. Adiponectin induces dendritic cell activation via PLCγ/JNK/NF-κB pathways, leading to Th1 and Th17 polarization. J. Immunol. 188, 2592–2601.
Klevorn, L.E., and Teague, R.M., 2016. Adapting cancer immunotherapy models for the real world. Trends Immunol. 37, 354–363.
Lauby-Secretan, B., Scoccianti, C., Loomis, D., et al., 2016. Body fatness and cancer—viewpoint of the IARC Working Group. N. Engl. J. Med. 375, 794–798.
Lee, B.C., Kim, M.S., Pae, M., et al., 2016. Adipose natural killer cells regulate adipose tissue macrophages to promote insulin resistance in obesity. Cell Metab. 23, 685–698.
Lumeng, C.N., Bodzin, J.L., and Saltiel, A.R., 2007. Obesity induces a phenotypic switch in adipose tissue macrophage polarization. J. Clin. Invest. 117, 175–184.
Lynch, L., Nowak, M., Varghese, B., et al., 2012. Adipose tissue invariant NKT cells protect against diet-induced obesity and metabolic disorder through regulatory cytokine production. Immunity 37, 574–587.
Mano, R., Hakimi, A.A., Zabor, E.C., et al., 2014. Association between visceral and subcutaneous adiposity and clinicopathological outcomes in non-metastatic clear cell renal cell carcinoma. Can. Urol. Assoc. J. 8, E675–E680.
Markowitz, J., Wesolowski, R., Papenfuss, T., et al., 2013. Myeloid-derived suppressor cells in breast cancer. Breast Cancer Res. Treat. 140, 13–21.
Mattioli, B., Straface, E., Quaranta, M.G., et al., 2005. Leptin promotes differentiation and survival of human dendritic cells and licenses them for Th1 priming. J. Immunol. 174, 6820–6828.
Maude, S.L., Frey, N., Shaw, P.A., et al., 2014. Chimeric antigen receptor T cells for sustained remissions in leukemia. N. Engl. J. Med. 371, 1507–1517.
McDermott, D.F., Drake, C.G., Sznol, M., et al., 2015. Survival, durable response, and long-term safety in patients with previously treated advanced renal cell carcinoma receiving nivolumab. J. Clin. Oncol. 33, 2013–2020.
Mirsoian, A., Bouchlaka, M.N., Sckisel, G.D., et al., 2014. Adiposity induces lethal cytokine storm after systemic administration of stimulatory immunotherapy regimens in aged mice. J. Exp. Med. 211, 2373–2383.
Mitchell, D.A., Batich, K.A., Gunn, M.D., et al., 2015. Tetanus toxoid and CCL3 improve dendritic cell vaccines in mice and glioblastoma patients. Nature 519, 366–369.
Motzer, R.J., Escudier, B., McDermott, D.F., et al., 2015. Nivolumab versus everolimus in advanced renal-cell carcinoma. N. Engl. J. Med. 373, 1803–1813.
Nagareddy, P.R., Kraakman, M., Masters, S.L., et al., 2014. Adipose tissue macrophages promote myelopoiesis and monocytosis in obesity. Cell Metab. 19, 821–835.
Naylor, C., and Petri, W.A., 2016. Leptin regulation of immune responses. Trends Mol. Med. 22, 88–98.
Nishimura, S., Manabe, I., Nagasaki, M., et al., 2009. $CD8^+$ effector T cells contribute to macrophage recruitment and adipose tissue inflammation in obesity. Nat. Med. 15, 914–920.
Norian, L.A., Rodriguez, P.C., O'Mara, L.A., et al., 2009. Tumor-infiltrating regulatory dendritic cells inhibit $CD8^+$ T cell function via L-arginine metabolism. Cancer Res. 69, 3086–3094.
Norian, L.A., Kresowik, T.P., Rosevear, H.M., et al., 2012. Eradication of metastatic renal cell carcinoma after adenovirus-encoded TNF-related apoptosis-inducing ligand (TRAIL)/CpG immunotherapy. PLoS One 7, e31085.
Ohashi, K., Parker, J.L., Ouchi, N., et al., 2010. Adiponectin promotes macrophage polarization toward an anti-inflammatory phenotype. J. Biol. Chem. 285, 6153–6160.
Okwan-Duodu, D., Umpierrez, G.E., Brawley, O.W., Diaz, R., 2013. Obesity-driven inflammation and cancer risk: role of myeloid derived suppressor cells and alternately activated macrophages. Am. J. Cancer Res 3, 21–33.
O'Shea, D., Corrigan, M., Dunne, M.R., et al., 2013. Changes in human dendritic cell number and function in severe obesity may contribute to increased susceptibility to viral infection. Int. J. Obes. (London) 37, 1510–1513.
Ostrand-Rosenberg, S., 2010. Myeloid-derived suppressor cells: more mechanisms for inhibiting antitumor immunity. Cancer Immunol. Immunother. 59, 1593–1600.
Reeves, G.K., Pirie, K., Beral, V., et al., 2007. Cancer incidence and mortality in relation to body mass index in the Million Women Study: cohort study. BMJ 335, 1134.

Schietinger, A., and Greenberg, P.D., 2014. Tolerance and exhaustion: defining mechanisms of T cell dysfunction. Trends Immunol. 35, 51–60.

Schietinger, A., Philip, M., Krisnawan, V.E., et al., 2016. Tumor-specific T cell dysfunction is a dynamic antigen-driven differentiation program initiated early during tumorigenesis. Immunity 45, 389–401.

Secord, A.A., Hasselblad, V., Von Gruenigen, V.E., et al., 2016. Body mass index and mortality in endometrial cancer: a systematic review and meta-analysis. Gynecol. Oncol. 140, 184–190.

Siegel, R.L., Miller, K.D., and Jemal, A., 2016. Cancer Statistics, 2016. CA Cancer J. Clin. 66, 7–30.

Smith, K.B., and Smith, M.S., 2016. Obesity statistics. Prim. Care 43, 121–135.

Topalian, S.L., Hodi, F.S., Brahmer, J.R., et al., 2012. Safety, activity, and immune correlates of anti-PD-1 antibody in cancer. N. Engl. J. Med. 366, 2443–2454.

Topalian, S.L., Drake, C.G., and Pardoll, D.M., 2015. Immune checkpoint blockade: a common denominator approach to cancer therapy. Cancer Cell 27, 450–461.

van Mierlo, G.J., Boonman, Z.F., Dumortier, H.M., et al., 2004. Activation of dendritic cells that cross-present tumor-derived antigen licenses $CD8^+$ CTL to cause tumor eradication. J. Immunol. 173, 6753–6759.

VanOosten, R.L., and Griffith, T.S., 2007. Activation of tumor-specific $CD8^+$ T Cells after intratumoral Ad5-TRAIL/CpG oligodeoxynucleotide combination therapy. Cancer Res. 67, 11980–11990.

Weitman, E.S., Aschen, S.Z., Farias-Eisner, G., et al., 2013. Obesity impairs lymphatic fluid transport and dendritic cell migration to lymph nodes. PLoS One 8, e70703.

Wilson, W.R., and Hay, M.P., 2011. Targeting hypoxia in cancer therapy. Nat. Rev. Cancer 11, 393–410.

Wolin, K.Y., Carson, K., Colditz, G.A., et al., 2010. Obesity and cancer. Oncologist 15, 556–565.

Yeung, O.W., Lo, C.M., Ling, C.C., et al., 2015. Alternatively activated (M2) macrophages promote tumour growth and invasiveness in hepatocellular carcinoma. J. Hepatol. 62, 607–616.

Zarkesh-Esfahani, H., Pockley, G., Metcalfe, R.A., et al., 2001. High-dose leptin activates human leukocytes via receptor expression on monocytes. J. Immunol. 167, 4593–4599.

Zitvogel, L., Apetoh, L., Ghiringhelli, F., Kroemer, G., 2008. Immunological aspects of cancer chemotherapy. Nat. Rev. Immunol. 8, 59–73.

CHAPTER

13

Redox-Dependent Circuits Regulating B Lymphocyte Physiology

Milena Bertolotti and Roberto Sitia

OUTLINE

Abstract

Redox chemistry has broad and important consequences for health and disease: we are now beginning to understand how the numerous redox-based circuits are integrated with other signaling pathways to regulate biological systems. In this essay, we summarize recent new findings that emerge from studies on the immune system, in particular B lymphocytes.

M.A. Hayat (ed): Immunology, Volume 1.
DOI: http://dx.doi.org/10.1016/B978-0-12-809819-6.00013-7

© 2018 Elsevier Inc. All rights reserved.

INTRODUCTION

Redox-based circuits in and between cells impact a variety of key functions and processes, including signaling, protein synthesis, proliferation, and tumor progression. Major modulators of redox homeostasis are reactive oxygen species (ROS), which are "radical" molecules containing an unpaired electron formed from reduction reactions involving oxygen. ROS overproduction and/or mismanagement can lead to oxidative stress, which results in direct or indirect damage of nucleic acids, proteins, and lipids, and has been implicated in cell death and aging. However, accumulating evidence indicates that ROS are not only foes, but also good friends, serving as critical signaling molecules in cell proliferation and survival (Finkel, 2000).

In particular, the effects of ROS and reactive nitrogen species (RNS) in innate and adaptive immunity have gained increasing scientific prominence. Thus, studies on how phagocytes kill engulfed pathogens led to the discovery of PhOx (phagocytic oxidase) also called NADPH oxidase 2 (Nox2). Besides playing a key role in innate immunity, Nox2 turned out to be a prototypic member of a family of key modulators of tyrosine phosphorylation-dependent signaling circuits, controlling tumorigenesis, B cell reactivity, and autoimmunity (Singel and Segal, 2016).

Each B cell is equipped with a multisubunit receptor (BCR) that recognizes an antigen with exquisite specificity. Upon encounter with antigen, BCR activation starts a multidimensional program entailing B cell proliferation, somatic mutation-affinity maturation, and terminal differentiation into antibody secreting plasma cells. It goes without saying that this response requires the integration of diverse, cell autonomous, and intercellular signaling pathways. These include contacts with regulatory T and dendritic cells and signals emanating from a network of cytokines and chemokines, which are progressively being dissected at the molecular level. A key yet still poorly recognized role is played in these processes by the physicochemical features of the B cell microenvironment, including metabolism, oxygen tension, and redox state (Angelini et al., 2002; Garcia-Manteiga et al., 2011).

Understanding the choreography of B cell responses is a formidable challenge. At the same time, the richness of robust in vitro, ex vivo, and in vivo experimental models, that have become available over the last decades, make cells of the B lineage a powerful system to tackle questions pertaining to the general theme of redox regulation.

A PECULIAR ROS: HYDROGEN PEROXIDE (H_2O_2)

Despite being useful for global descriptions and concision, the term ROS encompasses a vast family of heterogeneous molecules with diverse reactivities and half lives. To reconstruct the pathophysiological relevance of redox circuits, however, information on the molecular identity of each ROS is of critical importance. Indeed, it is only in defining the discrete chemical entities at work that we can dissect the downstream biological responses (Herrmann and Dick, 2012).

The main ROS encountered in living systems are superoxide ($O_2^{-\bullet\bullet}$), hydrogen peroxide (H_2O_2), hypochlorus acid (HOCl), singlet oxygen ($^1O_2$), lipid peroxides (ROOH), ozone (O_3), and hydroxyl radical ($OH^{\bullet}$). Amongst these, $O_2^{-\bullet}$, H_2O_2, and $OH^{-\bullet}$ are those with higher biological relevance. As already outlined above, RNS such as nitric oxide ($NO^{\bullet}$) are also important for cellular physiology as they can react with ROS to form oxidants, like peroxynitrite. However, apart from raising awareness of their huge pathophysiologic potentials, a discussion on RNS would go beyond the purposes of this chapter, which focuses mainly on the generation, transport, clearance, and signaling roles of H_2O_2 in B cell activation and differentiation.

In comparison to other ROS, H_2O_2 has the lowest reactivity, the highest stability, and the highest intracellular concentration. In particular, what qualifies H_2O_2 as a suitable second messenger are its enzymatic production and degradation, that provide spatiotemporal specificity, and its interesting chemistry, that gives specificity for oxidation of thiols (Forman et al., 2004). There are several sources of H_2O_2 production in mammalian cells: amongst these, the one for which there is strong evidence for a signaling function is $O_2^{-\bullet}$ generated by NADPH oxidases, followed by dismutation. The lifetime and potential diffusion distances for $O_2^{-\bullet}$ are very short (Mikkelsen and Wardman, 2003). Therefore, it seems likely that a reaction product of superoxide (i.e., H_2O_2), rather than the anion itself, is responsible for initiating the signaling response.

Support for H_2O_2 as the mediator of signal transduction has been gained from experiments in which scavengers of ROS were added to the extracellular milieu (Fisher, 2009). In fact the signaling cascade is relatively unaffected by the presence of superoxide dismutase, which converts $O_2^{-\bullet}$ to H_2O_2, but is abrogated by the presence of catalase or other H_2O_2 scavengers (Milovanova et al., 2008). Possible signal transducers include other products formed from $O_2^{-\bullet}$, such as $^{\bullet}OH$, peroxynitrite, or oxidized membrane components including lipid hydroperoxides or protein carbonyls. However, most of these more likely represent potentially toxic by-products of superoxide generation rather than physiologic signaling species. That H_2O_2 is the principal signal transducer for extracellularly generated $O_2^{-\bullet}$ is supported by the phenotypes observed after addition of exogenous H_2O_2 to living cells (Stone and Yang, 2006). The main limitation of studies using exogenous H_2O_2 lies in the concentrations that are normally necessary to obtain detectable phenotypes. Owing to its Janus like function, the high amounts of H_2O_2 used might simultaneously activate growth-promoting signals and exert toxicity.

Moreover, we have recently demonstrated that H_2O_2 needs proteinaceous channels to enter live cells, whose conductance is inhibited by diverse cellular stresses (Bertolotti et al., 2013, 2016; Medrano-Fernandez et al., 2016). It is also possible that the plasma membrane, as well as underlying cytosolic areas, are equipped to control the extent and duration of H_2O_2 diffusion. However, although it is clear that exogenously added H_2O_2 might not precisely mimic physiological signaling patterns, many results obtained so far were confirmed with H_2O_2 generated endogenously, mainly by NOX activation.

REDOX-DEPENDENCY OF B CELL ACTIVATION AND DIFFERENTIATION

The ultimate goal of B cells is to produce sufficient quantities of the desired, high-affinity antibodies. Many steps are necessary for the development of hematopoietic stem cells in the bone marrow into plasma cells in the periphery and the return of a few of them to the bone

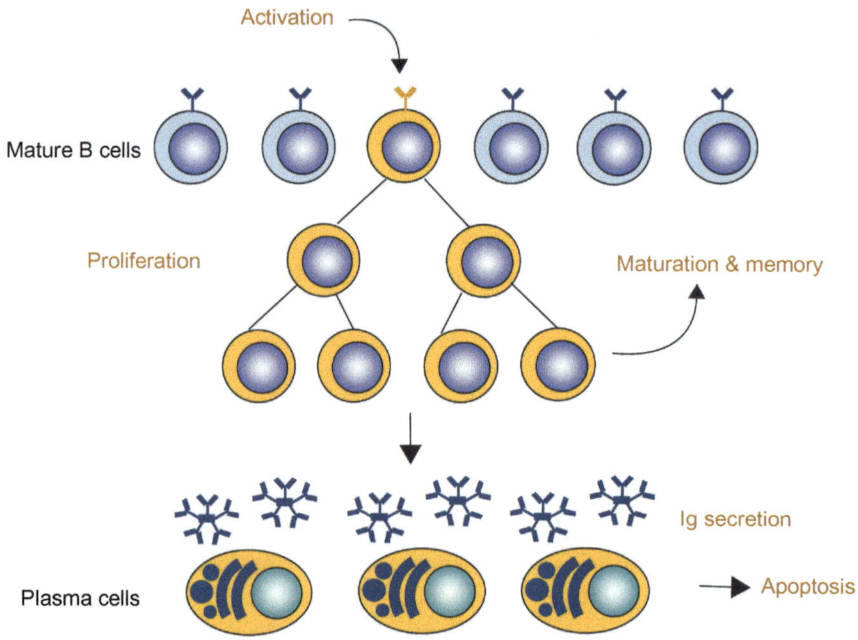

FIGURE 13.1 **B to plasma cell differentiation.** B cells are generated from lymphoid precursors in the bone marrow. If they succeed in producing a functional, nonautoreactive BCR, they differentiate into mature, naïve B cells. These cells express antigen-specific BCR of IgM and IgD classes and recirculate through the body. Upon encounter with antigens, mature B lymphocytes are activated and start a proliferative burst, followed by affinity maturation and development of memory cells, which are long-lived and provide an enhanced and improved immune reaction upon reencounter of cognate antigen. Some B cells undergo terminal differentiation towards antibody secreting plasma cells. Most plasma cells are short-lived and undergo apoptosis (Cenci et al., 2006). A few return to the bone marrow and survive for long periods producing lower levels of antibodies.

marrow. Far from describing this spectacular program in detail, we'll highlight here a few steps in which redox circuits play important roles, starting from mature B cells.

Activated B cells form germinal centers in lymphoid follicles, where proliferation and differentiation occur rapidly. Some of them swiftly differentiate into IgM producing plasma cells, to provide immediate protection against the pathogen. Some instead differentiate into memory B cells after affinity maturation (Fig. 13.1). Memory B cells divide very slowly if at all; they express surface Ig receptors, but do not secrete antibodies at a high rate. They have been selected for production of antibodies of increased affinity or binding kinetics (Murphy and Weaver, 2016).

As outlined earlier, other germinal center cells differentiate first into plasmablasts and then into plasma cells (Fig. 13.1). Plasmablasts begin to specialize in antibody secretion; their fate is to become nondividing, Ig secreting plasma cells. The majority of the plasma cells die after a few days of antibody production (Shapiro-Shelef and Calame, 2005). This limited survival of plasma cells, in addition to many crucial steps in terminal B lymphocyte differentiation, like activation and antibody secretion, need to be tightly controlled

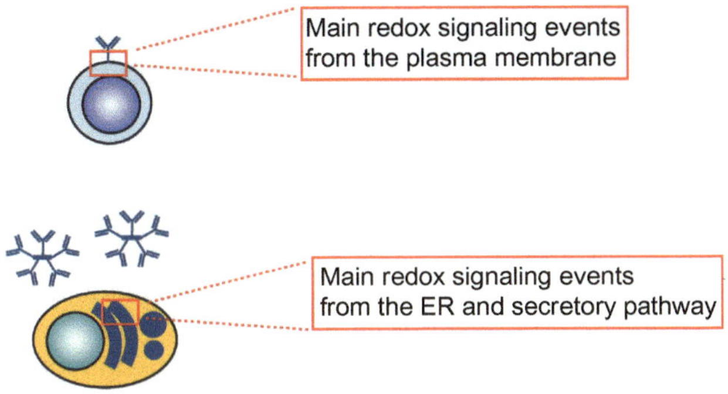

FIGURE 13.2 Sources of redox signals change during B cell differentiation.

and depend on finely tuned intra- and extracellular redox reactions (Bertolotti et al., 2010; Polikowsky et al., 2015; Reth, 2002; Wheeler and Defranco, 2012). The hormetic nature of oxygen and its reactive species, essential for survival but toxic when in excess, seems well suited to regulate these crucial life/death decisions. Indeed, evidence is accumulating for a role of H_2O_2 and redox remodeling at many levels of B cell existence and progression.

The pioneering work of the Finkel, Tonks, Rhee, and other labs showed that intracellular H_2O_2 impacts tyrosine kinase signaling, mainly by transiently inactivating phosphatases (Rhee, 2006; Tonks, 2005). It soon became clear that H_2O_2 is essential also for Toll-like receptors (TLRs) and B cell receptors (BCR) to amplify their signals upon binding of cognate ligands. We have recently confirmed that H_2O_2 molecules produced extracellularly by Nox2 enter cells via aquaporin 8 (AQP8) (Bertolotti et al., 2016). The transient oxidative wave generated in the proximity of ligand bound BCR and TLR boosts signaling (Fig. 13.2, upper part).

B to plasma cell differentiation requires the activation of Ire1-Xbp1 (Iwakoshi et al., 2003), a key unfolded protein response (UPR) arm (Ron and Walter, 2007). That differentiation entails endoplasmic reticulum (ER) stress is not surprising considering the massive increase in the size and activity of this organelle (van Anken et al., 2003). Differentiating B cells undergo also a profound reshaping of their redoxtatic machineries, with a dramatic increase in the intracellular H_2O_2 levels (Bertolotti et al., 2010). We have recently shown that different cellular stresses, ER stress included, inhibit AQP8-dependent H_2O_2 transport (Medrano-Fernandez et al., 2016). It is noteworthy that differentiating B cells are essentially unable to uptake exogenous H_2O_2, likely owing to their stressed state (this issue will be addressed in further details below—see "Membrane Permeability to H_2O_2 Is Impaired During Cell Stress in a Redox-Dependent Way" section). The combined result of these modifications is that H_2O_2 no longer originates from NOX stimulation in the extracellular space, but possibly from intracellular sources, mitochondria, and the ER being primary suspects (Fig. 13.2, lower part). The intense utilization of the antibody factory, forced to produce thousands of Ig molecules per second at the plasma cell level, entails in fact massive H_2O_2 production that could also contribute in limiting their lifespan (Cenci and Sitia, 2007).

B CELL ANTIGEN RECEPTOR SIGNALING IS AMPLIFIED BY H_2O_2

Naïve B cell activation is initiated by binding of antigen to the B cell antigen receptor (BCR) complex. Triggered BCR initiates a signaling cascade that leads to up-regulation of costimulatory molecules and mediates efficient antigen internalization followed by processing and presentation of antigenic peptides for recognition by specific T cells. TLRs stimulation provides a third signal that synergizes with BCR triggering (signal one) and T cell help (signal two), resulting in the induction of sustained proliferation and differentiation of human naïve B cells (Ruprecht and Lanzavecchia, 2006).

In the first step, BCR triggering, membrane-bound Ig molecules and Ig-αβ heterodimers function as antigen-binding and signaling subunits, respectively (Schamel and Reth, 2000). At least three PTKs (Lyn, Syk, and Btk) and a protein tyrosine phosphatase (PTP), SHP-1, are involved in signal transduction from the BCR. The negative regulator SHP-1 is inhibited by H_2O_2, which oxidizes a cysteine residue in the catalytic site of the enzyme. This explains why suitable doses of H_2O_2 can mimic antigen binding, triggering the same phosphorylation cascade as BCR cross-linking (Reth, 2002). Moreover, H_2O_2-treated B cells exhibit PTK dependent inositol 1,4,5-trisphosphate (IP3) generation and Ca^{2+} release (Qin et al., 2000). The initial, low-intensity activation of BCR signaling is amplified by the Ca^{2+}-dependent, self-sustaining feedback loop (Singh et al., 2005). Thus, oxidants and Ca^{2+} cooperatively regulate the strength and duration of BCR signaling, crucial parameters for determining the subtype of B cell generated from transitional, immature B cells (Casola, 2007). This point will be discussed in more detail below.

Surprisingly it was found that activation of effectors like Erk, JNK, p38, and Akt by the BCR was relatively insensitive to H_2O_2 depletion. Also chemokine receptor type 4 (CXCR4), which binds chemokines involved in B cell development and trafficking, activates MAP kinases via ROS-independent pathways, but its ability to activate Akt depends on ROS (Lee et al., 2007). In contrast, the surface receptor CD40, important to receive T cell costimulatory signals, activates JNK, p38, and Akt for the most part via H_2O_2-dependent pathways that are sensitive to scavengers such as *N*-acetyl-cysteine or ebselen (Lee et al., 2007). An important aspect of ligand-dependent BCR activation may be the rapid translocation of the receptor in the proximity of a source of H_2O_2 production (like NOX) or, vice versa, translocation of the source to the receptor. The synergic role of HVCN1, a proton channel that supports NOX activity, confirms that oxidation is needed, the activation of Syk and Akt being impaired in HVCN1-deficient cells (Capasso et al., 2010). However, not all signaling pathways downstream of Syk were equally impaired, removing the proton flux that sustains electrogenic H_2O_2 production. Ca^{2+} mobilization and Erk activation were not impaired in the absence of HVCN1, possibly indicating different activation thresholds (Capasso et al., 2010). These results imply that a complex interplay between distinct pathways regulates the outcome of a signal.

Furthermore, it is important to recall here that B cell activation is accompanied by an increase in overall cysteine sulfenylation, in both the cytosol and nucleus (Crump et al., 2012). Incubation of B cells with dimedone, a compound that covalently reacts with sulfenic acid to prevent its further oxidation or reduction, resulted in a block in anti-IgM-induced cell division, together with a failure to induce capacitative calcium entry and to maintain tyrosine phosphorylation (Crump et al., 2012). This provided further evidence that proteins necessary for B cell activation have to pass through cysteine sulfenylation to exert their functions.

How B Lymphocytes Control H_2O_2 (1): Crosstalks With Calcium

Calcium communicates with a number of other systems and pathways, including ROS. Indeed a reciprocal interaction exists between Ca^{2+} modulated ROS production and ROS modulated Ca^{2+} signaling (Gorlach et al., 2015) (Fig. 13.3A and B). In B lymphocytes, calcium entry likely influences mitochondrial ROS production, depending on their metabolic state. Moreover, it affects NOX activity.

Ca^{2+} primarily promotes ATP synthesis by stimulating enzymes of the Krebs cycle and oxidative phosphorylation in mitochondria: it has been hence suggested that increased metabolic rates consume more oxygen, resulting in increased H_2O_2 leakage from the respiratory chain (Brookes et al., 2004). Mitochondrial ROS were also reported to suppress BCR-dependent Lyn, Btk, and PLCγ2 phosphorylation and CD19 expression (Ogura et al., 2017), thus resulting in a general reduction of humoral immune responses. Other studies have demonstrated that mitochondrial ROS suppress plasma cell differentiation by attenuating heme synthesis and proliferation of B cells by negatively impacting mitochondrial metabolism (Jang et al., 2015).

Stimulation of B cells with anti-Ig normally leads to an increase in the number of detectable phosphoproteins and in the mobilization of free calcium ions in the cytoplasm. In the presence of calcium chelators or flux inhibitors, dampening in BCR signaling is observed (Singh et al., 2005). Lyn phosphorylation is particularly sensitive (see Fig. 13.3A), suggesting that the effects of Ca^{2+} deprivation are likely exercised at early steps in the BCR signaling pathway (Singh et al., 2005). Furthermore, H_2O_2 scavengers are equally able to inhibit BCR-dependent Lyn phosphorylation, but the addition of ionomycin has no effect, suggesting a redundancy of Ca^{2+} signals in this process. In other words, while both the BCR-dependent Ca^{2+} and ROS pathways influenced each other, it is ROS action that ultimately affects BCR signaling (Singh et al., 2005). Lyn acts also as a redox sensor, being activated through direct oxidation of C466 (Yoo et al., 2011): for this and the above-cited characteristics, it has an important role at the interphase between redox and Ca^{2+} signaling circuits.

A crosstalk between calcium and ROS is also crucial at the ER–mitochondrial interface. For instance, small increases in H_2O_2 can initiate Ca^{2+} release from ER stores (Hempel and Trebak, 2017). This would eventually influence a number of proteins involved in the TCA cycle and electron transport chain that rely on Ca^{2+} for their activity, hence controlling cellular metabolic flux and redox signaling in return. This has important implications for differentiating B lymphocytes, where the maintenance of energy production by mitochondria is essential in ensuring proliferation and sustained immunoglobulin production (van Anken et al., 2003). On the other hand, increased flux of Ca^{2+} into mitochondria is a driver of mitochondrial ROS production and an integral component of apoptosis. Altogether, these studies highlight how the mutual interplay between calcium and ROS is crucial for fine-tuning signaling networks.

How B Lymphocytes Control ROS (2): Modulation of NOX

NOX are complex enzymes that generate superoxide or directly H_2O_2. Their catalytic center is the membrane-associated protein gp91^{phox} complexed with p22^{phox}. Activation requires association with a phosphorylated form of p47 (also known as Ncf1), p40^{phox},

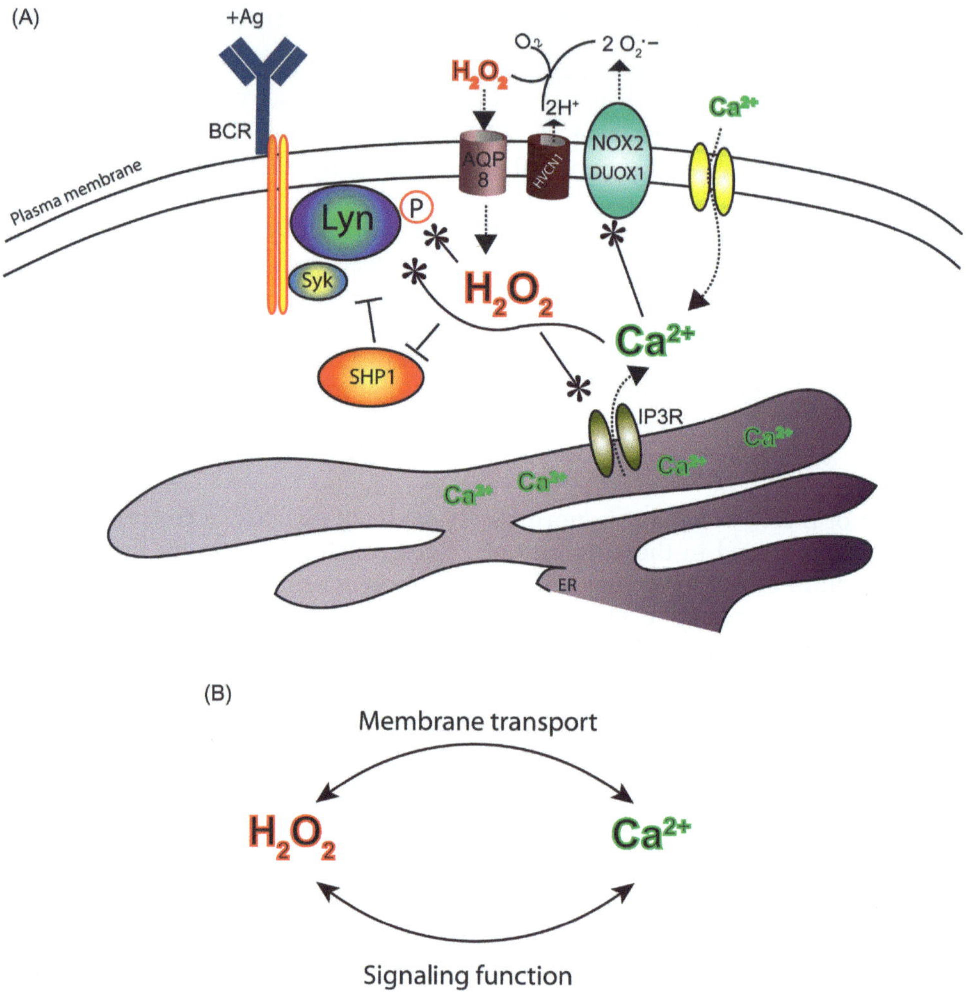

FIGURE 13.3 **Calcium and ROS signaling networks are closely related.** (A) Antigen binding to the BCR results in phosphorylation of ITAMs in the Igα/β heterodimer, creating docking sites for Lyn and Syk. H_2O_2 is generated extracellularly by NOX, its activity being sustained by the proton channel HVCN1. Upon crossing the plasma membrane through AQP8, H_2O_2 inhibits SHP-1, thus amplifying tyrosine kinase signaling paths. BCR activation also leads to the release of Ca^{2+} from ER stores, which is further stimulated by H_2O_2. Ca^{2+} is in turn able to modulate NOX activity. A second Ca^{2+} influx comes instead from the extracellular milieu (capacitative Ca^{2+} entry). H_2O_2 maintains high intracellular levels of Ca^{2+}, inhibiting its extrusion from the plasma membrane; in this way, H_2O_2 and Ca^{2+} synergize in amplifying BCR-dependent signaling. (B) Positive feedback loops between Ca^{2+} and H_2O_2, rely on reciprocal regulation of transporters and effectors.

p67^{phox}, and the small GTPase Rac (Bedard and Krause, 2007). $O_2^{-\bullet}$ and H_2O_2 are produced by these enzymes on the outer leaflet on the plasma membrane, where peroxidases generate more powerful oxidant species than H_2O_2 suitable for oxidizing extracellular substrates. An important NOX feature is also their inducibility by stress or inflammatory stimuli (Bedard and Krause, 2007). p47^{phox} and gp91^{phox} deficiency in humans leads to neutrophil dysfunction and chronic granulomatous disease (CGD), a primary immunodeficiency characterized by the inability to eradicate bacterial infections (Pollock et al., 1995).

First discovered in phagocytes, isoforms of NOX and dual oxidase (DUOX) have been described to generate H_2O_2 in a regulated manner also in other cell types, with important consequences upon signaling (Bedard and Krause, 2007). However, phagocytic activity was shown also in peritoneal B cells, which can kill engulfed bacteria through Nox2-produced ROS (Kovacs et al., 2015). It is important to recall here that $O_2^{-\bullet}$ and H_2O_2 are produced on the outer leaflet on the plasma membrane, where peroxidases generate more powerful oxidant species than H_2O_2 suitable for oxidizing extracellular substrates.

B lymphocytes express Nox2 and DUOX1 (Bedard and Krause, 2007; Reth, 2002; Singh et al., 2005). Intriguingly, we detected increased expression of NOXA2 ($p67^{phox}$) transcripts in lipopolysaccharide (LPS)-stimulated murine B lymphocytes (Masciarelli et al., unpublished observations). These preliminary findings deserve further analyses, especially since previous reports failed to detect NOXA2 in plasma cells (Inanami et al., 1998). DPI, a NOX inhibitor, does not diminish basal ROS levels but decreases ROS generation upon LPS or anti-IgM B cell stimulation (Capasso et al., 2010; Vene et al., 2010), suggesting that NOX mediate BCR-triggered ROS production. B cells from mice deficient in $gp91^{phox}$ are unable to generate ROS after BCR ligation (Richards and Clark, 2009). Experiments employing siRNA revealed that DUOX1 constitutes at least one of the components responsible for the early phase (from seconds to minutes) of the BCR-dependent ROS response, and thus is critical for establishing the BCR-dependent Ca^{2+}-ROS feedback loop above mentioned, ensuring optimal phosphorylation of Lyn following BCR activation (Singh et al., 2005).

Previous studies revealed significant changes in the intracellular ROS levels during B cell differentiation (Bertolotti et al., 2010; Vene et al., 2010). H_2O_2 increases sharply in the first days after LPS activation and again at the end of the culture period, before massive apoptosis ensues (Bertolotti et al., 2010). The initial rise in H_2O_2 seemed to correlate with proliferation, which is more rapid in B cells lacking H_2O_2 scavengers like Prx1 or Prx3 (Bertolotti et al., 2010). However, despite their lower H_2O_2 content during the initial phase, B cells lacking the antioxidant defense systems master regulator NF-E2-related factor 2 (Nrf2) proliferate similarly, suggesting the involvement of additional regulatory pathways (Bertolotti et al., 2010).

A formidable hurdle for the experimentalists is that ROS can be produced by various systems and in multiple locations and are counterbalanced by redundant antioxidant defense mechanisms. This complexity likely reflects the need for fine-tuning redoxtasis in differentiating B cells. The correlation between ROS generation and B cell activation/proliferation remains a key issue: the function of NOX, DUOX, and HVCN1 in regulating BCR-dependent proliferation and differentiation clearly deserves further investigation. NOX are also able to modulate major histocompatibility complex (MHC) class II antigen presentation by B cells (Crotzer et al., 2012): in the absence of functional $gp91^{phox}$ and $p40^{phox}$ subunits, human B cells displayed a reduced capacity for cytoplasmic antigen presentation and this could be related to the role of $p40^{phox}$ in promoting NOX assembly on endosomal membranes, that could eventually affect MHC II Ag processing and presentation within these organelles.

How B Lymphocytes Control ROS (3): Antioxidant Defense Systems

As anticipated earlier, elaborate antioxidant defense mechanisms—largely coordinated by transcription factor Nrf2—allow cells to neutralize excess ROS and handle oxidative stress. The complex web of antioxidant molecules includes low molecular weight molecules such as glutathione, as well as an array of protein antioxidants with specific subcellular localization and chemical reactivities. Among protein antioxidants, superoxide dismutases metabolize superoxide into H_2O_2, while catalases and peroxiredoxins rapidly catabolize H_2O_2. In addition, the thioredoxin and glutaredoxin families play a significant role in maintaining redox homeostasis in the cytosol and mitochondria, by the reduction of disulfide bonds in various target proteins. Mice or plants lacking individual factors display no or mild phenotype, implying a high degree of redundancy (Bertolotti et al., 2010; Blokhina and Fagerstedt, 2010), likely reflecting an evolutionary pressure to cope with oxidative stress.

In B cells, an important antioxidant defense is provided by ER-resident peroxiredoxin 4 (Prx4) and other peroxidases residing in the secretory pathway. The localization of many antioxidants in the ER is not accidental in cells whose main function is to guarantee high rates of protein folding to produce huge levels of Ig molecules; this point will be further discussed later.

In addition, a crucial role in B cell responses and other pathophysiologic processes, including inflammation and tumor development and progression, is played by the cysteine/cystine redox cycle (Conrad and Sato, 2012; Lo et al., 2008). The cycle is basically composed by system xc-, the redox couple cystine/cysteine, and intracellular oxido-reductases including thioredoxin. System xc- is a Nrf2-regulated, oxidative stress-inducible, cystine/glutamate antiporter which imports oxidized cystine releasing equimolar glutamate into the extracellular space. It consists of two protein components, the 4F2 heavy chain, necessary for membrane targeting of the heterodimer, and the xCT protein, responsible for transport activity (Conrad and Sato, 2012). While cysteine is more abundant in the reducing milieu of the cytosol, cystine prevails extracellularly. Once internalized, cystine is reduced to cysteine by thioredoxin. Internalized cysteine was long thought to be merely used for protein and GSH synthesis. However, increasing evidence indicates that some cysteine is actively released extracellularly via the neutral amino acid transport systems. There, cysteine is rapidly oxidized to cystine, and consequently taken up by system xc-. If this cystine/cysteine cycle persists, a steady state is reached, and cysteine remains reduced due a switch of the extracellular redox state from an oxidizing to a reduced state.

Since resting T, pre-B and B lymphocytes do not express xc-, cysteine must be made available to sustain their activation and proliferation (Angelini et al., 2002). The extracellular redox hence provides a rheostat for immune responses. In the case of T lymphocytes the "wet-nurse" has been shown to be the dendritic cell (Angelini et al., 2002) (Fig. 13.4). Which cell/s provide cysteine to B lymphocytes during their differentiation/activation remains to be established and can potentially represent a crucial druggable target/s for many diseases (Vene et al., 2011). Furthermore, it was recently discovered that xCT protects cancer cells from drug-induced oxidative stress by mediating detoxification and extrusion of chemotherapeutic drugs via interactions with multidrug resistance proteins (Dai et al., 2015). This would explain why this system is highly expressed by lymphomas and other malignant tumors (Gout et al., 2001) (see also Fig. 13.4).

A fascinating question is whether the antioxidant responses to enhanced ROS production reflect only a defense mechanism aimed at restoring redoxtasis or also exert functional effects on B lymphocytes differentiation. An active role for the antioxidant response has been established in various processes of the innate and adaptive immunity, including IL-1β

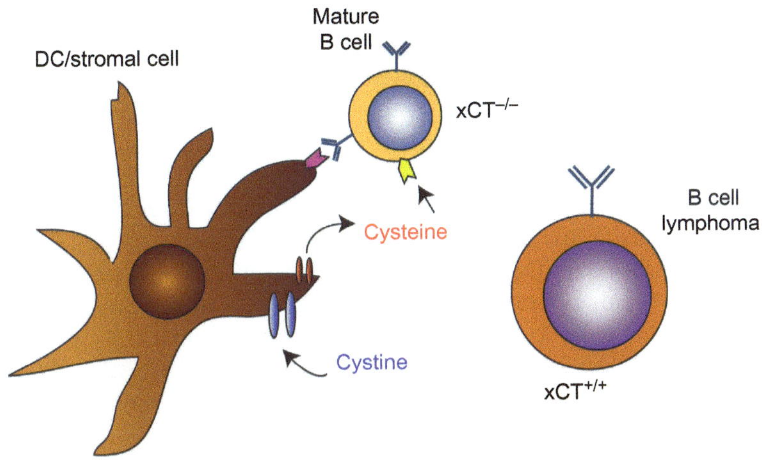

FIGURE 13.4 **The cystine/cysteine redox cycle in normal and malignant B cells.** A. The cycle is based on the initial uptake of cystine, its reduction to cysteine by the powerful cytosolic reductase activities, and the secretion of cysteine into the extracellular space. Cystine (Cys-*S*-*S*-Cys), the oxidized form of cysteine that prevails extracellularly, is imported via the glutamate/cystine antiporter system xc-. This system is composed of a specific light chain, xCT, and a heavy chain, 4F2, linked by a disulfide bridge. 4F2 is required for targeting the heterodimer to the plasma membrane, whereas the xCT protein is responsible for transport activity. Resting lymphocytes lack the xCT antiporter and are therefore unable to uptake oxidized cystine from the extracellular fluids. In the early phases of immune responses, DCs, which express high levels of system xc-, nurse B cells by taking up oxidized cystine and releasing cysteine that is easily internalized by them, allowing their survival and proliferation. xCT is expressed following BCR stimulation, so that activated B cells emancipate from cysteine requirement, as it happens for cancer cells, expressing high levels of the transporter.

secretion (Tassi et al., 2009, 2010) and T or B cell activation (Angelini et al., 2002). Similarly, in vitro exposure of B lymphocytes to exogenous antioxidants revealed that highly reducing conditions are detrimental in the early phases of B cell activation, though they promoted IgM secretion at later stages (Vene et al., 2010). Therefore also the antioxidant response plays a role in B cell differentiation, as well as ROS.

How B Lymphocytes Control ROS (4): Membrane Transport

So far we mentioned three different ways to exploit the positive side of ROS limiting their toxicity and oxidative stress.

1. Calcium-mediated control
2. NOX regulated activation
3. Antioxidant defenses involvement

Another form of physiological regulation concerns their membrane permeability and sequestration. H_2O_2 tends now to be generally accepted as a signaling molecule and specifically as the second messenger related to NOX: how can an extracellular molecule result in intracellular signaling events? H_2O_2 has often been believed to freely cross membranes but, contrary to this, recent studies point out that some membranes are rather poorly permeable to H_2O_2 (Antunes and Cadenas, 2000). This implies that transport of H_2O_2 may be regulated and that this regulation constitutes a major factor in the determination of cellular H_2O_2 concentration.

As shown by us and others, H_2O_2 transport often occurs through specific aquaporin channels, similar to what happens for H_2O (Bertolotti et al., 2013, 2016; Bienert et al., 2007; Miller et al., 2010). In particular, AQP8 is one of the ways through which H_2O_2 can enter cells and promote growth factor-dependent signaling (Bertolotti et al., 2013). Accordingly, interfering with H_2O_2 producers (Nox2) or transporters (AQP8) has profound effects on key downstream events in B cell physiology, such as IgM production and secretion (Bertolotti et al., 2016). Altogether, these studies suggest that individual cells and tissue types can potentially be tuned for their susceptibility to H_2O_2-mediated cellular signaling or stress, depending on which aquaporins or related channels are displayed on their cell surfaces (Dickinson and Chang, 2011).

MEMBRANE PERMEABILITY TO H_2O_2 IS IMPAIRED DURING CELL STRESS IN A REDOX-DEPENDENT WAY

Like most other secretory proteins, Ig are rich in disulfide bonds; they are cotranslationally translocated into the ER, and attain their proper quaternary structure in the early secretory pathway. Not surprisingly, therefore, the most dramatic feature of plasma cell differentiation is the expansion of the ER and other secretory organelles aimed at accommodating exuberant Ig synthesis. Cells that face an increased secretory demand generally activate the UPR, a multidimensional signaling cascade finalized to maintain ER homeostasis. In mammals, this is based on three main branches, initiated by PERK, ATF6, and Ire1, respectively (Ron and Walter, 2007). Ire1 triggers the nonconventional splicing of xbp1 transcripts. As a result, the active form of the XBP1 transcription factor (sXBP1) is produced to overall increase the secretory capacity (Sriburi et al., 2007). Different cells can selectively exploit the UPR branches (Wu and Kaufman, 2006): for instance terminal differentiating B cells require Ire1 and XBP1, but not PERK (Gass et al., 2008; Iwakoshi et al., 2003).

Interestingly, a novel redox-regulated cytoplasmic signaling function has recently been shown for Ire1 (Hourihan et al., 2016). Upon sulfenylation induced by cytosolic locally produced ROS, Ire1 initiates the Nrf2-mediated antioxidant response, clearly demonstrating how versatile redox regulation and signaling processes can be, by switching protein functions toward diverse directions. It will be important to elucidate the extent to which plasma cells discriminate between ER and cytosolic stress sensing.

As already mentioned, ER-resident Prx4 increases in parallel to the dramatic expansion of the ER (van Anken et al., 2003). Prx1 and 3 instead follow the pattern of other cytosolic and mitochondrial proteins and GPx1 rapidly decreases (Bertolotti et al., 2010). Thus, the changes observed reflect more than the simple architectural metamorphosis that hallmarks B to plasma cell differentiation. Intracellular H_2O_2 levels changed significantly during B cell differentiation, increasing from 24 to 48h after LPS activation, and again after about 4–5 days of mitogenic stimulation, before cells undergo apoptosis. Consistent with the sharp increase in H_2O_2 and in accordance to the new findings about Ire1, LPS-induced nuclear accumulation of Nrf2. Total Nrf2 levels increased as well in the first 2 days of differentiation, corresponding to the proliferative phase. This pattern correlated with the activation of two Nrf2-dependent genes, i.e., glutamate cysteine ligase-modifier involved in GSH biosynthesis and Heme oxygenase 1 (HMOX1) (Bertolotti et al., 2010). These results confirmed that the Nrf2 pathway is activated early upon LPS stimulation.

In general we can thus say that terminal B cell differentiation entails UPR activation (at least a branch of it), waves of H_2O_2 production, redox stress, and reshaping of the antioxidant responses, but the origins and the dynamics of these changes are still poorly understood. Altogether, these observations indicate that oxidative stress could contribute to limiting plasma cell lifespan, in a fashion possibly dependent on the rate and intensity of antibody secretion. Proteasomal insufficiency, a possible consequence of chronic UPR and ROS production, could hence cooperate in inducing death when enough antibodies have been produced.

The hormetic properties of H_2O_2—essential for cell survival, but toxic at high concentrations—imply tight regulation of its intracellular levels. We reasoned that, being more sensitive to oxidative toxicity (Cenci et al., 2006; Nerini-Molteni et al., 2008), cells undergoing stress might deploy mechanisms that limit H_2O_2 import and this was indeed shown by (Medrano-Fernandez et al., 2016). In B cells, if that is the case, the responses to exogenous H_2O_2 should vary depending on the day of LPS-induced differentiation, a process in which proteostasis and redoxtasis are challenged. In accordance to these predictions, we could indeed observe that at day 2, corresponding to high levels of stress (mostly of oxidative and proteostatic origin) (Bertolotti et al., 2010), the entry of exogenous H_2O_2—monitored by the H_2O_2-specific sensor HyPer (Belousov et al., 2006)—is impaired, in comparison to the other days. This is visualized both by plotting fluorescence ratio variations throughout time in time-course experiments (Fig. 13.5A) and by taking pictures at fixed timepoints (Fig. 13.5B).

INTRA- AND INTERCELLULAR REDOX SIGNALING

The ability to become insensitive to redox signal amplification coming from the extracellular environment could be a mechanism through which B lymphocytes overwhelmed with stress protect themselves from further damage. When stress overcomes certain thresholds, cells seem to suppress the entry of novel amplifying signals. However, in normal conditions, similar exogenous signals are necessary to proper B cell function, as demonstrated by the severe impairment in tyrosine phosphorylation showed by $Nox2^{-/-}$ cells upon BCR stimulation (Bertolotti et al., 2016) (Fig. 13.6, left). Surprisingly, although CGD patients have a clear defect in macrophage and neutrophil function, they do not display a general block in lymphocyte development as one would expect from molecules playing a crucial role in lymphocyte activation.

A possible explanation is that accessory cells can provide in vivo sufficient levels of H_2O_2 to $Nox2^{-/-}$ B cells (Fig. 13.6, right), thus sustaining IgM production and secretion. It would be of interest to decipher the Nox2 dependent regulatory circuits that are responsible for the increased IgM production.

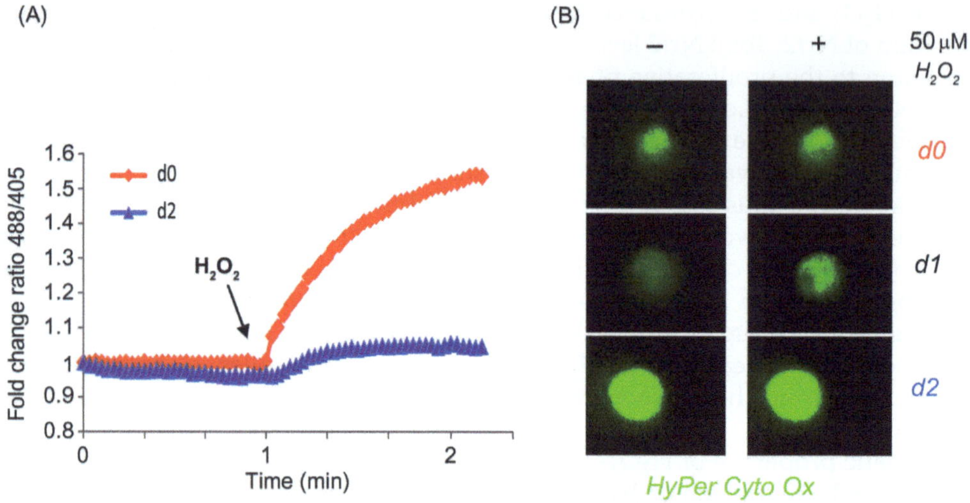

FIGURE 13.5 **Proteostatic insult is a prooxidative event affecting H_2O_2 membrane permeability.** (A) The effects of LPS-elicited oxidative stress on H_2O_2 uptake are displayed. Cells adhered onto poly-L-lysine were analyzed by immunofluorescence as in A. Images were collected every 2s for 3 or 5min. Postacquisition picture analysis was performed using ImageJ software and the 488/405-nm fluorescence excitation ratio of HyPer was calculated after background fluorescence subtraction. (B) Confocal representative images of the same acquisition experiment performed in A. The changes in HyPer fluorescence emission are used to measure H_2O_2 concentration, before and 2min after 50μM H_2O_2 addition. Note that cells increase in size upon LPS stimulation. However, concomitant with activation of IgM synthesis and ER stress, entry of exogenous H_2O_2 is inhibited.

REDOX-DEPENDENT CHEMOTAXIS IN B LYMPHOCYTES

A key aspect of lymphocyte physiology is their ability to traffic through secondary lymphoid organs. Whilst signaling by chemokine receptors has been characterized to a significant extent in T cells, the field of homing receptors pathways and regulators in B cells is still under development and its relationship with redox still controversial. For instance, the uptake of H_2O_2 through AQP3, an isoform with peroxiporin properties similar to AQP8 (Miller et al., 2010), was reported to have a role in chemokine-dependent T cell migration in immune responses (Hara-Chikuma et al., 2012).

As for T cells, a variety of chemokines regulate also B cell development and activation; still evidences for redox signaling involvement in their pathways are few. As an example, chemokine receptor type 4 (CXCR4), that binds chemokines involved in B cell development and trafficking, like CXCL12, was reported to activate Akt in a ROS-dependent manner (Lee et al., 2007). Negative regulation of B cell responses to chemokines was instead reported to depend on some specific Shc (Src homology 2 domain containing) adaptors (Patrussi et al., 2014). In particular, p66Shc isoform acts as a negative regulator of all steps of the chemotactic responses triggered by CXCR4 and CXCR5, by inhibiting Vav-dependent actin cytoskeleton reorganization. This effect on chemotaxis seems, however, independent of

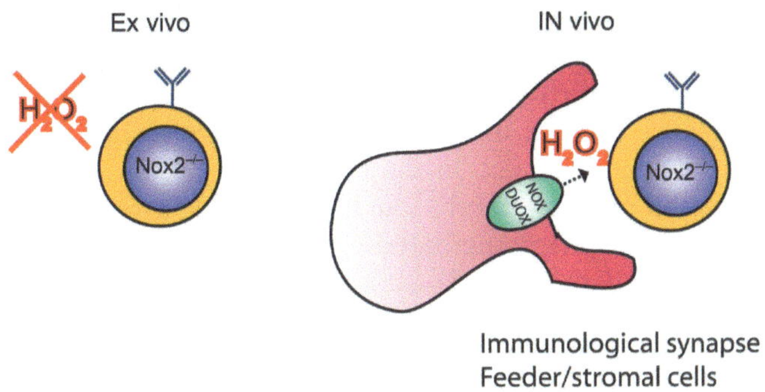

FIGURE 13.6 Difference between ex vivo/in vitro and in vivo redox circuits.

the ROS-accumulation ability of p66Shc, which is carried out through inhibition of fork-head transcription factors that control the expression of antioxidant enzymes (Nemoto and Finkel, 2002) and interrupting the mitochondrial respiratory chain (Giorgio et al., 2005). It is then not surprising that chronic lymphocytic leukemia (CLL) B cells have a defect in p66Shc expression and that this contributes to their extended survival (Capitani et al., 2012).

The homing-involved CXCL12 and the adhesion-modulator CCL21 are crucial chemokines that participate in naive T and B cell recruitment to the extra-follicular area in secondary lymphoid organs (Gunn et al., 1999). Work by Al Ghamdi et al. (2015) recently showed that supplementation with a well-known antioxidant, propolis, increased CCL21- and CXCL12-mediated B and T cell chemotaxis in diabetic mice, thus attenuating the abnormal lipid profiles, oxidative stress, inflammation, and impaired lymphocyte proliferation and migration towards chemokines that characterize immune system in diabetic disease (Al Ghamdi et al., 2015). The implication from these studies is that prooxidant and antioxidant molecules can exert immunomodulatory activity that could be crucial for the treatment of some pathological conditions, as better explained below in the last section of this chapter.

REDOX AND B CELL DISEASES

Subversion of redoxstasis could be important in cancer development and this explains why several ROS-modulating agents are undergoing clinical trials in oncology. The effectiveness of Bortezomib (a proteasome inhibitor) against multiple myeloma correlates with proteasomal overload (Bianchi et al., 2009; Cenci et al., 2006, 2011) as well as excessive ROS production (Hou et al., 2015). However, the latter increases also Nrf2-regulated HO-1 activity, thus conferring resistance to chemotherapy and genetic instability in multiple myeloma cells (Tibullo et al., 2016). Moreover, the Nox2-AQP8 axis could emerge as a target in plasma or B cell-mediated diseases, including CLL (dependent on continued BCR signaling), myelomas, systemic amyloidosis, or severe autoimmune disorders.

In this context, it was proposed that thymoquinone, a phytochemical compound with antioxidant properties, could be effective against multiple myeloma and alleviate diabetic complications, restoring the T cell immune response in diabetic offspring (Badr et al., 2011). In contrast, ellagic acid, a polyphenolic compound, showed anticancer activity by increasing ROS formation, mitochondria swelling, mitochondrial membrane potential decrease and cytochrome c release in CLL but not in healthy lymphocytes (Salimi et al., 2015). Prooxidant activity is cytotoxic for cancer cells also in the case of two PI3K inhibitors, which can effectively inhibit lymphoma cell proliferation by causing cell cycle arrest and can lead to cell death by inducing apoptosis and autophagy mediated by ROS accumulation (Civallero et al., 2015).

A contribution of ROS and redoxstasis alteration is also reported for autoimmune diseases. Two conditions are considered necessary for neo-antigenic determinants generated by free radicals to lead to autoimmunity: failure to remove or repair ROS damaged biomolecules, and an associated defect, probably in T cell signal processing and/or antigen presentation (Griffiths, 2008).

In conclusion, the relevance of redox circuitry in B cell pathophysiology is gaining momentum. However, considering the strict interplays between oxidative signals/stress and antioxidant responses, and the often unpredictable outcomes of disturbing the equilibrium, care should be taken in proposing treatments before we learn more about the underlying principles of redox regulation in B and other cell types.

Acknowledgments

We thank Iria Medraño-Fernandez for invaluable suggestions and critical reading of the manuscript, Simone Cenci, Max Cooper, Luca Rampoldi, and Michael Reth for helpful discussions. This work was made possible by grants from (AIRC (IG14559)), Fondazione Cariplo (2015 - 0591), Ministero della Salute (PE-2011-02352286), and Telethon (GGP15059) to RS; MB thanks the support of the ERC-grant 322972, the Excellence Initiative of the German Federal and State Governments (EXC294) and the Deutsche Forschungsgemeinschaft (TRR130).

References

Al Ghamdi, A.A., Badr, G., Hozzein, W.N., et al., 2015. Oral supplementation of diabetic mice with propolis restores the proliferation capacity and chemotaxis of B and T lymphocytes towards CCL21 and CXCL12 by modulating the lipid profile, the pro-inflammatory cytokine levels and oxidative stress. BMC Immunol. 16, 54.

Angelini, G., Gardella, S., Ardy, M., et al., 2002. Antigen-presenting dendritic cells provide the reducing extracellular microenvironment required for T lymphocyte activation. Proc. Natl. Acad. Sci. U.S.A. 99, 1491–1496.

Antunes, F., and Cadenas, E., 2000. Estimation of H_2O_2 gradients across biomembranes. FEBS Lett. 475, 121–126.

Badr, G., Lefevre, E.A., and Mohany, M., 2011. Thymoquinone inhibits the CXCL12-induced chemotaxis of multiple myeloma cells and increases their susceptibility to Fas-mediated apoptosis. PLoS One 6, e23741.

Bedard, K., and Krause, K.H., 2007. The NOX family of ROS-generating NADPH oxidases: physiology and pathophysiology. Physiol. Rev. 87, 245–313.

Belousov, V.V., Fradkov, A.F., Lukyanov, K.A., et al., 2006. Genetically encoded fluorescent indicator for intracellular hydrogen peroxide. Nat. Methods 3, 281–286.

Bertolotti, M., Yim, S.H., Garcia-Manteiga, J.M., et al., 2010. B- to plasma-cell terminal differentiation entails oxidative stress and profound reshaping of the antioxidant responses. Antioxid. Redox Signal. 13, 1133–1144.

Bertolotti, M., Bestetti, S., Garcia-Manteiga, J.M., et al., 2013. Tyrosine kinase signal modulation: a matter of H_2O_2 membrane permeability? Antioxid. Redox Signal. 19, 1447–1451.

Bertolotti, M., Farinelli, G., Galli, M., et al., 2016. AQP8 transports NOX2-generated H_2O_2 across the plasma membrane to promote signaling in B cells. J. Leukocyte Biol. 100, 1071–1079.

Bianchi, G., Oliva, L., Cascio, P., et al., 2009. The proteasome load versus capacity balance determines apoptotic sensitivity of multiple myeloma cells to proteasome inhibition. Blood 113, 3040–3049.

Bienert, G.P., Moller, A.L., Kristiansen, K.A., et al., 2007. Specific aquaporins facilitate the diffusion of hydrogen peroxide across membranes. J. Biol. Chem. 282, 1183–1192.

Blokhina, O., and Fagerstedt, K.V., 2010. Reactive oxygen species and nitric oxide in plant mitochondria: origin and redundant regulatory systems. Physiol. Plant 138, 447–462.

Brookes, P.S., Yoon, Y., Robotham, J.L., et al., 2004. Calcium, ATP, and ROS: a mitochondrial love-hate triangle. American journal of physiology. Cell Physiol. 287, C817–833.

Capasso, M., Bhamrah, M.K., Henley, T., et al., 2010. HVCN1 modulates BCR signal strength via regulation of BCR-dependent generation of reactive oxygen species. Nat. Immunol. 11, 265–272.

Capitani, N., Patrussi, L., Trentin, L., et al., 2012. S1P1 expression is controlled by the pro-oxidant activity of p66Shc and is impaired in B-CLL patients with unfavorable prognosis. Blood 120, 4391–4399.

Casola, S., 2007. Control of peripheral B-cell development. Curr. Opin. Immunol. 19, 143–149.

Cenci, S., and Sitia, R., 2007. Managing and exploiting stress in the antibody factory. FEBS Lett. 581, 3652–3657.

Cenci, S., Mezghrani, A., Cascio, P., et al., 2006. Progressively impaired proteasomal capacity during terminal plasma cell differentiation. EMBO J. 25, 1104–1113.

Cenci, S., van Anken, E., and Sitia, R., 2011. Proteostenosis and plasma cell pathophysiology. Curr. Opin. Cell Biol. 23, 216–222.

Civallero, M., Cosenza, M., Pozzi, S., et al., 2015. Activity of BKM120 and BEZ235 against Lymphoma Cells. BioMed Res. Int. 2015, 870918.

Conrad, M., and Sato, H., 2012. The oxidative stress-inducible cystine/glutamate antiporter, system x(c)-cystine supplier and beyond. Amino Acids 42, 231–246.

Crotzer, V.L., Matute, J.D., Arias, A.A., et al., 2012. Cutting edge: NADPH oxidase modulates MHC class II antigen presentation by B cells. J. Immunol. 189, 3800–3804.

Crump, K.E., Juneau, D.G., Poole, L.B., et al., 2012. The reversible formation of cysteine sulfenic acid promotes B-cell activation and proliferation. Eur. J. Immunol. 42, 2152–2164.

Dai, L., Noverr, M.C., Parsons, C., et al., 2015. xCT, not just an amino-acid transporter: a multi-functional regulator of microbial infection and associated diseases. Front. Microbiol. 6, 120.

Dickinson, B.C., and Chang, C.J., 2011. Chemistry and biology of reactive oxygen species in signaling or stress responses. Nat. Chem. Biol. 7, 504–511.

Finkel, T., 2000. Redox-dependent signal transduction. FEBS Lett. 476, 52–54.

Fisher, A.B., 2009. Redox signaling across cell membranes. Antioxid. Redox Signal. 11, 1349–1356.

Forman, H.J., Fukuto, J.M., and Torres, M., 2004. Redox signaling: thiol chemistry defines which reactive oxygen and nitrogen species can act as second messengers. Am. J. Physiol. Cell Physiol. 287, C246–256.

Garcia-Manteiga, J.M., Mari, S., Godejohann, M., et al., 2011. Metabolomics of B to plasma cell differentiation. J. Proteome Res. 10, 4165–4176.

Gass, J.N., Jiang, H.Y., Wek, R.C., et al., 2008. The unfolded protein response of B-lymphocytes: PERK-independent development of antibody-secreting cells. Mol. Immunol. 45, 1035–1043.

Giorgio, M., Migliaccio, E., Orsini, F., et al., 2005. Electron transfer between cytochrome c and p66Shc generates reactive oxygen species that trigger mitochondrial apoptosis. Cell 122, 221–233.

Gorlach, A., Bertram, K., Hudecova, S., et al., 2015. Calcium and ROS: a mutual interplay. Redox Biol. 6, 260–271.

Gout, P.W., Buckley, A.R., Simms, C.R., et al., 2001. Sulfasalazine, a potent suppressor of lymphoma growth by inhibition of the x(c)- cystine transporter: a new action for an old drug. Leukemia 15, 1633–1640.

Griffiths, H.R., 2008. Is the generation of neo-antigenic determinants by free radicals central to the development of autoimmune rheumatoid disease? Autoimmun. Rev. 7, 544–549.

Gunn, M.D., Kyuwa, S., Tam, C., et al., 1999. Mice lacking expression of secondary lymphoid organ chemokine have defects in lymphocyte homing and dendritic cell localization. J. Exp. Med. 189, 451–460.

Hara-Chikuma, M., Chikuma, S., Sugiyama, Y., et al., 2012. Chemokine-dependent T cell migration requires aquaporin-3-mediated hydrogen peroxide uptake. J. Exp. Med. 209, 1743–1752.

Hempel, N., and Trebak, M., 2017. Crosstalk between calcium and reactive oxygen species signaling in cancer. Cell Calcium 63, 70–96.

Herrmann, J.M., and Dick, T.P., 2012. Redox biology on the rise. Biol. Chem. 393, 999–1004.

Hou, J., Cui, A., Song, P., et al., 2015. Reactive oxygen species-mediated activation of the Src-epidermal growth factor receptor-Akt signaling cascade prevents bortezomib-induced apoptosis in hepatocellular carcinoma cells. Mol. Med. Rep. 11, 712–718.

Hourihan, J.M., Moronetti Mazzeo, L.E., Fernandez-Cardenas, L.P., et al., 2016. Cysteine sulfenylation directs IRE-1 to activate the SKN-1/Nrf2 antioxidant response. Mol. Cell 63, 553–566.

Inanami, O., Johnson, J.L., McAdara, J.K., et al., 1998. Activation of the leukocyte NADPH oxidase by phorbol ester requires the phosphorylation of p47PHOX on serine 303 or 304. J. Biol. Chem. 273, 9539–9543.

Iwakoshi, N.N., Lee, A.H., and Glimcher, L.H., 2003. The X-box binding protein-1 transcription factor is required for plasma cell differentiation and the unfolded protein response. Immunol. Rev. 194, 29–38.

Jang, K.J., Mano, H., Aoki, K., et al., 2015. Mitochondrial function provides instructive signals for activation-induced B-cell fates. Nat. Commun. 6, 6750.

Kovacs, I., Horvath, M., Lanyi, A., et al., 2015. Reactive oxygen species-mediated bacterial killing by B lymphocytes. J. Leukocyte Biol. 97, 1133–1137.

Lee, R.L., Westendorf, J., and Gold, M.R., 2007. Differential role of reactive oxygen species in the activation of mitogen-activated protein kinases and Akt by key receptors on B-lymphocytes: CD40, the B cell antigen receptor, and CXCR4. J. Cell Commun. Signal. 1, 33–43.

Lo, M., Wang, Y.Z., and Gout, P.W., 2008. The x(c)- cystine/glutamate antiporter: a potential target for therapy of cancer and other diseases. J. Cell. Physiol. 215, 593–602.

Medrano-Fernandez, I., Bestetti, S., Bertolotti, M., et al., 2016. Stress regulates aquaporin-8 permeability to impact cell growth and survival. Antioxid. Redox Signal. 24, 1031–1044.

Mikkelsen, R.B., and Wardman, P., 2003. Biological chemistry of reactive oxygen and nitrogen and radiation-induced signal transduction mechanisms. Oncogene 22, 5734–5754.

Miller, E.W., Dickinson, B.C., and Chang, C.J., 2010. Aquaporin-3 mediates hydrogen peroxide uptake to regulate downstream intracellular signaling. Proc. Natl. Acad. Sci. U.S.A. 107, 15681–15686.

Milovanova, T., Chatterjee, S., Hawkins, B.J., et al., 2008. Caveolae are an essential component of the pathway for endothelial cell signaling associated with abrupt reduction of shear stress. Biochim. Biophys. Acta 1783, 1866–1875.

Murphy, K., and Weaver, C., 2016. Janeway's Immunobiology, ninth ed. New York, NY: Garland Science.

Nemoto, S., and Finkel, T., 2002. Redox regulation of forkhead proteins through a p66shc-dependent signaling pathway. Science 295, 2450–2452.

Nerini-Molteni, S., Ferrarini, M., Cozza, S., et al., 2008. Redox homeostasis modulates the sensitivity of myeloma cells to bortezomib. Br. J. Haematol. 141, 494–503.

Ogura, M., Inoue, T., Yamaki, J., et al., 2017. Mitochondrial reactive oxygen species suppress humoral immune response through reduction of CD19 expression in B cells in mice. Eur. J. Immunol. 47, 406–418.

Patrussi, L., Capitani, N., Cannizzaro, E., et al., 2014. Negative regulation of chemokine receptor signaling and B-cell chemotaxis by p66Shc. Cell Death Dis. 5, e1068.

Polikowsky, H.G., Wogsland, C.E., Diggins, K.E., et al., 2015. Cutting edge: redox signaling hypersensitivity distinguishes human germinal center B cells. J. Immunol. 195, 1364–1367.

Pollock, J.D., Williams, D.A., Gifford, M.A., et al., 1995. Mouse model of X-linked chronic granulomatous disease, an inherited defect in phagocyte superoxide production. Nat. Genet. 9, 202–209.

Qin, S., Stadtman, E.R., and Chock, P.B., 2000. Regulation of oxidative stress-induced calcium release by phosphatidylinositol 3-kinase and Bruton's tyrosine kinase in B cells. Proc. Natl. Acad. Sci. U.S.A. 97, 7118–7123.

Reth, M., 2002. Hydrogen peroxide as second messenger in lymphocyte activation. Nat. Immunol. 3, 1129–1134.

Rhee, S.G., 2006. Cell signaling. H_2O_2, a necessary evil for cell signaling. Science 312, 1882–1883.

Richards, S.M., and Clark, E.A., 2009. BCR-induced superoxide negatively regulates B-cell proliferation and T-cell-independent type 2 Ab responses. Eur. J. Immunol. 39, 3395–3403.

Ron, D., and Walter, P., 2007. Signal integration in the endoplasmic reticulum unfolded protein response. Nat. Rev. Mol. Cell Biol. 8, 519–529.

Ruprecht, C.R., and Lanzavecchia, A., 2006. Toll-like receptor stimulation as a third signal required for activation of human naive B cells. Eur. J. Immunol. 36, 810–816.

Salimi, A., Roudkenar, M.H., Sadeghi, L., et al., 2015. Ellagic acid, a polyphenolic compound, selectively induces ROS-mediated apoptosis in cancerous B-lymphocytes of CLL patients by directly targeting mitochondria. Redox Biol. 6, 461–471.

Schamel, W.W., and Reth, M., 2000. Monomeric and oligomeric complexes of the B cell antigen receptor. Immunity 13, 5–14.

Shapiro-Shelef, M., and Calame, K., 2005. Regulation of plasma-cell development. Nat. Rev. Immunol. 5, 230–242.

Singel, K.L., and Segal, B.H., 2016. NOX2-dependent regulation of inflammation. Clin. Sci. (Lond.) 130, 479–490.
Singh, D.K., Kumar, D., Siddiqui, Z., et al., 2005. The strength of receptor signaling is centrally controlled through a cooperative loop between Ca^{2+} and an oxidant signal. Cell 121, 281–293.
Sriburi, R., Bommiasamy, H., Buldak, G.L., et al., 2007. Coordinate regulation of phospholipid biosynthesis and secretory pathway gene expression in XBP-1(S)-induced endoplasmic reticulum biogenesis. J. Biol. Chem. 282, 7024–7034.
Stone, J.R., and Yang, S., 2006. Hydrogen peroxide: a signaling messenger. Antioxid. Redox Signal. 8, 243–270.
Tassi, S., Carta, S., Vene, R., et al., 2009. Pathogen-induced interleukin-1beta processing and secretion is regulated by a biphasic redox response. J. Immunol. 183, 1456–1462.
Tassi, S., Carta, S., Delfino, L., et al., 2010. Altered redox state of monocytes from cryopyrin-associated periodic syndromes causes accelerated IL-1beta secretion. Proc. Natl. Acad. Sci. U.S.A. 107, 9789–9794.
Tibullo, D., Barbagallo, I., Giallongo, C., et al., 2016. Heme oxygenase-1 nuclear translocation regulates bortezomibinduced cytotoxicity and mediates genomic instability in myeloma cells. Oncotarget 7, 28868–28880.
Tonks, N.K., 2005. Redox redux: revisiting PTPs and the control of cell signaling. Cell 121, 667–670.
van Anken, E., Romijn, E.P., Maggioni, C., et al., 2003. Sequential waves of functionally related proteins are expressed when B cells prepare for antibody secretion. Immunity 18, 243–253.
Vene, R., Delfino, L., Castellani, P., et al., 2010. Redox remodeling allows and controls B-cell activation and differentiation. Antioxid. Redox Signal. 13, 1145–1155.
Vene, R., Castellani, P., Delfino, L., et al., 2011. The cystine/cysteine cycle and GSH are independent and crucial antioxidant systems in malignant melanoma cells and represent druggable targets. Antioxid. Redox Signal. 15, 2439–2453.
Wheeler, M.L., and Defranco, A.L., 2012. Prolonged production of reactive oxygen species in response to B cell receptor stimulation promotes B cell activation and proliferation. J. Immunol. 189, 4405–4416.
Wu, J., and Kaufman, R.J., 2006. From acute ER stress to physiological roles of the Unfolded Protein Response. Cell Death Differ. 13, 374–384.
Yoo, S.K., Starnes, T.W., Deng, Q., et al., 2011. Lyn is a redox sensor that mediates leukocyte wound attraction in vivo. Nature 480, 109–112.

CHAPTER

14

Immunomodulatory Attributes of Aged Garlic Extract and Its Components

Yeldur P. Venkatesh

OUTLINE

Abstract

For more than five centuries, garlic (*Allium sativum* L.) has acquired a reputation as a prophylactic and therapeutic agent in folklore and traditional medicine. Besides possessing various pharmacological activities, garlic has been shown to be a biological response modifier. Several laboratory and animal studies have shown that garlic stimulates a variety of immune-responder cells (dendritic cells, lymphocytes, macrophages, and natural killer cells). In the past, the immunomodulatory properties of garlic have been ascribed to its organosulfur (OS) compounds, the principal of which is allicin in crushed garlic. Based on recent findings, the immunomodulatory properties are also contributed by other major constituents of raw garlic, viz., the

M.A. Hayat (ed): *Immunology, Volume 1.*
DOI: http://dx.doi.org/10.1016/B978-0-12-809819-6.00014-9

© 2018 Elsevier Inc. All rights reserved.

lectins or agglutinins and water-soluble fructans. Garlic has also shown promise as an antitumor agent against various tumors in animal and limited human studies. An odorless preparation of garlic, aged garlic extract (AGE), which shows as much potency as raw garlic, has been investigated intensively during the last three decades. In this review, the immunomodulatory attributes of garlic extract and its components are described first, followed by the immunomodulatory attributes of AGE and its components.

The principal transformed OS compounds in AGE are *S*-allyl cysteine and *S*-allyl methylcysteine; the effects of these and other OS compounds in producing immunomodulation are covered in detail. Although AGE contains fructans, it represents only a small fraction (0.22%) of the total amount of fructans (~14 g/100 g) in raw garlic. In comparison with raw garlic extract, AGE contains mainly garlic lectins (ASA I and ASA II). The yield of immunomodulatory proteins from raw garlic and AGE is 28.5 and 25 mg/100 g fresh garlic, respectively. Lectins and fructans from AGE display lymphoproliferative activity and activation of macrophages including phagocytosis. In conclusion, it appears that the complete immunomodulatory potential of garlic or AGE is contributed by OS compounds, immunomodulatory proteins (mainly lectins), fructans, and fructosyl-arginine.

INTRODUCTION

Nonspecific immunostimulation appears to represent a practical modality to counteract immunosuppressions or a dysfunctional immune system, which may result from overuse of chemotherapeutic medicines or prolonged use of immunosuppressive drugs. Nonspecific immunostimulation constitutes either an alternative to or an adjuvant for conventional chemotherapy and prophylaxis of infections, for tumor as well as autoimmune diseases, especially when the host's immune system is impaired (Wagner et al., 1999; Barbosa, 2006). Immunostimulants or immunopotentiators are drugs leading predominantly to a nonspecific stimulation of immunological defense mechanisms. Most of them are not real antigens but are antigenomimetics or so-called "mitogens." Some immunostimulants may also stimulate suppressor T cells and thereby reduce immune resistance, hence the term "immunomodulation or immunoregulation," denoting any effect on, or change of, immune responsiveness (Wagner et al., 1999).

The therapeutic success obtained with recombinant forms of the body's own immune regulators (cytokines) such as interferons (IFN), tumor necrosis factor (TNF), colony stimulating factors, and interleukins (IL) is very limited, hence the underlying necessity to search for new immunostimulants from natural sources. In recent years, thymic peptides, biological response modifiers, synthetic low molecular weight compounds (levamisole), chemically modified nucleotides, polysaccharides (lentinan, β-glucan, inulin), and certain plant extracts (mistletoe extract) are also being used for achieving immunostimulation in various immunocomprised states (Hart et al., 1998; Elluru et al., 2006).

In the search for plant constituents with immunostimulating potency, plants used in traditional medicine would be ideal choices. Plants described for their antibacterial, antiviral, antifungal, or antitumor activities make ideal candidates for screening of immunomodulatory properties. From ancient literature (Codex Ebers of 1550 BC), garlic (*Allium sativum* L.) is one such plant prescribed as a preventive as well as curative medicine for several ailments. Garlic is a perennial plant composed of a compound bulb consisting of individual bulbs, the latter generally known as cloves which are enclosed together in a white or light pink skin. *Allium* is the largest and most important representative genus of the Amaryllidaceae family and includes garlic, onion (*Allium cepa* L.), shallot

(*Allium ascalonicum*), leek (*Allium ampeloprasum* L.), and chive (*Allium schoenoprasum* L.); these vegetables are popular in cuisines worldwide and are valued for their potent medicinal properties. The term garlic originates from the Anglo-Saxon "gar-leac" or spear plant. The genus *Allium* is derived from the Celtic word *all*, meaning pungent or burning. Garlic and onion are the two species widely grown for culinary use (Block, 2010).

Garlic contains water (62–68%), carbohydrate (26–30%), protein (1.5–2.1%), amino acids (1–1.5%), organosulfur (OS) compounds (1.1–3.5%), and fiber (1.5%), all based on fresh weight (Koch and Lawson, 1996). Carbohydrates are the most abundant class of compounds present in garlic bulbs and account for ~77% of the dry weight. The majority of the carbohydrate material in garlic and other *Allium* species consist of water-soluble fructose polymers called fructans or fructosans (Block, 2010). It has been established that ~65% of the dry weight of garlic consists of fructans, a water-soluble dietary fiber; hence, fructans constitute ~84% of the carbohydrate content of garlic and have been structurally characterized (Losso and Nakai, 1997; Baumgartner et al., 2000; Tsukamoto et al., 2008).

Garlic ranks highly among foods that help to prevent disease, largely due to its high content of OS compounds and antioxidant activity as revealed from extensive studies from laboratory findings and animal models. However, fresh garlic is not suitable for everyone; it can cause gastric irritation and its pungent odor that lingers on the breath and skin is a social deterrent; these disagreeable effects of fresh garlic are due to allicin, an oxidant released upon cutting or crushing the clove (Koch and Lawson, 1996). The presence and potency of garlic OS compounds vary with respect to the mode of garlic preparation and extraction. Cooked or processed garlic products show different kinds of garlic OS compounds, some of which are highly unstable and instantly decomposed. These factors are probably responsible for the observed paradoxical results of garlic's effects. Trio et al. (2014) have summarized the biotransformation processes of garlic alliin (*S*-allyl-1-cysteine sulfoxide) into different garlic OS compounds as well as the garlic OS compounds' compositions from different garlic preparations and reviewed the chemopreventive functions and molecular mechanisms focusing on the antiinflammation, antioxidation, antidiabetes, and anticancer activity behind different garlic OS compounds. It is generally believed that garlic modulates cytokine secretion and that such modulation may provide a mechanism of action for many of their therapeutic effects (Arreola et al., 2015).

Fructans are resistant to hydrolysis by human digestive enzymes but can be fermented by the colonic microbiota to produce short chain fatty acids, metabolic by-products that possess immunomodulatory activity. The indirect role of fructans in stimulating probiotic growth is one of the mechanisms through which fructans exert their prebiotic activity and improve health. Recently, a more direct mechanism for fructan activity has been suggested; fructans may interact with immune cells in the intestinal lumen to modulate immune responses in the body (Di Bartolomeo and Van den Ende, 2015). Fructans are currently being investigated for their potential as "ROS (reactive oxygen species) scavengers" that benefit intestinal epithelial cells by improving their redox environment (Franco-Robles and López, 2015).

Certain lectins are known to have immunomodulatory activity; one of the most notable activities is mitogenicity, i.e., the triggering of quiescent, nondividing lymphocytes into a state of growth and proliferation. After the discovery of the mitogenic lectin, phytohemagglutinin from kidney bean (*Phaseolus vulgaris*) (Nowell, 1960), there have been many mitogenic lectins discovered from various other sources. The presence of lectins in garlic was first studied by Van Damme et al. (1991), although occurrence of lectin activity in garlic was initially reported by Nachbar and Oppenheim (1980). The proteins of garlic largely remain a less investigated area when compared to the vast amount of research work carried out on the OS compounds of garlic (Tapiero et al., 2004; Iciek et al., 2009; Kuttan and Thejass, 2009).

The majority of the protein component of garlic is constituted by the enzyme alliinase (cysteine sulfoxide lyase or alliin lyase; 110 kDa of two 55 kDa subunits) and the 25 kDa lectins, *A. sativum* agglutinin I and II (ASA_{25}; ASA I and ASA II) (Koch and Lawson, 1996; Van Damme et al., 2008). Alliinase is the enzyme that catalyzes the conversion of alliin to allicin (diallyl thiosulfinate), which is responsible for the typical mercaptan odor of garlic. Alliinase is a glycoprotein containing 5.5–6% neutral sugars, and forms a complex with concanavalin A (Con A) and garlic mannose-specific lectin, ASA I (Van Damme et al., 1992a; Rabinkov et al., 1994, 1995; Dam et al., 1998). In addition to alliinase, the presence of a high molecular weight glycoprotein, *A. sativum* agglutinin (110 kDa; ASA_{110} with two identical subunits of ~50 kDa) showing mitogenic activity and mannose-specificity has been reported (Gupta and Sandhu, 1996, 1997).

The two 25 kDa mannose-specific garlic bulb lectins, ASA I and ASA II, are the only proteins other than alliinase that have been well characterized. Garlic lectins have similar binding specificities (towards mannose sugars in $\alpha 1 \rightarrow 3$ and $\alpha 1 \rightarrow 6$ linkages) as snowdrop lectin (*Galanthus nivalis* agglutinin, GNA), the first lectin of the monocot mannose-binding lectin family to be isolated and characterized. Hence, garlic lectins belong to the GNA superfamily of lectins; however, garlic lectins have only two subunits, unlike GNA, which has four. Although the exact physiological role of the GNA-related lectins is not known, they are considered to be vegetative storage or defense proteins (Van Damme et al., 1998, 2008).

The cDNA clones for ASA I and ASA II have been characterized. While ASA I is a heterodimer composed of two different subunits of 11.5 and 12.5 kDa, ASA II is a homodimer of two identical 12 kDa subunits (Van Damme et al., 1991; Kaku et al., 1992). ASA I is translated as a precursor containing a signal peptide, and the two mature polypeptides joined by a linker region (Van Damme et al., 1992b). The initial lectin precursor is processed co- and posttranslationally to remove the signal peptide, the linker sequence, and a C-terminal extension. The coding sequence of ASA II is almost identical to that of the 12.5 kDa subunit of ASA I, with the addition of a signal peptide. The processing of the ASA II homodimer is thought to be similar to that of the Amaryllidaceae lectins, and only involves removal of the signal peptide and a C-terminal extension (Van Damme et al., 1992b).

IMMUNOMODULATION BY GARLIC

Several studies demonstrate a direct toxic effect of garlic to sarcoma and gastric, colon, bladder, and prostate cancer cells in tissue culture, but these effects cannot explain the inhibition of growth of transplanted cancer in animal models; the most likely explanation of this effect is immune stimulation (Lamm and Riggs, 2001). Garlic stimulates proliferation of lymphocytes and macrophage phagocytosis, induces the infiltration of macrophages and lymphocytes in transplanted tumors, induces splenic hypertrophy, stimulates release of IL-2, TNF-α and IFN-γ, enhances natural killer (NK) cell, killer cell and lymphokine-activated killer cell activity; all these activities represent effective stimulation of the immune response. Thus, garlic may be useful in preventing the suppression of immune response that is associated with increased risk of malignancy.

Charron et al. (2015) conducted a randomized crossover feeding trial in which 17 volunteers consumed a garlic-containing meal (100 g white bread, 15 g butter, and 5 g raw, crushed garlic) or a garlic-free control meal (100 g white bread and 15 g butter) after 10 days of consuming a controlled, garlic-free diet. They found that seven genes (including nuclear factor of activated T cells (NFAT) activating protein with immunoreceptor tyrosine-based activation motif 1 (NFAM1) and oncostatin M (OSM)) were upregulated by intake of a single garlic meal. OSM is a pleiotropic cytokine that inhibits several tumor cell lines in culture. The authors indicate that the bioactivity of garlic is multifaceted and includes activation of genes related to immunity, apoptosis, and xenobiotic metabolism in humans and Mono Mac 6 cells.

Epidemiologic studies, while limited in their abilities to assess *Allium* consumption, indicate some associations of *Allium* vegetable consumption with decreased risk of cancer, particularly cancers of the gastrointestinal tract. The majority of supportive evidence on *Allium* vegetables' cancer-preventive effects comes from mechanistic studies; these studies highlight potential mechanisms of individual OS compounds and of various preparations/extracts of these vegetables, including decreased bioactivation of carcinogens, antimicrobial activities, and redox modification. Nicastro et al. (2015) postulate that *Allium* vegetables and their components produce effects at each stage of carcinogenesis and affect many biologic processes that modify cancer risk.

Immunomodulatory Properties of Organosulfur (OS) Compounds

The immunomodulatory effects of diallyl sulfide (DAS), diallyl disulfide (DADS), and allyl methyl sulfide (AMS) were studied in BALB/c mice by Kuttan (2000). After treatment with five doses (20 mg/dose), the total white blood cell (WBC) count was enhanced significantly. DADS showed the maximum number of WBCs on the 6th day, and highest antibody titer of 516 on the 12th day. Administration of DAS and DADS enhanced the weight of spleen and thymus. DADS, AMS, and DAS administration enhanced the number of plaque-forming cells (PFC) in the spleen; maximum numbers of PFC were observed in DADS-treated animals. Bone-marrow cellularity also increased significantly in DADS- and AMS-treated animals. These results are suggestive of an immunostimulating effect of OS compounds.

The effects of DADS, an oil-soluble OS compound on the growth of estrogen receptor (ER)-positive (KPL-1 and MCF-7) and ER-negative (MDA-MB-231 and MKL-F) human breast cancer cell lines, were examined by Nakagawa et al. (2001). In MDA-MB-231 cells, DADS antagonized the effect of linoleic acid (LA), a potent breast cancer cell stimulator (at DADS = 1.8 μM and LA $\geq 6.5 \times 10^2$ μM concentration), and synergized the effect of eicosapentaenoic acid (EPA), a potent breast cancer cell suppressor (at DADS $> 3 \times 10^{-3}$ μM and EPA $> 6.3 \times 10^{-1}$ μM concentration). The results suggest that DADS could be a promising anticancer agent for hormone-dependent/independent breast cancers, and may harmonize with polyunsaturated fatty acids (known modulators of breast cancer cell growth).

Ajoene is one of the main compounds formed from heating crushed garlic, as a mixture of E- and Z-isomers (E- and Z-4,5,9-trithiadodeca-1,6,11-triene 9-oxide). Ajoene possesses a broad spectrum of biological activities including anticancer activity. Structure-activity studies on ajoene and ajoene analogues have revealed that the Z-isomer is moderately more active than the E-isomer. Its cytotoxicity towards cancer cells is postulated to occur via an apoptotic mechanism involving activation of the mitochondrial-dependent caspase cascade (Kaschula et al., 2010). Upon administration of Ajoene to mice for 4 weeks, Washiya et al. (2013) observed increased fecal IgA and intestinal IgA levels.

Foroutan-Rad et al. (2017) reviewed the effectiveness, toxicities, and mechanisms of pharmaceutical actions of different garlic extracts and OS compounds isolated from garlic against *Leishmania* spp. in a variety of in vitro, in vivo, and clinical trials reports, and concluded that garlic with immunomodulatory effects and apoptosis induction contributes to the treatment of leishmaniasis, which is caused by an obligate intracellular protozoa belonging to *Leishmania* genus.

Immunomodulation by Proteins in Garlic

The mitogenic effect of whole garlic extracts and ASA_{110} has been studied on T-lymphocytes (Colic and Savic, 2000; Colic et al., 2002). Apart from the characterization of ASA_{110} (Gupta and Sandhu, 1996, 1997), the immunomodulatory proteins (ImPs) of garlic have only recently been defined. An immunomodulatory glycoprotein of ~14 kDa purified from raw garlic was found to augment a delayed-type hypersensitivity (DTH) response (Ghazanfari et al., 2002) and NK-cell activity in BALB/c mice when administered intraperitoneally. Administration of fresh garlic extract (20 mg/kg; i.p.) produced an optimum augmentation of NK activity in BALB/c mice (Hassan et al., 2003). Further, they showed that a glycoprotein of ~14 kDa isolated from garlic extract could induce NK augmentation against K562 tumor cell line. In vivo studies also confirmed that this protein could induce a resistance to the growth of spontaneous mammary carcinoma in BALB/c mice (Hassan et al., 2003).

The protein fraction of garlic was found to augment parasite (*Leishmania major*) engulfment and destruction of intracellular parasites by peritoneal macrophages in vivo (Ghazanfari et al., 2006). In another study, a 10–14 kDa protein fraction of garlic was identified as responsible for induction of nitric oxide (NO) production, thus promoting the activation of reduced macrophage activity (Gamboa-León et al., 2007). In all these studies, the active components of the protein fraction(s) of garlic possessing immunomodulatory properties have not been clearly defined or identified.

Recently, ImPs from garlic, displaying lymphoproliferation and macrophage activation in vitro, have been isolated from raw garlic extract by a two-step process (ultrafiltration and anion-exchange chromatography). The flowchart for the preparation of ImPs from raw garlic is shown in Fig. 14.1. The major ImPs have been identified as corresponding to the abundant garlic lectins or agglutinins, viz., ASA I and ASA II occurring in a ratio of 4:1 by weight in raw garlic (Clement et al., 2010). The stability of the garlic ImPs in highly acidic or alkaline conditions, or at higher temperatures, as observed by Clement and Venkatesh (2010), reinforces the immunomodulatory potential of garlic as a therapeutic food even in cooked or processed forms. Phagocytosis of yeast cells by macrophages in the presence of the garlic

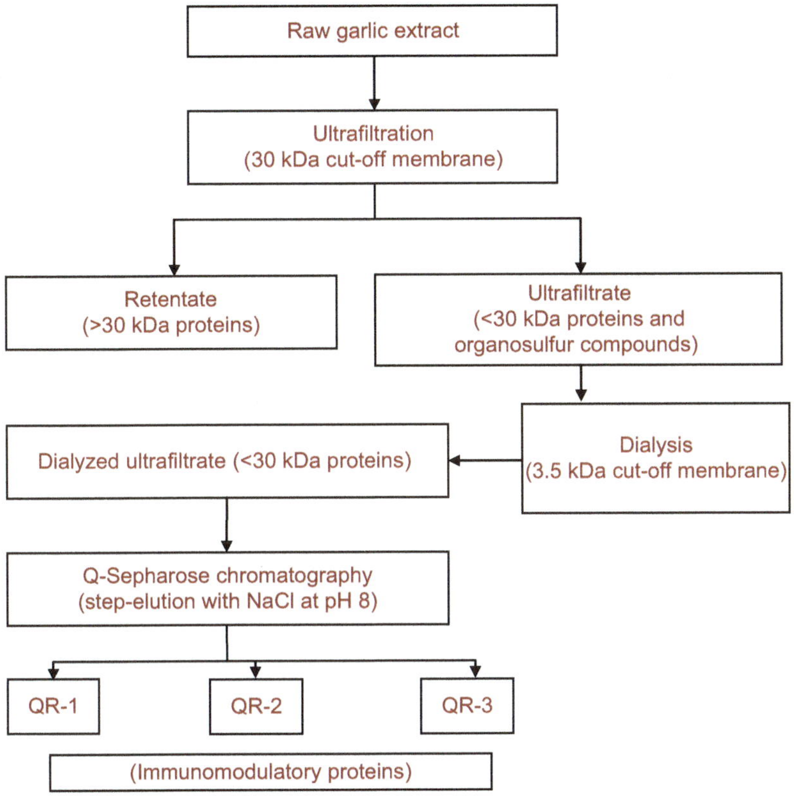

FIGURE 14.1 Flowchart for the purification of immunomodulatory proteins (ImPs) from raw garlic. QR-2 and QR-1 represent ASA I and ASA II, respectively, as characterized by Clement et al. (2010). QR-3 which lacks one subunit of ASA I is similar to ASA I in immunomodulatory properties, but is devoid of hemagglutination activity. The amounts of proteins isolated are 5.7 (QR-1), 22.0 (QR-2) and 0.8 (QR-3) mg for 100 g raw garlic. Total amount of ImPs from raw garlic is 28.5 mg/100 g raw garlic (fresh weight basis).

lectin ASA I is enhanced (Fig. 14.2). Both ASA I and ASA II have the property of intrinsic immunogenicity in BALB/c mice upon intradermal, intranasal (mucosal), and oral administrations; however, ASA I is more potent than ASA II. Although raw garlic extract (10 µg/mL) showed a significant increase in ROS generation, garlic lectins failed to induce ROS from rat peritoneal macrophages (Pramod et al., 2015). Garlic lectins are mitogenic and comitogenic for human peripheral blood lymphocytes.

Natural antibodies to avidin, peanut agglutinin, wheat germ agglutinin, soybean agglutinin, and garlic lectins have been reported in normal human sera and found to be polyreactive (Bubb et al., 1993; Tchernychev et al., 1995; Tchernychev and Wilchek, 1996). Many polyreactive antibodies have been shown to possess broad antibacterial and antiviral activities (Baumgarth et al., 2005; Zhou et al., 2007a,b). All these studies highlight the usefulness of garlic ImPs for stimulation of the immune system by promoting natural antibodies in immunocompromised and/or immune dysfunction clinical situations (Venkatesh and

Prasanna Kumar, 2013). Further, they also emphasize the utility of garlic ImPs as potential adjuvants in vaccine preparations or for boosting the natural immune system by way of processed foods containing such ImPs. The ImPs of garlic could represent one of the critical components responsible for an array of beneficiary immunomodulatory responses described for garlic till date.

In an animal model, Ebrahimi et al. (2013) showed that intratumor injection of the protein fraction purified from fresh garlic extract provoked a significant increase and infiltration of $CD8^+$ subpopulation of T-lymphocytes into the breast tumor site, as well as a decrease in tumor size in BALB/c mice. The protein fraction increased peripheral blood $CD8^+$ T-lymphocytes in treated animals. Further, the protein fraction augmented the DTH response compared to the control group. However, it should be noted that the protein fraction was composed of three single polypeptides, with molecular masses of ~10–13 kDa and different isoelectric points; these proteins have not been characterized, and their identities are not known at present.

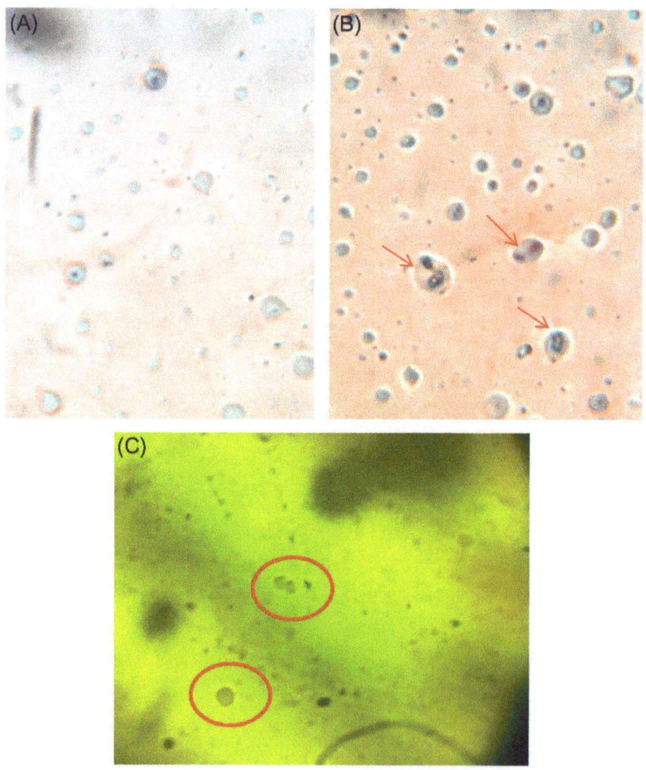

FIGURE 14.2 Phagocytosis of yeast cells by LPS-stimulated rat peritoneal macrophages in the presence of garlic lectins (magnification: 3600×). Panel (A): without garlic lectin, ASA I; panels (B) and (C): with 2 μg of garlic lectin, ASA I. *Arrows* point to Giemsa stained rat peritoneal macrophages containing engulfed yeast cells. Panel (A) with control cells shows fewer macrophages with engulfed yeast cells when compared to cells incubated with ASA I for 90 min (panel B). Panel (C) shows macrophages (shown as circled) adhering to yeast cells.

Onion, belonging to the same genus as garlic, also contains a monocot mannose-specific lectin (Van Damme et al., 1991, 2008). Recently, onion lectin (*A. cepa* agglutinin; ACA) has been isolated from onion extract in a yield of ~1 mg/kg (Prasanna and Venkatesh, 2015). ACA is nonglycosylated and a noncovalent tetramer of 50 kDa (subunit molecular mass is ~12 kDa). ACA (0.1 µg/well) enhanced the proliferation of murine thymocytes at 24 h; however, it does not proliferate B cell-enriched rat splenocytes. Overall, purified ACA induces a Th1-type immune response in vitro. Though present in low amounts, ACA may contribute to the immune-boosting potential of the popular spice onion since considerable amounts are consumed on a daily basis universally.

Immunomodulation by Fructans in Garlic

Recently, fructans have been isolated from the flow-through pool of Q-Sepharose chromatography of raw garlic extract; the yield of raw garlic fructans (RGFs) is 13.8 g/100 g (fresh weight) garlic (Chandrashekara and Venkatesh, 2016). RGF stimulated murine splenocytes to a similar extent (four- to fivefold vs untreated cells) as seen in the case of established polysaccharide immunostimulators such as galactan, arabinogalactan, mannan, and zymosan. Further, RGF activated macrophages present in rat peritoneal exudates to release NO, which was significant at both 24 and 48 h. Yeast cell phagocytosis by activated macrophages was also significantly enhanced by RGF. Since fructans are present in abundance in raw garlic (14–17% by fresh weight), it appears to have potential in the development of functional foods for immunomodulation, by inclusion in foods during processing.

Fructo-oligosaccharides (FOS) was isolated from onion, belonging to the same genus as garlic, by hot 80% ethanol extraction in a yield of ~4.5 g/100 g fresh weight (Kumar et al., 2015). Structural analyses of fractionated FOS showed that onion FOS ranged from mainly trisaccharides to hexasaccharides. Onion FOS (50 µg/mL) significantly increased the proliferation of mouse splenocytes/thymocytes, the production of NO by peritoneal exudates cells (PECs) from Wistar rats, and augmentation of intracellular free radicals production and phagocytic activity of isolated murine PECs. Onion FOS belongs to the group of inulin-type fructans, and possesses immunostimulatory activities in vitro (Kumar et al., 2015).

From the foregoing account on immunomodulation by garlic, it is clear that several components in garlic appear to work synergistically to produce various effects, but, because of garlic's chemical complexity, processing methods yield preparations with differing efficacy and safety. The efficacy and safety of these preparations in preparing dietary supplements based on garlic are also contingent on the processing methods employed. Although there are many garlic supplements commercially available, they fall into one of four categories, namely: (1) dehydrated garlic powder, (2) garlic oil, (3) garlic oil macerate, and (4) aged garlic extract (AGE). Highly unstable thiosulfinates, such as allicin, disappear during processing and are quickly transformed into a variety of OS components. Garlic and garlic supplements are consumed in many cultures for their hypolipidemic, antiplatelet, and pro-circulatory effects. In addition to these beneficial effects, some garlic preparations also appear to possess hepatoprotective, immune-enhancing, anticancer, and chemopreventive activities (Amagase et al., 2001). It is generally believed that garlic prevents cold and flu symptoms through immune enhancement.

Saponins in Garlic and Other *Allium* Species

The steroidal saponins are important biologically active molecules that are considered to be responsible for the observed bioactivity of *Allium* species, including antifungal, cytotoxic, enzyme-inhibitory, and others. Furthermore, nonsulfur compounds like saponins may contribute to the essential biological activities of garlic. Recently, Sobolewska et al. (2016) have reviewed the literature data concerning chemistry and biological activity of several steroidal saponins from the *Allium* genus. However, their immunomodulatory activities have not yet been investigated, with the exception of a new steroidal saponin isolated from the bulbs of *A. ampeloprasum* L. var. porrum (Adão et al., 2012). Results of this study indicated that it exhibited hemolytic activity in the in vitro assays, and immunological adjuvant activity on the cellular immune response against ovalbumin (OVA) antigen.

The immunomodulatory attributes of the commonly used dietary garlic supplement AGE and its several isolated components are addressed henceforth in this review.

AGED GARLIC EXTRACT

An alternate source of garlic that is odorless and richer in antioxidants than fresh garlic bulbs is AGE, one of the best known garlic supplements. AGE is a concentrated form of processed garlic that has been shown in a number of scientific studies to be safe and effective in providing health benefits. The additional biological effects such as hepatoprotective, immune-enhancing, anticancer, and chemopreventive activities attributed to AGE may be due to compounds such as *S*-allylcysteine (SAC), *S*-allylmercaptocysteine (SAMC), *N*(α)-fructosyl arginine (FruArg), saponins, and others, formed during the long-term extraction process. It is not necessary for garlic preparations to contain odorous compounds like thiosulfinates (e.g., allicin) to be effective; they decompose and disappear during any processing. Although allicin-like transient components are not directly active, available data suggests that an allicin-free garlic preparation that is standardized with a bioavailable component such as SAC is active and various effects of garlic may be attributed to it (Amagase, 2006).

AGE has manifold biological activities including immunomodulative and antioxidative effects. It is used as a major component of nonprescription tonics and cold-prevention medicines or dietary supplements. AGE is marketed under the trade name Kyolic by *Wakunaga Company* (Japan). AGE is, as the name suggests, produced by aging of garlic; fresh garlic is sliced, macerated, and kept in 15–25% ethanol at ambient temperature for 18–20 months (Lawson, 1993). The incubation medium or extract is then filtered and concentrated to dryness and is stored at −20°C for further studies. Chemical changes occur when garlic extracts are aged for long time (Lawson and Wang, 1995). Compounds like allicin, γ-glutamyl-*S*-1-propenylcysteine, are found to decrease during the course of ageing. AGE is richer in antioxidants than other commercial garlic preparations and fresh garlic (Wang et al., 2015), and it also boosts cellular antioxidants, including glutathione that helps in maintaining a healthy immune system and prevents drug toxicity, and peroxidases that eliminate toxic peroxides. The OS compounds, SAC and SAMC, which are produced only during the ageing process, are responsible for the antioxidant activity of AGE.

Kyo et al. (1997) examined the effect of AGE on the function of mast cells and activated T-lymphocytes adopting the in vitro histamine release system, the in vivo IgE-mediated skin reaction system and the in vivo late phase reaction system. The observed results suggest that AGE application could modify, directly or indirectly, the function of mast cells, basophils, and activated T-lymphocytes which play a leading role in allergic cascade reactions including inflammation.

In a study on the effect of AGE on NO production (measured as NO metabolites nitrite and nitrate) in mice plasma, AGE (2.86 g/kg, p.o.) temporarily increased NO production by 30–40% from 15 to 60 min after administration (Morihara et al., 2002). These authors conclude that AGE increased NO production by activating constitutive NO synthase, but not inducible NO synthase and may be a useful supplement for the prevention of cardiovascular disease.

AGE, extracted for more than 10 months, is less irritating and less toxic (Lawson and Wang, 1995). Several experimental studies have demonstrated that AGE possesses antioxidative, antistress, immunomodulating, cardiovascular, and hepatoprotective properties (Kasuga et al., 2001). Kyo et al. (1998) found that AGE could be a significant immune modifier, which maintains the homeostasis of immune functions, and could exhibit antitumor activities through immune modulation. Consequently, AGE stimulated the proliferation of mouse spleen cells and the release of cytokines (IL-2, TNF-α and IFN-γ), increased NK activities, and enhanced phagocytosis by peritoneal macrophages. AGE treatment also stimulated the reactivity of lymphocytes in response to cytokines or mitogens.

In the psychological stress model, AGE significantly prevented the decrease in spleen weight and restored the reduction of anti-sheep red blood cell (SRBC) hemolytic PFCs caused by the electrical stress; these results indicated that psychological stress qualitatively and quantitatively impairs immune function, and that AGE is extremely useful for preventing psychologically induced damage (Kyo et al., 1999). In the IgE-mediated allergic mouse model, AGE significantly decreased the antigen-specific ear swelling induced by picryl chloride application to the ear and i.v. administration of antitrinitrophenyl antibody (Kyo et al., 2001). In the transplanted carcinoma cell model, AGE significantly inhibited the growth of Sarcoma-180 (allogenic) and LL/2 lung carcinoma (syngenic) cells transplanted into mice. Concomitantly, increases in NK and killer activities of spleen cells were observed in Sarcoma-180 bearing mice administered AGE.

In order to confirm the effects of AGE on colorectal adenomas (n = 51), Tanaka et al. (2004) conducted a double-blind randomized study using high-AGE (AGE 2.4 mL/day) and low-AGE (AGE 0.16 mL/day) dose groups. The number and size of adenomas before intake (0 month) and 6 and 12 months after intake indicated a potential suppressive effect on colorectal adenomas. The results suggest the possibility of preventive and therapeutic effects of AGE on colorectal adenomas, though it is necessary to investigate these in larger-scale and longer-term trials (Tanaka et al., 2004, 2006).

In a randomized double-blind trial of patients (n = 50) with inoperable colorectal, liver, or pancreatic cancer, AGE was administered to one group and a placebo was administered to another for 6 months. Although no difference was observed in the quality-of-life, both the number of NK cells and the NK-cell activity increased significantly in the AGE group (Ishikawa et al., 2006); this particular study showed that administering AGE to patients with advanced cancer of the digestive system improved NK-cell activity. AGE suppressed the proliferation of three colorectal cancer cell lines (HT29, SW480, and SW620) in the same way, but the invasive activities of only SW480 and SW620 cells were inhibited by AGE (Matsuura et al., 2006). These results indicated that AGE could prevent tumor formation by inhibiting angiogenesis through the suppression of endothelial cell motility, proliferation, and tube formation. Matsuura et al. (2006) suggest that AGE would be a good chemopreventive agent for colorectal cancer because of its antiproliferative action on colorectal carcinoma cells and inhibitory activity on angiogenesis.

The chemopreventive effect of AGE on colon carcinogenesis and cell proliferation in 1,2-dimethylhydrazine (DMH)-induced colon neoplastic rats has been studied (Katsuki et al., 2006). Rats were given weekly injections of DMH (20 mg/kg, s.c.) for 20 weeks, and fed either a basal diet or diet containing 4% AGE. Serum from AGE-treated rats contained detectable SAC. The AGE diet significantly reduced the number of colon tumors and aberrant crypt foci (ACF) compared to the basal diet; these findings suggest that AGE has a chemopreventive effect on colon carcinogenesis through suppression of cell proliferation.

The effect of i.p. administration of AGE on an established allergic airway inflammation in BALB/c mice was investigated (Zare et al., 2008). The results indicated that AGE caused a significant decrease in the hallmark criteria of allergic airway inflammation levels which included eosinophil percentage in lavage, peribronchial lung eosinophils, IgG1 level in lavage and serum, grade of mucous-producing goblet cells, and peribronchial and perivascular inflammation. Zare et al. (2008) conclude that AGE has the potential for attenuation of inflammatory features of allergic airway inflammation in a murine model.

Expression of CD36 scavenger receptors on macrophages is involved in oxidized low-density lipoprotein uptake and foam cell formation during atherosclerotic lesion development. Morihara et al. (2010) examined the effects of AGE, which increases cellular total thiols and glutathione concentrations, on CD36 expression in human monocytes or macrophages (THP-1 cells and primary human monocytes). Their data indicate that AGE inhibits CD36 expression by modulating the peroxisome proliferator activator receptor gamma (PPARgamma) pathway in human macrophages and monocytes differentiation into macrophages.

In Kunming mice bearing murine foregastric carcinoma, it was shown that upon administration of aged black garlic extract (ABGE) intraperitoneally for 2 weeks, significant antitumor effects of ABGE were observed, such as growth inhibition of inoculated tumors (Wang et al., 2012). Further investigation of serum superoxide dismutases, glutathione peroxidase, IL-2 and the increased indices of spleen and thymus indicated that the anticancer action of ABGE may be due to its antioxidant and immunomodulative effects.

In recent studies it has been shown that administration of AGE (100 mg/kg, i.p.) resulted in improved immune responses against subcutaneously implanted WEHI-164 fibrosarcoma tumors in BALB/c mice (Fallah-Rostami et al., 2013; Tabari and Ebrahimpour, 2014). Mice which received AGE have significantly longer survival time compared to control mice. The inhibitory effect on tumor growth was seen in AGE-treated mice. The $CD4^+/CD8^+$ ratio and in vitro IFN-γ production of splenocytes were significantly increased in AGE group. WEHI-164 specific cytotoxicity of splenocytes from AGE mice was also significantly increased.

Naltrexone (NTX), an opioid receptor antagonist, has immunomodulatory and antitumor effects. Recently, it was found that AGE (100 mg/kg, i.p.) showed synergistic effects with NTX (0.5 mg/kg, i.p.) on inhibition of WEHI-164 fibrosarcoma tumor growth and increment of survival times (Ebrahimpour et al., 2013). Mice who received AGE + NTX had significantly longer survival times compared with the mice treated with AGE or NTX alone; an enhanced inhibitory effect on tumor growth was seen in the combination therapy group. The $CD4^+/CD8^+$ ratio and in vitro IFN-γ production of splenocytes were significantly increased in AGE + NTX and NTX groups. WEHI-164 specific cytotoxicity of splenocytes was also significantly increased in AGE + NTX treated mice.

In the group of F344 rats with DMH-induced carcinogenesis receiving a basal diet containing 3% w/w AGE, there was a decrease in the number of ACF without any changes in gross tumor pathology (Jikihara et al., 2015). AGE showed a lower number of adenoma and adenocarcinoma lesions by histological analysis; immunohistochemical staining indicated that AGE suppressed the proliferative activity in adenoma and adenocarcinoma lesions, but showed no effect on normal colon mucosa. Further, the authors demonstrated that AGE delayed cell cycle progression by down regulating cyclin B1 and cdk1 expression via inactivation of NF-κB in human colorectal cancer cells, but did not induce apoptosis.

Innate immune cells are responsible for the inflammation necessary to kill pathogens; two innate lymphocytes—γδ-T and NK cells—appear to be susceptible to diet modification. Nantz et al. (2012) made a study of the effect of AGE on cell proliferation and activation and inflammation, and whether those changes affected the occurrence and severity of colds and flu. After 45 days of AGE consumption (2.56 g AGE/day) or placebo supplements, γδ-T and NK cells proliferated better and were more activated than cells from the placebo group in healthy human participants ($n = 120$; 21–50 years) during the cold and flu season. After 90 days, although the number of illnesses was not significantly different, the AGE group showed reduced cold and flu severity, with a reduction in the number of: (1) symptoms, (2) days participants functioned suboptimally, and (3) work/school days missed. These results suggested that AGE supplementation may enhance immune cell function and may be partly responsible for the reduced severity of colds and flu reported, perhaps with less accompanying inflammation (Nantz et al., 2012; Percival, 2016).

Immunomodulatory Attributes of OS Compounds in Aged Garlic Extract (AGE)

Generally AGE is prepared by storing sliced garlic in 15–20% aqueous ethanol for 20 months, followed by filtration and concentration (Hirao et al., 1987). The final liquid extract has been reported to contain 10% (w/v) ethanol and small amounts of several water-soluble OS compounds (Lawson, 1993) which are listed in Table 14.1. Recently, a rapid postcolumn HPLC method using hexaiodoplatinate reagent for qualitative and quantitative analyses of OS compounds in AGE was developed (Matsutomo and Kodera, 2016). The method has several advantages: less interference from nonsulfur compounds, high sensitivity, good correlation coefficients, and high resolution that can separate >20 OS compounds, including several isomers, in garlic preparations in a single run. Identification of *cis*-*S*-1-propenylcysteine (*cis*-S1PC) and γ-glutamyl-*S*-allyl-mercaptocysteine was possible in AGE.

Lee et al. (1994) showed that thioallyl compounds, natural constituents of garlic and/or AGE and known to inhibit malignant cells, can also reduce the proliferation of normal cells. Garlic OS compounds inhibit carcinogen activation, boost phase 2 detoxifying processes, cause cell cycle arrest in G2/M phase, stimulate the mitochondrial apoptotic pathway, and

TABLE 14.1 Content of Organosulfur (OS) Compounds in Aged Garlic Extract (Lawson, 1993)

Organosulfur Compound	Amount (mg/g)
S-Allyl cysteine (SAC)	0.30
S-Propenyl cysteine (SPC or S1PC)	0.15
S-Methyl cysteine (SMC)	0.11
γ-Glutamyl-*S*-allylcysteine	0.28
γ-Glutamyl-*S*-1-propenylcysteine	0.17
S-Allyl mercaptocysteine (SAMC)	0.04
Cysteine	0.01
Alliin	<0.02

increase acetylation of histones (Iciek et al., 2009). Further, these researchers have presented other little known aspects of molecular action of OS compounds, like modulation of cellular redox state, involvement in signal transduction, and posttranslational modification of proteins by sulfane sulfur or by formation of mixed disulfides (*S*-thiolation reactions).

The ability of OS compounds to hinder cancer cell proliferation and viability closely correlates with the length of the sulfur chain. Available data support a mechanism of mitotic arrest of cancer cells due to the alteration of the microtubule network, probably as a consequence of the high reactivity of sulfur atoms against the thiol groups of different cellular macromolecules controlling crucial regulatory functions (Cerella et al., 2011); these findings indicate a promising potential for the use of garlic OS compounds in chemoprevention and chemotherapy.

ATP-binding cassette transporter A1 (ABCA1) is a key mediator of cholesterol efflux to apoA-I in lipid-loaded macrophages, which is the first step of reverse cholesterol transport in vivo and a critical step in preventing atherosclerosis; enhanced ABCA1 expression may inhibit foam cell formation and consequently reduce atherogenic risk. In one study, it was shown that SAC can increase ABCA1 expression in human THP-1 macrophages and may be beneficial in promoting reverse cholesterol efflux (Malekpour-Dehkordi et al., 2013).

Schäfer and Kaschula (2014) have recently proposed a link between the immunomodulating activity of garlic OS compounds and the prevention of cancer. They hypothesize that garlic elicits antiinflammatory and antioxidative responses that aid in priming the organism towards eradication of an emerging tumor. In a recent study, Yoo et al. (2014) investigated the effect of aged black garlic (ABG) on the inhibition of IgE-mediated allergic response in RBL-2H3 cells and in vivo passive cutaneous anaphylaxis. These authors conclude that ABG suppresses the allergic response, and the mechanism for its antiallergic action may involve suppressions of Syk, cytosolic phospholipase A_2 (cPLA2), 5-lipoxygenase (5-LO), and cyclooxygenase-2 (COX-2). The antiallergic actions of ABG, ethyl acetate extract or an active fraction of ABG suggest that they may be useful as functional foods for allergic diseases (Yoo et al., 2014).

In a recent study, the effect of SAC and S1PC on intestinal IgA production was investigated (Suzuki et al., 2016). S1PC enhanced IgA production in mouse splenic lymphocytes in

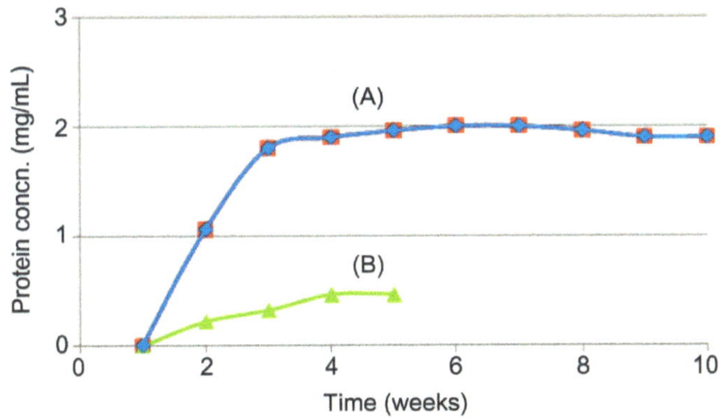

FIGURE 14.3 Protein content of supernatant during ageing or reextraction of garlic in 25% ethanol at 25°C. Protein was quantified by Bradford assay. Curve (A): Quantitation of protein during ageing of garlic in 25% ethanol at 25°C; *x*-axis is shown in weeks. Curve (B): Quantitation of protein in reextracted residual garlic; *x*-axis should be noted as months instead of weeks.

culture; however, SAC was ineffective. In addition, oral administration of S1PC to mice for 5 days increased the IgA level in the intestinal lavage fluids and number of IgA-producing cells in Peyer's patches. Furthermore, S1PC induced the expression of X-box binding protein 1 (Xbp1) mRNA, an inducer of plasma cell differentiation, in Peyer's patches. These results suggest that S1PC increases IgA-producing cells via the enhancement of Erk1/2-mediated Xbp1 expression in the intestine (Suzuki et al., 2016).

Immunomodulatory Properties of Proteins in AGE

Morioka et al. (1993) reported that the protein fraction of AGE enhanced NK-cell activity and cytotoxicity of macrophages towards tumor cells; proliferation of human lymphocytes mediated by IL-2 and con A was also enhanced.

An investigation of the changes occurring in the protein fraction during ageing of garlic shows that protein is gradually released into the supernatant up to a period of 4 weeks; thereafter, the protein concentration stabilizes at ~2 mg/mL (Fig. 14.3, curve A). Upon reextraction of the residual garlic following filtration and air-drying, it was found that protein is again released, albeit more slowly, and the protein concentration stabilizes at ~0.5 mg/mL at 5 months (Figure 14.3, curve B). Upon SDS-PAGE analysis of AGE at various times, it was found that only three major proteins in the 12–14 kDa regions were seen; typically, ~25 mg protein was obtained from AGE, starting with 100 g of raw garlic (Chandrashekar and Venkatesh, 2009). Three proteins were separated by Q-Sepharose chromatography at pH 8, viz., QA-1, QA-2, and QA-3, all of which displayed immunomodulatory and mannose-binding activities; however, QA-2 showed the highest mitogenic activity. The identity of QA-2 and QA-1 ImPs with the garlic lectins ASA I and ASA II, respectively, has been confirmed by specific hemagglutination activity (Fig. 14.4). Though QA-3 exhibits mitogenic activity, it is devoid of hemagglutination activity, probably due to the lack of one subunit of the homodimer (Chandrashekar and Venkatesh, 2009).

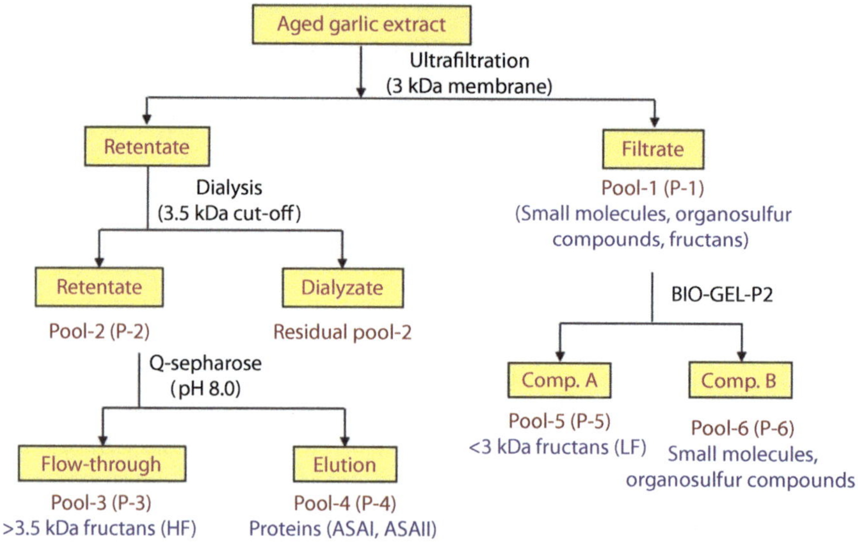

FIGURE 14.4 Flowchart for the isolation of low molecular weight fructans (<3 kDa; LF) and high molecular weight fructans (>3.5 kDa; HF) from aged garlic extract. ASA I and ASA II denote garlic lectins (*Allium sativum* agglutinins I and II) present in both aged garlic extract (Chandrashekar and Venkatesh, 2009) and raw garlic (Clement et al., 2010). *Source: Reproduced with permission from Chandrashekar, P.M., Prashanth, K.V., Venkatesh, Y.P., 2011. Isolation, structural elucidation and immunomodulatory activity of fructans from aged garlic extract. Phytochemistry 72, 255–264. © 2010 Elsevier Ltd.*

Ahmadabad et al. (2011) showed that the semipurified 14 kDa protein isolated from AGE increased the expression of CD40 molecule on dendritic cells (DCs) isolated from the spleen of BALB/c mice, but did not influence CD86 and MHC-II molecules. Furthermore, no significant differences were noticed in the pulsed-DCs with 14 kDa protein and nonpulsed DCs on the allogenic mixed lymphocyte reaction test (Ahmadabad et al., 2011). Daneshmandi et al. (2011) assessed the effect of protein fractions of garlic on peritoneal macrophages and found that the purified 14- and 47-kDa protein fractions do not show significant proliferation of stimulated macrophages. Both 14- and 47-kDa protein fractions significantly suppressed production of NO from macrophages. Cytotoxicity of macrophages' supernatant on WEHI-164 fibrosarcoma cells was not affected by garlic protein fractions. It appears that both 14- and 47-kDa fractions of AGE are able to suppress NO production from macrophages, show no cytotoxic effect on macrophages and do not increase the tumoricidal property of macrophages.

In another study, the effect of the purified 47 kDa protein extracted from AGE on the expression of DC surface markers was evaluated (Hasan et al., 2012). The 47 kDa protein-treated DCs lowered the expression of DC maturation markers including CD40, CD86, and MHC-II in comparison with nontreated DCs, but no statistical difference between the two groups was observed. Upon treatment of DCs with 47 kDa protein, DCs down regulated the expression of costimulatory and MHC-II surface molecules, which is similar to tolerogenic DC phenotype. Based on these results, Hasan et al. (2012) suggest that the 47 kDa protein can be utilized as a potential candidate to generate tolerogenic DCs in vitro.

Immunomodulation by Fructans in AGE

The amount of fructans in AGE represents a very small fraction (0.22%) of the total fructans in raw garlic. Both high molecular weight (>3.5 kDa; HF) and low molecular weight (<3 kDa; LF, oligosaccharide with 10 units) fructans have been isolated from AGE (Fig. 14.4); NMR structural analysis revealed that both have (2→1) β-D-fructofuranosyl bonds linked to a terminal glucose at the nonreducing end and β-D-fructofuranosyl branching on its backbone (Chandrashekar et al., 2011). Both HF and LF displayed mitogenic activity and activation of macrophages including phagocytosis; these in vitro activities were comparable to that of known polysaccharide immunostimulators such as zymosan and mannan. This was the first evidence to show that fructans derived from AGE possess immunomodulatory properties.

Garlic fructans isolated from AGE (termed aged garlic fructans or AGF) produce a significant humoral (serum IgG) response to OVA (30 μg; an experimental antigen) in BALB/c mice administered mucosally by either intranasal or oral route—a delayed response appearing on 50th day at a dose of 30 μg AGF by intranasal route. However, the serum IgG response was seen earlier on 35th day at a dose of 100 μg AGF by oral route. Higher concentrations of AGF (>50 μg) were inhibitory for adjuvant activity by intranasal administration. These observations indicate that AGF display immunoadjuvant activity for a test antigen though the humoral immune response is delayed (Chandrashekar and Venkatesh, 2012).

Immunomodulation Exerted by Fructosyl-arginine (FruArg) in AGE

The amino-carbonyl (Maillard) reaction of amino acids with sugars is a nonenzymatic browning reaction that takes place during the processing, cooking, and storage of foods. Maillard reaction products such as FruArg have been shown to possess interesting chemical and biological properties including antimutagenic and antioxidant activity (Ide et al., 1999). Zhou et al. (2014) demonstrated that both AGE and FruArg can significantly inhibit lipopolysaccharide (LPS)-induced NO production in LPS-activated murine BV-2 microglial cells; ~78% of the proteins responding to AGE and FruArg treatments are in common, suggesting that the proteins differentially affected by treatment with AGE and FruArg are involved in inflammatory responses and the Nrf2-mediated oxidative stress response. It appears that AGE and FruArg attenuate neuroinflammatory responses and promote resilience in BV-2 microglial cells by suppressing NO production and by regulating expression of multiple protein targets associated with oxidative stress.

CONCLUSIONS

Garlic and its components or its various processed forms (including AGE) have been proposed as promising candidates for maintaining the homeostasis of the immune system. Garlic appears to enhance the functioning of the immune system by stimulating certain cell types, such as macrophages, lymphocytes, NK cells, DCs, cytotoxic T-lymphocytes, and eosinophils, by mechanisms including modulation of cytokine secretion, immunoglobulin production, phagocytosis, and macrophage activation. Immunoregulation by garlic extracts and isolated compounds can contribute to the treatment and prevention of pathologies such as the common cold and flu, obesity, metabolic syndrome, cardiovascular disorders, gastric ulcer, autoimmune diseases, and certain cancers. Garlic modulates cytokine secretion and such modulation may provide a mechanism of action for many of its therapeutic effects. All the major components of AGE—transformed OS compounds, lectins (ASA I and ASA II), fructans, and FruArg—appear to contribute to the overall immunomodulatory attributes of AGE. Currently, there is an emerging consensus on the benefits of the use of fructans in functional foods, sustaining health via direct immunomodulatory and antioxidant effects or through indirect, prebiotic mechanisms. Garlic and AGE have shown great promise as antitumor agents; though not completely curative, they have the potential for use as an adjuvant immunotherapy in various immune dysfunctional states. The intrinsic immunogenicity of garlic lectins in producing natural antibodies and immunoadjuvant activity of fructans are of considerable significance, since garlic is a nontoxic food source. Considering the immense pharmacological benefits and bioactivities of garlic including immunomodulation, it is no wonder that the Indian traditional system of medicine, Ayurveda, describes garlic as a "complete food."

Acknowledgments

The research work conducted in the author's laboratory was supported by the Council of Scientific and Industrial Research (CSIR), New Delhi, Government of India, in the form of institutional research projects MLP-040 and MLP-057 during the years 2005–10.

References

Adão, C.R., Pereira da Silva, B., Tinoco, L.W., et al., 2012. Haemolytic activity and immunological adjuvant effect of a new steroidal saponin from *Allium ampeloprasum* var. porrum. Chem. Biodivers. 9, 58–67.

Ahmadabad, H.N., Hassan, Z.M., Safari, E., et al., 2011. Evaluation of the immunomodulatory effect of the 14kDa protein isolated from aged garlic extract on dendritic cells. Cell. Immunol. 269, 90–95.

Amagase, H., 2006. Clarifying the real bioactive constituents of garlic. J. Nutr. 136 (3 Suppl.), 716S–725S.

Amagase, H., Petesch, B.L., Matsuura, H., et al., 2001. Intake of garlic and its bioactive components. J. Nutr. 131 (3 Suppl.), 955S–962S.

Arreola, R., Quintero-Fabián, S., López-Roa, R.I., et al., 2015. Immunomodulation and anti-inflammatory effects of garlic compounds. J. Immunol. Res. 2015, 401630. http://dx.doi.org/10.1155/2015/401630.

Barbosa, I.L., 2006. Cellular therapy – adoptive immunotherapy/immunostimulation. Transfusion Apheresis Sci. 34, 98–99.

Baumgarth, N., Tung, J.W., and Herzenberg, L.A., 2005. Inherent specificities in natural antibodies: a key to immune defense against pathogen invasion. Springer Semin. Immunopathol. 26, 347–362.

Baumgartner, S., Dax, T.G., Praznik, W., et al., 2000. Characterization of the high-molecular weight fructans isolated from garlic (*Allium sativum* L.). Carbohydr. Res. 328, 177–183.

Block, E., 2010. Garlic and Other Alliums: The Lore and the Science. RSC Publishing, Royal Society of Chemistry, UK, Cambridge.

Bubb, M.O., Green, F., Conradie, J.D., et al., 1993. Natural antibodies to avidin in human serum. Immunol. Lett. 35, 277–280.

Cerella, C., Dicato, M., Jacob, C., et al., 2011. Chemical properties and mechanisms determining the anti-cancer action of garlic-derived organic sulfur compounds. Anticancer Agents Med. Chem. 11, 267–271.

Chandrashekar, P.M., and Venkatesh, Y.P., 2009. Identification of the protein components displaying immunomodulatory activity in aged garlic extract. J. Ethnopharmacol. 124, 384–390.

Chandrashekar, P.M., and Venkatesh, Y.P., 2012. Fructans from aged garlic extract produce a delayed immunoadjuvant response to ovalbumin antigen in BALB/c mice. Immunopharmacol. Immunotoxicol. 34, 174–180.

Chandrashekar, P.M., Prashanth, K.V., and Venkatesh, Y.P., 2011. Isolation, structural elucidation and immunomodulatory activity of fructans from aged garlic extract. Phytochemistry 72, 255–264.

Chandrashekara, P.M., and Venkatesh, Y.P., 2016. Immunostimulatory properties of fructans derived from raw garlic (*Allium sativum* L.). Bioact. Carbohydr. Dietary Fibre 8, 65–70.

Charron, C.S., Dawson, H.D., Albaugh, G.P., et al., 2015. A single meal containing raw, crushed garlic influences expression of immunity- and cancer-related genes in whole blood of humans. J. Nutr. 145, 2448–2455.

Clement, F., and Venkatesh, Y.P., 2010. Dietary garlic (*Allium sativum*) lectins, ASA I and ASA II, are highly stable and immunogenic. Int. Immunopharmacol. 10, 1161–1169.

Clement, F., Pramod, S.N., and Venkatesh, Y.P., 2010. Identity of the immunomodulatory proteins from garlic (*Allium sativum*) with the major garlic lectins or agglutinins. Int. Immunopharmacol. 10, 316–324.

Colic, M., and Savic, M., 2000. Garlic extracts stimulate proliferation of rat lymphocytes in vitro by increasing IL-2 and IL-4 production. Immunopharmacol. Immunotoxicol. 22, 163–181.

Colic, M., Vucevic, D., Kilibarda, V., et al., 2002. Modulatory effects of garlic extracts on proliferation of T-lymphocytes in vitro stimulated with concanavalin A. Phytomedicine 9, 117–124.

Dam, T.K., Bachhawat, K., Rani, P.G., et al., 1998. Garlic (*Allium sativum*) lectins bind to high mannose oligosaccharide chains. J. Biol. Chem. 273, 5528–5535.

Daneshmandi, S., Hajimoradi, M., Ahmadabad, H.N., et al., 2011. Effect of 14-kDa and 47-kDa protein molecules of age garlic extract on peritoneal macrophages. Immunopharmacol. Immunotoxicol. 33, 21–27.

Di Bartolomeo, F., and Van den Ende, W., 2015. Fructose and fructans: opposite effects on health? Plant Foods Hum. Nutr. 70, 227–237.

Ebrahimi, M., Mohammad Hassan, Z., Mostafaie, A., et al., 2013. Purified protein fraction of garlic extract modulates cellular immune response against breast transplanted tumors in BALB/c mice model. Cell J. 15, 65–75.

Ebrahimpour, S., Tabari, M.A., Youssefi, M.R., et al., 2013. Synergistic effect of aged garlic extract and naltrexone on improving immune responses to experimentally induced fibrosarcoma tumor in BALB/c mice. Pharmacognosy Res. 5, 189–194.

Elluru, S., Van Huyen, J.P., Delignat, S., et al., 2006. Molecular mechanisms underlying the immunomodulatory effects of mistletoe (*Viscum album* L.) extracts Iscador. Arzneimittelforschung 56, 461–466.

Fallah-Rostami, F., Tabari, M.A., Esfandiari, B., et al., 2013. Immunomodulatory activity of aged garlic extract against implanted fibrosarcoma tumor in mice. N. Am. J. Med. Sci. 5, 207–212.

Foroutan-Rad, M., Hazrati Tappeh, K., and Khademvatan, S., 2017. Antileishmanial and immunomodulatory activity of *Allium sativum* (garlic): a review. J. Evid. Based Complementary Altern. Med. 22, 141–155.

Franco-Robles, E., and López, M.G., 2015. Implication of fructans in health: immunomodulatory and antioxidant mechanisms. ScientificWorldJournal 2015, 289267. http://dx.doi.org/10.1155/2015/289267.

Gamboa-León, M.R., Aranda-González, I., Mut-Martín, M., et al., 2007. In vivo and in vitro control of *Leishmania mexicana* due to garlic-induced NO production. Scand. J. Immunol. 66, 508–514.

Ghazanfari, T., Hassan, Z.M., and Ebrahimi, M., 2002. Immunomodulatory activity of a protein isolated from garlic extract on delayed type hypersensitivity. Int. Immunopharmacol. 2, 1541–1549.

Ghazanfari, T., Hassan, Z.M., and Khamesipour, A.J., 2006. Enhancement of peritoneal macrophage phagocytic activity against *Leishmania major* by garlic (*Allium sativum*) treatment. J. Ethnopharmacol. 103, 333–337.

Gupta, A., and Sandhu, R.S., 1996. Mitogenic activity of high molecular weight mannose specific agglutinin. Indian J. Biochem. Biophys. 33, 325–327.

Gupta, A., and Sandhu, R.S., 1997. A new high molecular weight agglutinin from garlic (*Allium sativum*). Mol. Cell. Biochem. 166, 1–9.

Hart, T., Lamont, A., and Williams, D., 1998. New drug targets in inflammation and immunomodulation. Drug Discov. Today 3, 516–521.

Hasan, N.A., Zuhair, M.H., Safari, E., et al., 2012. Evaluation of the effect of the 47 kDa protein isolated from aged garlic extract on dendritic cells. Iran. J. Basic Med. Sci. 15, 745–751.

Hassan, Z.M., Yaraee, R., Zare, N., et al., 2003. Immunomodulatory affect of R10 fraction of garlic extract on natural killer activity. Int. Immunopharmacol. 3, 1483–1489.

Hirao, Y., Sumioka, I., Nakagami, S., et al., 1987. Activation of immunoresponder cells by the protein fraction from aged garlic extract. Phytother. Res. 1, 161–164.

Iciek, M., Kwiecień, I., and Włodek, L., 2009. Biological properties of garlic and garlic-derived organosulfur compounds. Environ. Mol. Mutagen. 50, 247–265.

Ide, N., Lau, B.H., Ryu, K., et al., 1999. Antioxidant effects of fructosyl arginine, a Maillard reaction product in aged garlic extract. J. Nutr. Biochem. 10, 372–376.

Ishikawa, H., Saeki, T., Otani, T., et al., 2006. Aged garlic extract prevents a decline of NK cell number and activity in patients with advanced cancer. J. Nutr. 136 (3 Suppl.), 816S–820S.

Jikihara, H., Qi, G., Nozoe, K., et al., 2015. Aged garlic extract inhibits 1,2-dimethylhydrazine-induced colon tumor development by suppressing cell proliferation. Oncol. Rep. 33, 1131–1140.

Kaku, H., Goldstein, I.J., Van Damme, E.J., et al., 1992. New mannose-specific lectins from garlic (*Allium sativum*) and ramsons (*Allium ursinum*) bulbs. Carbohydr. Res. 229, 347–353.

Kaschula, C.H., Hunter, R., and Parker, M.I., 2010. Garlic-derived anticancer agents: structure and biological activity of ajoene. Biofactors 36, 78–85.

Kasuga, S., Uda, N., Kyo, E., et al., 2001. Pharmacologic activities of aged garlic extract in comparison with other garlic preparations. J. Nutr. 131 (3 Suppl.), 1080S–1084S.

Katsuki, T., Hirata, K., Ishikawa, H., et al., 2006. Aged garlic extract has chemopreventative effects on 1,2-dimethylhydrazine-induced colon tumors in rats. J. Nutr. 136 (3 Suppl.), 847S–851S.

Koch, H.P., and Lawson, L.D. (Eds.), 1996. Garlic: The Science and Therapeutic Application of *Allium sativum* L. and Related Species, second ed. Williams & Wilkins, Baltimore, MD.

Kumar, P.V., Prashanth, K.V., and Venkatesh, Y.P., 2015. Structural analyses and immunomodulatory properties of fructo-oligosaccharides from onion (*Allium cepa*). Carbohydr. Polym. 117, 115–122.

Kuttan, G., 2000. Immunomodulatory effect of some naturally occurring sulphur-containing compounds. J. Ethnopharmacol. 72, 93–99.

Kuttan, G., and Thejass, P., 2009. Diallyl sulfide from garlic. In: Agarwal, B.B., Kunnumakkara, A.B. (Eds.), Molecular Targets and Therapeutic Uses of Spices: Modern Uses for Ancient Medicine. World Scientific Publishing Co. Pvt. Ltd., Singapore, pp. 197–224.

Kyo, E., Uda, N., Kakimoto, M., et al., 1997. Anti-allergic effects of aged garlic extract. Phytomedicine 4, 335–340.

Kyo, E., Uda, N., Suzuki, A., et al., 1998. Immunomodulation and antitumor activities of aged garlic extract. Phytomedicine 5, 259–267.

Kyo, E., Uda, N., Ushijima, M., et al., 1999. Prevention of psychological stress-induced immune suppression by aged garlic extract. Phytomedicine 6, 325–330.

Kyo, E., Uda, N., Kasuga, S., et al., 2001. Immunomodulatory effects of aged garlic extract. J. Nutr. 131 (3 Suppl.), 1075S–1079S.

Lamm, D.L., and Riggs, D.R., 2001. Enhanced immunocompetence by garlic: role in bladder cancer and other malignancies. J. Nutr. 131 (3 Suppl.), 1067S–1070S.

Lawson, L.D., 1993. Bioactive organosulfur compounds of garlic and garlic products. In: Kinghorn, A.D., Balandrin, M.F. (Eds.), Human Medical Agents from Plants. American Chemical Society, Washington, DC, pp. 306–330. ACS Symposium Series 534.

Lawson, S.D., and Wang, Z.Y.J., 1995. Changes in the organosulfur compounds released from garlic during aging in water, dilute ethanol, or dilute acetic acid. J. Toxicol. 14, 214.

Lee, E.S., Steiner, M., and Lin, R., 1994. Thioallyl compounds: potent inhibitors of cell proliferation. Biochim. Biophys. Acta 1221, 73–77.

Losso, J.N., and Nakai, S., 1997. Molecular size of garlic fructooligosaccharides and fructopolysaccharides by matrix-assisted laser desorption ionization mass spectrometry. J. Agric. Food Chem. 45, 4342–4346.

Malekpour-Dehkordi, Z., Javadi, E., Doosti, M., et al., 2013. S-Allylcysteine, a garlic compound, increases ABCA1 expression in human THP-1 macrophages. Phytother. Res. 27, 357–361.

Matsutomo, T., and Kodera, Y., 2016. Development of an analytic method for sulfur compounds in aged garlic extract with the use of a postcolumn high performance liquid chromatography method with sulfur-specific detection. J. Nutr. 146, 450S–455S.

Matsuura, N., Miyamae, Y., Yamane, K., et al., 2006. Aged garlic extract inhibits angiogenesis and proliferation of colorectal carcinoma cells. J. Nutr. 136 (3 Suppl.), 842S–846S.

Morihara, N., Sumioka, I., Moriguchi, T., et al., 2002. Aged garlic extract enhances production of nitric oxide. Life Sci. 71, 509–517.

Morihara, N., Ide, N., and Weiss, N., 2010. Aged garlic extract inhibits CD36 expression in human macrophages via modulation of the PPARgamma pathway. Phytother. Res. 24, 602–608.

Morioka, N., Sze, L.L., Morton, D.L., et al., 1993. A protein fraction from aged garlic extract enhances cytotoxicity and proliferation of human lymphocytes mediated by interleukin-2 and concanavalin A. Cancer Immunol. Immunother. 37, 316–322.

Nachbar, M.S., and Oppenheim, J.D., 1980. Lectins in the United States diet: a survey of lectins in commonly consumed foods and a review of the literature. Am. J. Clin. Nutr. 33, 2338–2345.

Nakagawa, H., Tsuta, K., Kiuchi, K., et al., 2001. Growth inhibitory effects of diallyl disulfide on human breast cancer cell lines. Carcinogenesis 22, 891–897.

Nantz, M.P., Rowe, C.A., Muller, et al., 2012. Supplementation with aged garlic extract improves both NK and γδ-T cell function and reduces the severity of cold and flu symptoms: a randomized, double-blind, placebo-controlled nutrition intervention. Clin. Nutr. 31, 337–344.

Nicastro, H.L., Ross, S.A., and Milner, J.A., 2015. Garlic and onions: their cancer prevention properties. Cancer Prev. Res. (Phila.) 8, 181–189.

Nowell, P.C., 1960. Phytohemagglutinin: an initiator of mitosis in cultures of normal human leucocytes. Cancer Res. 20, 462–464.

Percival, S.S., 2016. Aged garlic extract modifies human immunity. J. Nutr. 146, 433S–436S.

Pramod, S.N., Vigneshwaran, V., and Venkatesh, Y.P., 2015. Immunomodulatory effects of hemagglutinating lectins from potato (*Solanum tuberosum*) and garlic (*Allium sativum*) on human and murine lymphocytes. Int. J. Pharm. Pharm. Sci. 7 (Suppl. 1), 147–153.

Prasanna, V.K., and Venkatesh, Y.P., 2015. Characterization of onion lectin (*Allium cepa* agglutinin) as an immunomodulatory protein inducing Th1-type immune response *in vitro*. Int. Immunopharmacol. 26, 304–313.

Rabinkov, A., Zhu, X.Z., Grafi, G., et al., 1994. Alliin lyase (alliinase) from garlic (*Allium sativum*): biochemical characterization and cDNA cloning. Appl. Biochem. Biotechnol. 48, 149–171.

Rabinkov, A., Wilchek, M., and Mirelman, D., 1995. Alliinase (alliin lyase) from garlic (*Allium sativum*) is glycosylated at ASN146 and forms a complex with a garlic mannose-specific lectin. Glycoconj. J. 12, 690–698.

Schäfer, G., and Kaschula, C.H., 2014. The immunomodulation and anti-inflammatory effects of garlic organosulfur compounds in cancer chemoprevention. Anticancer Agents Med. Chem. 14, 233–240.

Sobolewska, D., Michalska, K., Podolak, I., et al., 2016. Steroidal saponins from the genus *Allium*. Phytochem. Rev. 15, 1–35.

Suzuki, J., Yamaguchi, T., Matsutomo, T., et al., 2016. S-1-Propenylcysteine promotes the differentiation of B cells into IgA-producing cells by the induction of Erk1/2-dependent Xbp1 expression in Peyer's patches. Nutrition 32, 884–889.

Tabari, M.A., and Ebrahimpour, S., 2014. Effect of aged garlic extract on immune responses to fibrosarcoma tumor in BALB/c mice. Indian J. Cancer 51, 609–613.

Tanaka, S., Haruma, K., Kunihiro, M., et al., 2004. Effects of aged garlic extract (AGE) on colorectal adenomas: a double-blinded study. Hiroshima J. Med. Sci. 53, 39–45.

Tanaka, S., Haruma, K., Yoshihara, M., et al., 2006. Aged garlic extract has potential suppressive effect on colorectal adenomas in humans. J. Nutr. 136 (3 Suppl.), 821S–826S.

Tapiero, H., Townsend, D.M., and Tew, K.D., 2004. Organosulfur compounds from alliaceae in the prevention of human pathologies. Biomed. Biopharmacother. 58, 183–193.

Tchernychev, B., and Wilchek, M., 1996. Natural human antibodies to dietary lectins. FEBS Lett. 397, 139–142.

Tchernychev, B., Rabinkov, A., Mirelman, D., et al., 1995. Natural antibodies to dietary proteins: the existence of natural antibodies to alliinase (alliin lyase) and mannose-specific lectin from garlic (*Allium sativum*) in human serum. Immunol. Lett. 47, 53–57.

Trio, P.Z., You, S., He, X., et al., 2014. Chemopreventive functions and molecular mechanisms of garlic organosulfur compounds. Food Funct. 5, 833–844.

Tsukamoto, S., Okamoto, K., Inanaga, J., et al., 2008. Purification, characterization and biological activities of a garlic oligosaccharide. J. UOEH 30, 147–157.

Van Damme, E.J.M., Goldstein, I.J., and Peumans, W.J., 1991. A comparative study of mannose-binding lectins from the Amaryllidaceae and Alliaceae species. Phytochemistry 30, 509–514.

Van Damme, E.J.M., Smeets, K., Torrekens, S., et al., 1992a. Isolation and characterization of alliinase cDNA clones from garlic (*Allium sativum* L.) and related species. Eur. J. Biochem. 209, 751–757.

Van Damme, E.J.M., Smeets, K., Torrekens, S., et al., 1992b. The closely related homomeric and heterodimeric mannose-binding lectins from garlic are encoded by one-domain and two-domain lectin genes, respectively. Eur. J. Biochem. 206, 413–420.

Van Damme, E.J.M., Peumans, W.J., Barre, A., et al., 1998. Plant lectins: a composite of several distinct families of structurally and evolutionary related proteins with diverse biological roles. Crit. Rev. Plant Sci. 17, 575–692.

Van Damme, E.J.M., Lannoo, N., and Peumans, W.J., 2008. Plant lectins. Adv. Bot. Res. 48, 107–209.

Venkatesh, Y.P., and Prasanna Kumar, V., 2013. Natural antibodies to dietary components. Glob. J. Immunol. Allerg. Dis. 1, 25–34. http://dx.doi.org/10.14205/2310-6980.2013.01.01.4.

Wagner, H., Kraus, S., and Jurcic, K., 1999. Search for potent immunostimulating agents from plants and other natural sources. In: Wagner, H. (Ed.), Immunomodulatory Agents from Plants. Birkhauser Verlag, Basel, pp. 1–39.

Wang, X., Jiao, F., Wang, Q.W., et al., 2012. Aged black garlic extract induces inhibition of gastric cancer cell growth in vitro and in vivo. Mol. Med. Rep. 5, 66–72.

Wang, X., Liu, R., Yang, Y., et al., 2015. Isolation, purification and identification of antioxidants in an aqueous aged garlic extract. Food Chem. 187, 37–43.

Washiya, Y., Nishikawa, T., and Fujino, T., 2013. Enhancement of intestinal IgA production by Ajoene in mice. Biosci. Biotechnol. Biochem. 77, 2298–2301.

Yoo, J.M., Sok, D.E., and Kim, M.R., 2014. Anti-allergic action of aged black garlic extract in RBL-2H3 cells and passive cutaneous anaphylaxis reaction in mice. J. Med. Food 17, 92–102.

Zare, A., Farzaneh, P., Pourpak, Z., et al., 2008. Purified aged garlic extract modulates allergic airway inflammation in BALB/c mice. Iran. J. Allergy Asthma Immunol. 7, 133–141.

Zhou, Z.H., Tzioufas, A.G., and Notkins, A.L., 2007a. Properties and function of polyreactive antibodies and polyreactive antigen-binding B cells. J. Autoimmun. 29, 219–228.

Zhou, Z.H., Zhang, Y., Hu, Y.F., et al., 2007b. The broad antibacterial activity of the natural antibody repertoire is due to polyreactive antibodies. Cell Host & Microbe 1, 51–61.

Zhou, H., Qu, Z., Mossine, V.V., et al., 2014. Proteomic analysis of the effects of aged garlic extract and its FruArg component on lipopolysaccharide-induced neuroinflammatory response in microglial cells. PLoS One 9, e113531. http://dx.doi.org/10.1371/journal.pone.0113531. eCollection 2014.

CHAPTER

15

MUC1 and Cancer Immunotherapy

Chuanwei Yang, James L. Murray and Nuhad K. Ibrahim

OUTLINE

Abstract

MUC1 is a glycosylated transmembrane protein. In normal cells, it is expressed in the apical surface of epithelial cells. Cancer cells express up to 100-fold more MUC1 protein than normal cells. In addition, cancer cells usually have abnormal glycosylation of MUC1 protein. MUC1-associated antibody production and cellular immune responses have been shown to exert a positive effect on cancer patient outcomes. Therefore, boosting MUC1-associated immunity through the use of vaccines, MUC1-specific antibodies, or T cell transfers represents an important strategy in the fight against cancer.

Over the past three decades, different types of MUC1-based vaccines have been developed, both in preclinical experimentation and clinical trials. These vaccines include peptide/protein vaccines, vaccinia virus vaccine, dendritic cell based vaccine, and DNA vaccine. In the process of attempting to enhance the immune responses to the vaccines, vaccine carriers such as mannan, keyhole limpet hemocyanin, and adjuvants including Bacillus Calmette-Guerin, incomplete Freund's adjuvants, and saponin, were all tried along with the vaccines with varied successes. We reviewed the literature and presented findings from both preclinical studies and clinical trials related to MUC1 immunotherapy. We also discussed strategies to boost the immune responses, which included the pretreatment with low dose cyclophosphamide before vaccine administration as a Treg suppressive strategy.

New evidence and hypotheses are emerging on the effectiveness of MUC1 vaccines in treating subsets of cancer patient population. Those include that innate preimmune status of individual patients may predict

M.A. Hayat (ed): *Immunology, Volume 1*.
DOI: http://dx.doi.org/10.1016/B978-0-12-809819-6.00015-0

© 2018 Elsevier Inc. All rights reserved.

benefits of MUC1-based immunotherapy and that Tamoxifen, but not aromatase inhibitors, in combination with MUC1 immunotherapy benefits estrogen receptor positive breast cancer patients. Encouraging results from these hypotheses driven clinical studies added a breath of fresh air into the field of MUC1 cancer immunotherapy. Future success of MUC1 immunotherapy may also reside in the combination strategies of its use with chemotherapy, targeted therapy, and hormone therapy.

INTRODUCTION

MUC1 is a member of human mucin gene family which includes more than 16 different mucin species. MUC1 was first cloned in 1990, located on chromosome 1922, and its genetic sequence was identified by several independent research groups in the same year (Gendler et al., 1990; Ligtenberg et al., 1990; Wreschner et al., 1990). The gene contains seven exons among which exon two has a varied number of 60 bp unit tandem repeats (Lan et al., 1990). The 60 bp tandem repeat unit varies in number between individuals and codes for exactly 20 amino acid peptides or icosapeptides (Gendler et al., 1990; Hanisch and Muller, 2000). This region of variable numbers of tandem repeats (VNTR) may have from 20 to over 100 repeats and make up 40–80% of the whole gene sequence, depending on genetic polymorphism (Hanisch and Muller, 2000). MUC1 gene encodes for mucin1 or MUC1 protein, a glycosylated protein with a molecular weight of 300 kDa to over 1000 kDa (Baeckstrom et al., 1991). MUC1 protein is a transmembrane protein with an extracellular region, a transmembrane domain, and a cytoplasmic tail (Yang et al., 2007). The icosapeptide encoded from the VNTR has five potential *O*-glycosylation sites. Five other potential *N*-glycosylation sites of mucin1 exist near the C-terminal domain of the protein (Hanisch and Muller, 2000). MUC1 protein plays roles in epithelial protection, cell-cell adhesion, protein bindings with beta- and gamma-catenins and ICAM-1, and cellular signal transduction (Baldus et al., 2004). Identities of MUC1 in the literature are often confusing. So far over 20 alias names have been reported for MUC1. Those include ADMCKD1, Cancer Antigen 15-3 (CA 15-3), CD227 Antigen, DF3 antigen, Episialin, epithelial membrane antigen (EMA), epitectin, H23 Antigen, H23AG, KL-6, MUC1/SEC, MUC1/ZD, MCKD-1, MAM-6, MCD, NCRC-11 antigen, PEMT, Peanut-lectin reactive urinary mucin, polymorphic epithelial mucin, and tumor-associated epithelial mucin (Kinlough et al., 2004; Baldus et al., 2004). MUC1 is expressed in the apical surface of epithelial cells under physiological conditions in normal individuals (Hanisch and Muller, 2000). Its expression is upregulated in most carcinomas where the polarity of epithelial cells is lost and MUC1 is expressed on all sides of the cancer cell surface instead. Overexpressed MUC1 protein can be secreted or shed into the extracellular space surrounding the cancer cells. As a result, MUC1 protein can be detected in cancer cells supernatant in culture, and in the blood samples of patients (Ligtenberg et al., 1992; Hilkens et al., 1986; Magnani et al., 1983; Johnson et al., 1986). Cancer cells also display abnormal glycosylation of MUC1. In normal cells, MUC1 peptide epitopes are often masked by carbohydrate structures attached to the peptide, whereas in cancer cells, the peptide epitopes are often exposed, due to shorter and fewer branches of the carbohydrate structures, allowing specific antibodies to bind the epitopes (Girling et al., 1989; Lloyd et al., 1996). In addition, *O*-glycan chains are more sialylated in cancer, and inhibition of sialylation decreases the migratory ability of malignant cells (Lloyd et al., 1996), and therefore its malignant potential.

MUC1 and the Immune System

MUC1 purified from cancer patients had an inhibitory effect on antitumor immune response by interfering with T cell proliferation and function (Agrawal et al., 1998b). MUC1 secreted from cancer cells suppressed the lysis of cancer cells by natural killer cell-mediated killing (Zhang et al., 1997). However, MUC1 protein is expressed on the activated T cell surface, but not on the resting T cells, as revealed by antibody based assays such as flow cytometry, immunohistochemistry, and immunofluorescence staining (Agrawal et al., 1998a; Correa et al., 2003); as such, it seems to have a longer and more branched carbohydrate structure, which is rather different from the MUC1, with under glycosylation and fewer branched *O*-glycans seen on the cancer cells. The functional implications of MUC1 on T cell surface is not fully understood (Agrawal et al., 1998a).

Increased levels of MUC1 in blood and anti-MUC1 antibodies were noticed in normal individuals during pregnancy or after childbirth. MUC1 core peptide sequences can stimulate the proliferation of T cells isolated from pregnant or multiparous donors in vitro whereas T cells isolated from nulliparous women did not respond well in the presence of the same MUC1 core peptides in vitro (Agrawal et al., 1995, 1996). Epidemiological studies have shown that there were decreased breast cancer incidences in mothers with multiple pregnancies compared to women without pregnancy (Kalache et al., 1993; Lambe et al., 1998). This pregnancy-derived benefit of reduced breast cancer risk could well be from MUC1-mediated immune responses. It has been speculated that increased expression of MUC1 and the appearance of anti-MUC1 antibodies during pregnancy may prime T cells against future cancer cells (Agrawal et al., 1995, 1996).

Indeed, antibody production against MUC1 is an important part of MUC1-associated immune responses. There was correlation between the high level of anti-MUC1 antibodies in blood samples of cancer patients and the high number of MUC1-primed B lymphocytes in the lymph nodes of those patients (Petrarca et al., 1999). Indeed, breast cancer patients at early stages who were tested positive of anti-MUC1 antibodies in blood drawn before treatment had better prognosis compared to patients whose sera were negative for the antibodies (Von Mensdorff-Pouilly et al., 2000; Blixt et al., 2011). Breast cancer patients are less likely to develop cancer metastasis with the appearance of anti-MUC1 antibodies in their sera. In addition to pregnancies, recent observations have indicated that anti-MUC1 antibodies can develop in some healthy individuals who had abnormal tumor-like MUC1 expression from earlier infections or chronic inflammation. These individuals have decreased cancer incidences (Cramer et al., 2005). These naturally occurring antibodies against MUC1 may kill circulating tumor cells with MUC1 antigen expression on their surface by complement-mediated cell lysis or antibody-dependent killing mechanisms. Mounting evidence showing a protective effect of MUC1-associated immunity against cancer implies that inducing or boosting MUC1-associated immune responses through the use of vaccines, antibodies, or T cell transfers may have profound effects in cancer immunotherapy.

Vaccine-Based Strategy to Boost the Immune Response Against Cancer

There are different types of vaccines which have been designed and used in preclinical settings and clinical trials. Current vaccines include peptide/protein vaccine, DNA based vaccine, vaccinia virus vaccine, and dendritic cell (DC)-MUC1 positive cancer cell fusion vaccine.

Peptide/protein vaccine. Experiments were done to evaluate the efficacy of synthetic peptides, fusion protein, or native MUC1 protein expressed in tumor cell lines in generating immune response in mice in the 1990s (Apostolopoulos et al., 1994, 1995a,b). It was demonstrated that synthetic peptides with various lengths or fusion protein alone induced only antibody production without cytotoxic T lymphocyte (CTL) response, whereas tumor cell lines expressing MUC1 protein by transfection produced strong CTL response and a weak antibody production (Apostolopoulos et al., 1994, 1995b). This observation led to the search for ways to trigger and enhance CTL response in formulating peptide or fusion protein vaccines. For example, investigators experimented with the use of carriers and/or adjuvants in boosting the immune response (Bi et al., 2016). One such carrier is mannan, a polysaccharide composed of mannose residues. Further studies indicated that mannan can serve as a ligand for the mannose receptor present on the surface of DCs and macrophages (Apostolopoulos et al., 2000). Mannan can exist in two states: oxidized or reduced. The oxidized mannan, when conjugated to MUC1 fusion protein, was able to induce strong CTL, but weak antibody production in mice. On the other hand, the reduced mannan conjugated MUC1 fusion protein induced only minimal CTL, but high antibody production in mice (Apostolopoulos et al., 1995a). MUC1 transgenic mice immunized with oxidized manna-MUC1 fusion protein conjugate elicited strong CTL and, as a result, rejected the tumor (Lees et al., 2000).

Synthetic MUC1 peptide vaccine was first introduced in a phase I trial in 1993 (Goydos et al., 1996). Mannan-MUC1 fusion protein vaccine was first used in 25 patients with adenocarcinoma of the breast, colon, stomach, or rectum (Karanikas et al., 1997). It was reported that 13 out of 25 patients had antibody production against MUC1 and 2 out of 10 patients had CTL response (Karanikas et al., 1997). Antitumor activity was not evaluated in that trial. Mannan-MUC1 vaccine was evaluated in a phase I/II clinical trial in 12 patients with adenocarcinoma for its T cell response (Karanikas et al., 2000). The study demonstrated that $CD8^+CD69^+$ T cells from all 12 immunized patients produced 3 times or higher levels of TNF-alpha, interferon (IFN)-gamma than 12 normal healthy donors after in vitro stimulation with MUC1 peptide, indicative of T cell response to the immunization, although $CD4^+CD69^+$ T cells showed no difference in the production of TNF-alpha and IFN-gamma between immunized patients and normal subjects (Karanikas et al., 2000). Oxidized mannan-MUC1 was investigated in a clinical trial in 31 patients with stage II breast cancer. A 5.5-year follow-up after immunization demonstrated that 16 patients in the experimental group receiving the oxidized mannan-MUC1 vaccine had no recurrence while 4 out of 15 patients in the placebo group had recurrent tumors. Among patients in the experimental group, 9 out of 13 patients (69%) had MUC1-specific antibody production and 4 out of 10 patients had T cell response, but no patients in the placebo group showed an immune response to MUC1 (Apostolopoulos et al., 2006). A 15-year clinical follow-up indicated that the recurrent rate was significantly lower in the experimental vaccine group (2/16 patients, 12.5%) than the placebo group (9/15 patients, 60%). There was no sign of autoimmunity during the 15-year follow-up (Vassilaros et al., 2013).

In addition to mannan as a carrier, other carriers include liposomes which have adjuvant properties besides delivering antigens to antigen-presenting cells (APCs; Guan et al.,

1998). Both liposome-encapsulated and surface-bound MUC1 peptide vaccines were made and shown to trigger strong T cell responses. Surface-bound MUC1 peptide vaccine was also found to induce MUC1-specific antibody response (Guan et al., 1998). Liposome-encapsulated MUC1 peptide vaccine was administered in a mouse model of breast cancer and shown to produce MUC1-specific CTL response in vivo (Mukherjee et al., 2003). Liposome-encapsulated MUC1 peptide vaccine (BLP25) was also investigated in a Phase I clinical trial in a cohort of patients with advanced nonsmall-cell lung cancer (NSCLC). The trial showed that the vaccine was safe and induced CTL response in 5 out of 12 patients, but the vaccine did not induce MUC1-specific antibody production and did not inhibit tumor growth (Palmer et al., 2001). In a Phase IIb trial, patients with advanced NSCLC were divided into two groups with 83 patients in the control group and 88 patients in the experimental group. The control group received the best supportive care alone, while the experimental group received BLP25 vaccine plus best supportive care. Results from this trial indicated that there was a trend of increased survival in the treatment group, but the difference did not reach statistical significance (Butts et al., 2005). Longer follow-up study of these patients showed that 3-year survival rate was significantly higher in patients receiving BLP25 vaccine plus supportive care than patients receiving best supportive care alone (Butts et al., 2011). Encouraging results from phase IIb trial of BLP25 vaccine led to a large-scale, international, double-blind phase III trial for patients with stage III NSCLC after chemoradiotherapy (Butts et al., 2014). Patients in the phase III trial were assigned to either drug Tecemotide (L-BLP25) treatment group with 1006 patients or placebo group with 507 patients. The study team reported a median overall survival (OS) of 25.6 months in the drug treatment group and 22.3 months in the placebo group, with no statistically significant difference between the two groups (Butts et al., 2014). The safety and efficacy of Tecemotide (L-BLP25) was recently evaluated in 400 patients with HER2-negative early breast cancer in a prospective "neoadjuvant" phase II trial (Singer et al., 2017). The study team did not find significant difference in residual cancer burden (RCB0/1) rate or pathological complete response rate between the experimental group receiving Tecemotide plus standard chemotherapy, and the control group receiving standard chemotherapy alone, although the vaccine was safe for breast cancer patients (Singer et al., 2017).

Another carrier is keyhole limpet hemocyanin (KLH). KLH is a protein which has the ability to induce immune responses in a nonspecific manner. Conjugates of MUC1 peptide to KLH were shown to trigger immune response with strong antibody production and minimal CTL response in preclinical studies and clinical trials (Kim et al., 2000; Ragupathi et al., 2003; Sabbatini et al., 2007). In the preclinical studies, mice immunized with KLH-MUC1 peptide plus QS-21 adjuvant induced strong antibody production with weak or no CTL response (Kim et al., 1999, 2000). In one clinical trial, KLH-MUC1 peptide conjugate plus QS-21 was administered in nine high-risk breast cancer patients but without evidence of disease at the time of the trial. It was reported that all immunized patients produced high IgM and IgG antibody titers against MUC1 without evidence of T cell activation (Gilewski et al., 2000). The same study team evaluated the T cell response in another trial involving six patients with breast cancer. Again they did not observe any T cell response against MUC1 after MUC1-KLH conjugate plus QS-21 immunization (Musselli et al., 2002).

It is known that the immunogenic activities of the synthetic peptides as vaccines were further enhanced by the use of different adjuvants which include incomplete Freund's adjuvants (IFA), Bacillus Calmette-Guerin (BCG) (Li et al., 2000), monophosphoryl lipid A (MPLA), purified plant *Quillaja saponaria* extract (QS-21) (Zhu and Tuo, 2016), and (Polyriboinosinic-Polyribocytidylic acid stabilized with Poly-l-lysine and Carboxymethylcellulose poly-ICLC) (Ammi et al., 2015). This concept led to the experimentations of around 20 different adjuvants in a combined setting with MUC1 peptide-KLH conjugate (Kim et al., 1999).

BCG adjuvant was used with the MUC1 peptide in the early 1990s. In one trial, the MUC1 synthetic peptide used contained 105 amino acids spanning five tandem repeats in the extracellular VNTR domain. The study team recruited 63 patients with late-stage breast, colon, and pancreatic cancers. Although the peptide vaccine with BCG mix was tolerated well by participants, most of the patients failed to have strong immune responses (Goydos et al., 1996). In another trial, a different adjuvant SB-AS2 was used. A 100 amino acid synthetic peptide was mixed with SB-AS2 to give to 16 patients with early stage pancreatic cancer. In that trial, patients were evenly divided into four groups with a different single dose of the peptide mix for each group. Most of the patients did not show strong humoral or cellular responses (Ramanathan et al., 2005). But in a clinical trial using adjuvant polyinosinic-polycytidylic acid stabilized with poly-L-lysine and carboxymethylcellulose (Poly-ICLC), the peptide vaccine mix with the adjuvant was administered in a prophylactic setting to 39 patients who had premalignant lesions in their colon. The MUC1 peptide vaccine elicited strong anti-MUC1 antibody IgG production and long-term immune memory in 44% of participants (17 out of 39 patients). Results from this trial implies that the MUC1 peptide vaccine may be useful and important in cancer prevention in the premalignant stage (Kimura et al., 2013). It was recorded that QS-21 purified from plant extract was the strongest adjuvant in stimulating IgM and IgG antibody production against MUC1 in this setting (Kim et al., 1999). All of the adjuvants used in the experiments failed to induce significant CTLs against MUC1, although T cell responses against KLH were noted.

To improve MUC1 vaccine-induced T cell responses, scientists experimented with glycopeptides attaching sugar residues to tumor antigens. This formulation produced stronger CTLs than the nonglycosylated MUC1 peptide (Apostolopoulos et al., 2003). Another glycopeptides, sialyl-T(n)-MUC1 were also developed. Sialyl-T(n) or sialyl-TN is a carbohydrate epitope present on many glycoproteins including mucin1. One research group used tetanus toxoid as a carrier together with sialyl-T(n)-MUC1 for immunization. Results showed that strong immune responses were produced using this strategy (Kaiser et al., 2009; Hossain and Wall, 2016).

We have used sialyl-TN (STn)-KLH vaccine in clinical trials for the treatment of metastatic breast cancer (Miles et al., 2011; Ibrahim et al., 2013). In a large multicenter phase III trial involving 1028 patients with metastatic breast cancer, it was reported that no overall benefit in time to progression (TTP) or OS was seen, although STn-KLH vaccine was safe and well tolerated (Miles et al., 2011). Preplanned stratification and analysis of the patient population in the trial revealed that patients in STn-KLH group who received concomitant endocrine therapy had longer TTP and OS than patients in KLH only control group with concomitant endocrine therapy (Ibrahim et al., 2013).

Additional post hoc analysis of the data suggested that estrogen receptor (ER) positive patients who received, in addition, tamoxifen showed improved survival, but not those who received an aromatase inhibiter (Ibrahim et al., 2013). This hypothesis generating clinical observation may be partially explained by the MUC1–ER interaction. Indeed, interaction of MUC1 oncoprotein and ER has been shown to increase the stability of ER alpha and activate the ER mediated pathway, leading to overexpression of ER downstream targets such as Rab31 (Wei et al., 2006; Yang et al., 2006; Jin et al., 2012; Tecalco-Cruz et al., 2016). In addition, MUC1-C (C-terminal subunit) can interact and activate tyrosine kinase receptors such as EGFR and ErbB2. MUC1-C can also localize to the nucleus and activate Wnt/β-catenin and NF-kB/RelA pathways (Kufe, 2013). Activation of receptor tyrosine kinases can activate downstream effectors such as akt and MAPK, leading to phosphorylation of ER at specific residues, which in turn affecting the stability and activities of ER (Ali et al., 1993; Shou et al., 2004).

It was demonstrated over 30 years ago that pretreatment with cyclophosphamide potentiated both cellular and humoral immune responses to an antigen when administered days later both in animal models and in patients with metastatic cancer (Berd et al., 1982). Low dose cyclophosphamide augmented immune response to cancer vaccine by depleting the pool of suppressor T cells in cancer patients (Berd et al., 1988). In two phase I/II clinical trials involving 50 cancer patients with advanced disease, pretreatment of low dose cyclophosphamide by intravenous infusion generated both IgM and IgG antibody titer at significantly higher levels to STn-KLH vaccine than patients without cyclophosphamide infusion and with improved survival (Miles and Papazisis, 2003). In addition, low dose cyclophosphamide had the ability to reduce the number of regulatory T cells (Tregs) and inhibit its suppressive function on immune response against tumor (Ghiringhelli et al., 2004; Lutsiak et al., 2005). To improve MUC1 vaccine mediated immune response, a low dose of cyclophosphamide—an agent used for chemotherapy in clinic—was used pre-vaccination to reduce the number of Tregs and inhibit the suppressive function of Tregs on immune response against tumor (Ghiringhelli et al., 2004; Lutsiak et al., 2005; Miles et al., 2011). In other clinical studies, metronomic treatment of low dose cyclophosphamide reduced Tregs and induced tumor-specific CTL responses, leading to improved clinical outcome in metastatic breast cancer patients (Ge et al., 2012).

Globo H is a hexasaccharide with α(1,2)-linked fucose, which was isolated from MCF7 (Menard et al., 1983). Similar to MUC1, it is overexpressed in epithelial cell tumors such as colon, ovarian, gastric, pancreatic, endometrial, lung, prostate, and breast cancers (Bremer et al., 1984; Canevari et al., 1983). Active immunotherapy with Globo H coupled to KLH along with saponin as an adjuvant was studied in two phase I trials. The vaccine-induced Globo H-specific antibodies which mediated in vitro binding and cytotoxicity to Globo H expressing breast cancer cells.

Recently, a large, randomized, double blind, placebo-controlled, phase 2/3 trial was performed in patients with metastatic breast cancer who had <2 episodes of disease progression, and who achieved at least stable disease after >1 anticancer regimen (Huang et al., 2016). 349 patients—70% with hormone receptor positive breast cancer, 13% triple negative, and 62% who received hormone therapy—were randomized 2:1 to receive either vaccine (Globo H- KLH; OPT-822) or PBS control on weeks 1, 2, 3, 9, 13, 17, 25, and 37, or until progression, along with a low dose cyclophosphamide. Primary and secondary endpoints were progression-free survival (PFS) and OS. In summary, no difference was observed in PFS in the intent to treat analysis: however, both PFS and OS were significantly improved in the 50% of patients who developed early specific antibody responses to vaccine. These findings were observed irrespective of breast cancer subtype (i.e., ER+ or ER negative).

Vaccinia virus vaccine. Vaccinia virus-MUC1 was able to induce CTL response which was enhanced by vaccinia virus-B7-1 administered concurrently in mice (Akagi et al., 1997). This vaccinia virus-MUC1 inoculation prevented pulmonary metastasis after tumor challenge and the coadministration of vaccinia virus-B7 with Vaccinia virus-MUC had no additive effect on the prevention of murine pulmonary metastasis (Akagi et al., 1997). Recombinant vaccinia virus expressing human MUC1 and IL2 (TG1031) was evaluated in nine patients with inoperable advanced breast cancer. The vaccine was well tolerated. Two out of nine patients had detectable levels of CTL against MUC1, but none of the patients had detectable MUC1-specific antibody titers (Scroll et al., 2000). In another trial, two dose levels of TG1031 were evaluated in 28 metastatic breast cancer patients. Two patients had around 50% partial regression of their tumors, which lasted for 11 months, and 15 patients had stable disease (Scroll et al., 2003). A modified highly attenuated strain of vaccinia virus named Modified Vaccinia Ankara (MVA) incorporating MUC1 antigen and cytokine IL2 (MVA-MUC1-IL2 or TG4010) was used in preclinical studies and clinical trials. It was demonstrated that MVA-MUC1-IL2 vaccine worked synergistically with radiotherapy to induce CTL response specific for MUC1 tumor antigen in mice (Hillman et al., 2017). Two phase I studies were undertaken in 13 patients with adenocarcinoma. Three dose levels of the vaccine were administered intramuscularly (Rochlitz et al., 2003). Five out of 13 patients showed a weak CTL response against MUC1 and 4 out of 13 patients had stable disease for 6 months (Rochlitz et al., 2003). A phase II study demonstrated that the vaccine was feasible and induced MUC1-specific CTL response in patients with late-stage NSCLC (Ramlau et al., 2008). Another study showed that the same formulation improved PSA doubling time in prostate cancer patients (Dreicer et al., 2008). A phase 2B study involving 148 patients with advanced NSCLC indicated that TG4010 vaccine in combination with standard chemotherapy had a higher 6-month PFS rate than the chemotherapy only control group (Quoix et al., 2011). A follow-up confirmatory phase 2B/3 trial involving 222 patients with advanced NSCLC (TIME) showed that TG4010 plus chemotherapy seemed to increase PFS compared to placebo plus chemotherapy control group (Quoix et al., 2016). Vaccine using attenuated vaccinia virus and fowlpox vector with antigen CEA and MUC1 and immune molecules B7.1, ICAM-1, and LFA-3 (TRICOM) expression (also called PANVAC vaccine) was administered to 25 patients with metastatic carcinoma in a pilot study (Gulley et al., 2008). Nine out of 16 patients showed CTL response to MUC1 and/or CEA. One patient with breast cancer had decreased tumor size in the liver metastasis site and one patient with ovarian cancer had a favorable clinical response for 18 months after immunization (Gulley et al., 2008). PANVAC vaccine was also investigated in a phase 2 trial in 48 patients with metastatic breast cancer. 25/48 patients were assigned to arm A with treatment of PANVAC plus docetaxel while 23/48 patients were in arm B receiving docetaxel alone. Median PFS for patients in arm A was 7.9 months compared to PFS of 3.9 months for patients receiving docetaxel alone in arm B ($p = .09$). Results from this trial indicate that combined therapy of docetaxel with PANVAC may offer a better clinical outcome than docetaxel treatment alone (Heery et al., 2015).

Retrospective analyses of immune responsiveness in patients from two randomized trials, Globo H vaccine and PANVAC/PROSTVAC vaccine, have led to intriguing hypotheses which suggest that the innate preimmune status of individual patients may be responsible for improved survival in the vaccinated subset. For example, analyses of predefined lymphocyte subsets in peripheral blood in patients receiving PANVAC or PROSTVAC vaccinia virus vaccines for which a specific biologic function had been previously reported, showed no specific prognostic value in PFS in patients randomized to either docetaxel (breast cancer) or radiotherapy alone (prostate cancer); however, in patients receiving these agents along with their respective vaccines the peripheral immunoscore revealed significant differences in PFS (Farsaci et al., 2016). Similarly, half of the patients receiving Globo H who mounted significantly higher antibody responses to the vaccine had an improved PFS and OS compared to patients who demonstrated weak responses, or unvaccinated controls (Huang et al., 2016). These data argue for additional prospective clinical trials in which patients are stratified into groups where they would be most likely to benefit from vaccination based on preexisting immune parameters.

DC based vaccine. DCs are powerful Antigen Presenting Cells (APCs). Strategies such as pulsing DC with tumor antigenic peptides or tumor whole-cell lysates have been used to boost immunity against cancer (Brossart et al., 2000; Liu et al., 2008; Rainone et al., 2016; Chang et al., 2005). A phase I/II study demonstrated that immunization with in vitro grown autologous DC pulsed with MUC1 peptide increased survival for 12 patients with pancreatic and biliary tumors after surgery (Lepisto et al., 2008). Fusion of DCs with tumor cells containing different tumor antigens represents a similar strategy in DC based vaccine (Gong et al., 2008). Immunized mice with fusion cells showed protection from tumor challenge (Gong et al., 1997). The immune response produced was demonstrated to be effective against tumors in MUC1 transgenic mice bearing MUC1 positive tumors (Gong et al., 1998). In one study, half of the immunized mice with DC–MUC1 fusion vaccine did not develop tumors, and another half had delayed onset of tumor in a spontaneous breast cancer mouse model (Xia et al., 2003). Immunized mice had strong MUC1-specific CTL responses, especially CD8 positive T cell response (Koido et al., 2002). DC–cancer cell fusion vaccine has been applied in clinical trials. In a phase I study, the vaccine was used for patients with metastatic breast and kidney cancer. It was found that the vaccine was well tolerated and induced CTL response with two breast cancer patients experiencing disease regression and five kidney cancer patients having tumor stabilization (Avigan et al., 2004).

DNA vaccine. Naked DNA construct can be injected into animals through intramuscular or intradermal routes. Expressed proteins from the constructs can be processed by intracellular and extracellular pathways for antigen presentation (Pillai et al., 2015). Intramuscular injections of full-length MUC1 cDNA were able to induce both CTL and humoral responses against MUC1 in mice, which in turn protected immunized mice from subsequent tumor cell challenge. The protective effects were dependent on the presence of both CD4 and CD8 positive T cells (Plunkett et al., 2004). Low doses of intradermal injection of MUC1 were shown to prevent tumor metastasis into the lung (Kamata et al., 2002). When combined with plasmid DNA coding for IL-18, MUC1 cDNA was able to break tolerance in MUC1 transgenic mice and kill the MUC1-expressing tumor cell challenge (Snyder et al., 2006). Cytokine IL2 expression combined with MUC1 DNA immunization could trigger stronger immune response in the form of higher antibody production and more CTLs than MUC1 DNA alone (Zhang et al., 2008). The DNA vaccine has not yet been tested in clinical trials.

Passive anti-MUC1 antibody adoption for immunotherapy. Healthy individuals could have naturally occurring anti-MUC1 antibody present. These individuals often have lower risk of developing cancer (Pinheiro et al., 2010; Cramer et al., 2005). The production of naturally occurring anti-MUC1 antibodies in healthy people appears to be a response to the expression of abnormal MUC1 on the cell surface resulted from conditions unrelated to cancer such as infections or inflammations (Cramer et al., 2005). The presence of anti-MUC1 IgG and IgM antibodies in the serum of patients have a favorable predictive value in terms of survival and reduced incidence of distal metastasis in multiple cancers including pancreatic cancer (Hamanaka et al., 2003), breast cancer (Von Mensdorff-Pouilly et al., 2000; Blixt et al., 2011), and gastric cancer (Kurtenkov et al., 2007). Then the question arises: how to use anti-MUC1 antibody in cancer patients for immunotherapy? Efforts have been made to produce anti-MUC1 antibodies for clinical trials.

At least two MUC1-specific mouse antibodies (muHMFG-1, mAB-AR20.5) and two MUC1-specific humanized antibodies (hPAM4, AS1402) have been used in trials. In a phase III trial, muHMFG-1 antibodies labeled with yttrium-90 were injected into the peritoneal cavity of patients with ovarian cancer after surgery and chemotherapy. Results indicate a positive association between anti-MUC1 antibody administration and OS with endogenous anti-MUC1 antibody production in the recipients (Oei et al., 2008). Anti-MUC1 humanized antibody hPAM4 labeled with Yttrium-90 was tested in a phase I trial for 20 patients with advanced pancreatic cancer. One single dose was administered in this trial. Three out of 20 patients achieved partial response with tumor shrinkage and seven out of 20 patients remained stable for 1.5–5.6 months (Gulec et al., 2011). We have tested anti-MUC1 humanized AS1402 in two clinical trials. In a phase I trial involving 26 patients with advanced breast cancer, we demonstrated that AS1402 administration was well tolerated in patients (Pegram et al., 2009). This was followed by a phase II trial in which 110 hormone receptor positive patients with locally advanced or metastatic breast cancer were enrolled. Results from this trial showed that there was no difference between the patient group receiving letrozole plus AS1402 and the group receiving letrozole alone. The addition of AS1402 to letrozole offered no benefit to patients with hormone-positive breast cancer (Ibrahim et al., 2011). Recently a MUC1-specific human antibody was developed and showed tumor specificity and potential for therapy against epithelial cell cancer (Lohmueller et al., 2016).

MUC1-directed adoptive T cell immunotherapy. In one phase I/II study, four patients with previously treated metastatic breast cancer were recruited. Peripheral blood mononuclear cells (PBMCs) from patients were stimulated two times in vitro with MUC1 antigen before being infused back into the patients. Two patients had microscopic disease and the other two patients had no evidence of disease. Results from this trial demonstrated that tumor burden had an inverse effect on MUC1 antigen induced CTL response and cytokine production (Wright et al., 2009). Some investigators also experimented with using both MUC1 peptide pulsed DCs and cytotoxic lymphocytes sensitized with MUC1-expressing cancer cells in clinical trials. In one study, 20 patients with unresectable or recurrent pancreatic cancer were recruited. PBMC were separated into adherent cells for MUC1–DCs and floating cells for MUC1–CTLs. MUC1 ex vivo exposed DCs were injected intradermally and MUC1–CTLs were infused intravenously. One patient with lung metastasis had a complete response and cancer in five patients remained progression-free (Kondo et al., 2008). In another phase I/II trial involving four recurrent ovarian cancer patients, MUC1 peptide and IL2 stimulated generation of autologous CD4 effector cells was done ex vivo and the resulting IL-10 and IFN-gamma producing CD4 effector cells were infused into patients by intraperitoneal injection. One patient remained disease-free for over a year and another patient had prolonged survival for 16 months even with recurrent disease (Dobrzanski et al., 2012). Investigators also tried to use gemcitabine to inhibit Tregs to enhance antitumor immunity. In one clinical trial, 42 patients with unresectable or recurrent pancreatic cancer were recruited. MUC1-mRNA transfected DCs and CTLs plus gemcitabine were used in this adoptive immunotherapy study (Shindo et al., 2014). After the treatment, one patient had complete response and three patients had partial response. 22 patients remained stable. One-year survival rate was 51%. The authors concluded that this combination approach might be feasible and effective for pancreatic cancer. So far, clinical trials for adoptive T cell immunotherapy are small-scale with limited sample size, involving late-stage cancer patients. There could be differences between early stage and late-stage cancer patients in the effectiveness of adoptive T cell immunotherapy. This needs to be taken into consideration in the design of future clinical trials. In addition, emerging inhibitors of PD-1/PD-L1 and CTLA-4 could be combined with adoptive T cell immunotherapy. In the meantime, large-scale clinical trials are warranted to evaluate the efficacy of adoptive T cell immunotherapy for the treatment of different types of cancer in clinic.

SUMMARY

In summary, significant findings have been made in MUC1-based immunotherapy. Over the course of the past three decades, novel strategies were developed and utilized to boost both cellular and humoral responses to tumor antigen MUC1. Results from both preclinical studies and clinical trials indicate that MUC1 is a promising target for cancer immunotherapy, owing to its 100-fold increased expression in cancer cells than in normal tissues. MUC1-based vaccines are well tolerated in patients. Clinical trials using MUC1 vaccines have been carried out in patients with breast cancer, lung cancer, colorectal cancer, pancreatic cancer, prostate cancer, and ovarian cancer. Clinical trials involving MUC1-directed adoptive T cell immunotherapy have also been done in patients with breast cancer, ovarian cancer, and pancreatic cancer, with some successes. Interpretation of the results from these trials were limited due to their small sample sizes. Passive anti-MUC1 antibody adoptive immunotherapy is feasible and safe in patients. The combination of the anti-MUC1 (Theratope) with tamoxifen (but not with AI) seems to improve OS in metastatic breast cancer; however the combined use of the humanized anti-MUC1 antibody AS1402 with letrozole was not beneficial in hormone-positive breast cancer patients.

The success of MUC1-based immunotherapy in the future may reside in the strategies combining MUC1 immunotherapies with chemotherapy, SERM or with novel inhibitors of PD-1/PD-L1 and CTLA-4. The addition of low doses of cyclophosphamide may be part of the strategy to enhance the immune response, as well. Just as patients are selected for trials of targeted agents based on the expression of specific genetic mutations, the next decade should witness a host of immunotherapy trials in which those patient subsets most likely to benefit from immunotherapy based on specific immune reactivity de novo, will enjoy a more prolonged survival compared to those patients without such immunity.

References

Agrawal, B., Reddish, M.A., Krantz, M.J., et al., 1995. Does pregnancy immunize against breast cancer? Cancer Res. 55, 2257–2261.

Agrawal, B., Reddish, M.A., and Longenecker, B.M., 1996. In vitro induction of MUC-1 peptide-specific type 1 T lymphocyte and cytotoxic T lymphocyte responses from healthy multiparous donors. J. Immunol. 157, 2089–2095.

Agrawal, B., Krantz, M.J., Parker, J., et al., 1998a. Expression of MUC1 mucin on activated human T cells: implications for a role of MUC1 in normal immune regulation. Cancer Res. 58, 4079–4081.

Agrawal, B., Krantz, M.J., Reddish, M.A., et al., 1998b. Cancer-associated MUC1 mucin inhibits human T-cell proliferation, which is reversible by IL-2. Nat. Med. 4, 43–49.

Akagi, J., Hodge, J.W., McLaughlin, J.P., et al., 1997. Therapeutic antitumor response after immunization with an admixture of recombinant vaccinia viruses expressing a modified MUC1 gene and the murine T-cell costimulatory molecule B7. J. Immunother. 20 (1), 38–47.

Ali, S., Metzger, D., Bornert, J., Chambon, P., 1993. Modulation of transcriptional activation by ligand dependent phosphorylation of the human oestrogen receptor A/B region. EMBO J. 12 (3), 1153–1160.

Ammi, R., De Waele, J., Willemen, Y., et al., 2015. Poly(I:C) as cancer vaccine adjuvant: knocking on the door of medical breakthroughs. Pharmacol. Ther. 146, 120–131.

Apostolopoulos, V., Xing, P.X., and Mckenzie, I.F., 1994. Murine immune response to cells transfected with human MUC1: immunization with cellular and synthetic antigens. Cancer Res. 54 (19), 5186–5193.

Apostolopoulos, V., Pietersz, G.A., Loveland, B.E., et al., 1995a. Oxidative/reductive conjugation of mannan to antigen selects for T1 or T2 immune responses. Proc. Natl. Acad. Sci. 92 (22), 10128–10132.

Apostolopoulos, V., Pietersz, G.A., Xing, P.X., et al., 1995b. The immunogenicity of MUC1 peptides and fusion protein. Cancer Lett. 90 (1), 21–26.

Apostolopoulos, V., Barnes, N., Pietersz, G.A., et al., 2000. Ex vivo targeting of the macrophage mannose receptor generates anti-tumor CTL responses. Vaccine 18 (27), 3174–3184.

Apostolopoulos, V., Yuriev, E., Ramsland, P.A., et al., 2003. A glycopeptide in complex with MHC class I uses the GalNAc residue as an anchor. Proc. Natl. Acad. Sci. 100 (25), 15029–15034.

Apostolopoulos, V., Pietersz, G.A., Tsibanis, A., et al., 2006. Pilot phase III immunotherapy study in early-stage breast cancer patients using oxidized mannan MUC1[ISRCTN71711835]. Breast Cancer Res. 8 (3), R27.

Avigan, D., Vasir, B., Gong, J., et al., 2004. Fusion cell vaccination of patients with metastatic breast and renal cancer induces immunological and clinical responses. Clin. Cancer Res. 10 (14), 4699–4708.

Baeckstrom, D., Hansson, G.C., Nilsson, O., et al., 1991. Purification and characterization of a membrane-bound and a secreted mucin-type glycoprotein carrying the carcinoma-associated sialyl-Lea epitope on distinct core proteins. J. Biol. Chem. 266, 21537–21547.

Baldus, S.E., Engelmann, K., and Hanisch, F., 2004. MUC1 and the MUCs: a family of human mucins with impact in cancer biology. Crit. Rev. Clin. Lab. Sci. 41 (2), 189–231.

Berd, D., Mastrangelo, M.J., Engstrom, P.F., Paul, A., Maguire, H., 1982. Augmentation of the human immune response by cyclophosphamide. Cancer Res. 42, 4862–4866.

Berd, D., and Mastrangelo, M.J., 1988. Effect of low dose cyclophosphamide on the immune system of cancer patients: depletion of CD4+, 2H4+ suppressor-inducer T-cells. Cancer Res. 48, 1671–1675.

Bi, S., Bailey, W., and Brisson, C., 2016. Performance of keyhole limpet hemocyanin (KLH) as an antigen carrier for protein antigens depends on KLH property and conjugation route. J. Immunol. 196 (1 Suppl.) 76.16.

Blixt, O., Bueti, D., Burford, B., et al., 2011. Autoantibodies to aberrantly glycosylated MUC1 in early stage breast cancer are associated with a better prognosis. Breast Cancer Res. 13, R25.

Bremer, E.G., Levery, S.B., Sonnino, S., et al., 1984. Characterization of a glycosphingolipid antigen defined by the monoclonal antibody MBr1 expressed in normal and neoplastic epithelial cells of human mammary gland. J. Biol. Chem. 259 (23), 14773–14777.

Brossart, P., Wirths, S., Stuhler, G., et al., 2000. Induction of cytotoxic T-lymphocyte responses in vivo after vaccinations with peptide-pulsed dendritic cells. Blood 96 (9), 3102–3108.

Butts, C., Murray, N., Maksymiuk, A., et al., 2005. Randomized PhaseIIB trial of BLP25 liposome vaccine in stage IIIB and IV non-small-cell lung cancer. J. Clin. Oncol. 23 (27), 6674–6681.

Butts, C., Maksymiuk, A., Goss, G., et al., 2011. Updated survival analysis in patients with stage IIIB or IV non-small-cell lung cancer receiving BLP25 liposome vaccine(L BLP25): phase IIB randomized, multicenter, open-label trial. J. Cancer Res. Clin. Oncol. 137, 1337–1342.

Butts, C., Socinski, M.A., Mitchell, P.L., et al., 2014. Tecemotide (L-BLP25) versus placebo after chemoradiotherapy for stage III non-small-cell lung cancer (START): a randomized, double-blind, phase 3 trial. Lancet Oncol. 15 (1), 59–68.

Canevari, S., Fossati, G., Balsari, A., et al., 1983. Immunochemical analysis of the determinant recognized by a monoclonal antibody (MBr1) which specifically binds to human mammary epithelial cells. Cancer Res. 43 (3), 1301–1305.

Chang, G.C., Lan, H.C., Juang, S.H., et al., 2005. A pilot clinical trial of vaccination with dendritic cells pulsed with autologous tumor cells derived from malignant pleural effusion in patients with late-stage lung carcinoma. Cancer 103 (4), 763–771.

Correa, I., Plunkett, T., Vlad, A., et al., 2003. Form and pattern of MUC1 expression on T cells activated in vivo or in vitro suggests a function in T-cell migration. Immunology 108, 32–41.

Cramer, D.W., Titus-Ernstoff, L., McKolanis, J.R., et al., 2005. Conditions associated with antibodies against the tumor-associated antigen MUC1 and their relationship to risk for ovarian cancer. Cancer Epidemiol. Biomarkers Prev. 14, 1125–1131.

Dobrzanski, M.J., Rewers-Felkins, K.A., Samad, K.A., et al., 2012. Immunotherapy with IL-10- and IFN-r-producing CD4 effector cells modulate "natural" and "inducible" CD4 TReg cell subpopulation levels: observations in four cases of patients with ovarian cancer. Cancer Immunol. Immunother. 61 (6), 839–854.

Dreicer, R., Stadler, W.M., Ahmann, F.R., et al., 2008. Mva-Muc1-IL12 vaccine immunotherapy (TG4010) improves PSA doubling time in patients with prostate cancer with biochemical failure. Invest. New Drugs 27 (4), 379–386.

Farsaci, B., Donahue, R.N., Grenga, I., et al., 2016. Analyses of pretherapy peripheral immunoscore and response to vaccine therapy. Cancer Immunol. Res. 4 (9), 755–765.

Ge, Y., Domschke, C., Stoiber, N., et al., 2012. Metronomic cyclophosphamide treatment in metastasized breast cancer patients: immunological effects and clinical outcome. Cancer Immunol. Immunother. 61 (3), 353–362.

Gendler, S.J., Lancaster, C.A., Taylor-Papadimitriou, J., et al., 1990. Molecular cloning and expression of human tumor-associated polymorphic epithelial mucin. J. Biol. Chem. 265 (25), 15286–15293.

Ghiringhelli, F., Larmonier, N., Schmitt, E., et al., 2004. $CD4^+$ $CD25^+$ regulatory T cells suppress tumor immunity but are sensitive to cyclophosphamide which allows immunotherapy of established tumors to be curative. Eur. J. Immunol. 34, 336–344.

Gilewski, T., Adluri, S., Ragupathi, G., et al., 2000. Vaccination of high-risk breast cancer patients with mucin-1 (MUC1) keyhole limpet hemocyanin conjugate plus QS-21. Clin. Cancer Res. 6 (5), 1693–1701.

Girling, A., Bartkova, J., Burchell, J., et al., 1989. A core protein epitope of the polymorphic epithelial mucin detected by the monoclonal antibody SM-3 is selectively exposed in a range of primary carcinomas. Int. J. Cancer 43, 1072–1076.

Gong, J., Chen, D., Kashiwaba, M., et al., 1997. Induction of antitumor activity by immunization with fusions of dendritic and carcinoma cells. Nat. Med. 3 (5), 558–561.

Gong, J., Chen, D., Kashiwaba, M., et al., 1998. Reversal of tolerance to human MUC1 antigen in MUC1 transgenic mice immunized with fusions of dendritic and carcinoma cells. Proc. Natl. Acad. Sci. 95 (11), 6279–6283.

Gong, J., Koido, S., and Calderwood, S.K., 2008. Cell fusion: from hybridoma to dendritic cell-based vaccine. Expert Rev. Vaccines 7 (7), 1055–1068.

Goydos, J.S., Elder, E., Whiteside, T.L., et al., 1996. A phase I trial of a synthetic mucin peptide vaccine. Induction of specific immune reactivity in patients with adenocarcinoma. J. Surg. Res. 63, 298–304.

Guan, H.H., Budzynski, W., Koganty, R.R., et al., 1998. Liposomal formulations of synthetic MUC1 peptides: effects of encapsulation versus surface display of peptides on immune responses. Bioconjug. Chem. 9 (4), 451–458.

Gulec, S.A., Cohen, S.J., Pennington, K.L., et al., 2011. Treatment of advanced pancreatic carcinoma with 90Y-Clivatuzumab Tetraxetan: a phase I single-dose escalation trial. Clin. Cancer Res. 17, 4091–4100.

Gulley, J.L., Arlen, P.M., Tsang, K.Y., et al., 2008. Pilot study of vaccination with recombinant CEA-MUC-1-TRICOM poxviral-based vaccines in patients with metastatic carcinoma. Clin. Cancer Res. 14 (10), 3060–3069.

Hamanaka, Y., Suehiro, Y., Fukui, M., et al., 2003. Circulating anti-MUC1 IgG antibodies as a favorable prognostic factor for pancreatic cancer. Int. J. Cancer 103, 97–100.

Hanisch, F.G., and Muller, S., 2000. MUC1: the polymorphic appearance of a human mucin. Glycobiology 10, 439–449.

Heery, C.R., Ibrahim, N.K., Arlen, P.M., et al., 2015. Docetaxel alone or in combination with a therapeutic cancer vaccine (PANVAC) in patients with metastatic breast cancer a randomized clinical trial. JAMA Oncol. 1 (8), 1087–1095.

Hilkens, J., Kroezen, V., Bonfrer, J.M., et al., 1986. MAM-6 antigen, a new serum marker for breast cancer monitoring. Cancer Res. 46, 2582–2587.

Hillman, G.G., Reich, L.A., Rothstein, S.E., et al., 2017. Radiotherapy and MTA-MUC1-IL-2 vaccine act synergistically for inducing specific immunity to MUC-1 tumor antigen. J. Immunother. Cancer 5, 4. http://dx.doi.org/10.1186/s40425-016-0204-3.

Hossain, M.K., and Wall, K.A., 2016. Immunological evaluation of recent MUC1 glycopeptide cancer vaccines. Vaccines 4 (25). http://dx.doi.org/10.3390/vaccines4030025.

Huang, C.S., Yu, A.L., Tseng, L.M., et al., 2016. Randomized phase II/III trial of active immunotherapy with OPT-822/OPT-821 in patients with metastatic breast cancer J. Clin. Oncol., ASCO, 2016.

Ibrahim, N.K., Yariz, K.O., Bondarenko, I., et al., 2011. Randomized phase II trial of letrozole plus anti-MUC1 antibody AS1402 in hormone receptor-positive locally advanced or metastatic breast cancer. Clin. Cancer Res. 17 (21), 6822–6830.

Ibrahim, N.K., Murray, J.L., Zhou, D., et al., 2013. Survival advantage in patients with metastatic breast cancer receiving endocrine therapy plus sialyl Tn-KLH vaccine: post hoc analysis of a large randomized trial. J. Cancer 4 (7), 577–584. http://dx.doi.org/10.7150/jca.7028.

Jin, C., Rajabi, H., Pitroda, S., et al., 2012. Cooperative interaction between the MUC1-C oncoprotein and the Rab31 GTPase in estrogen receptor-positive breast cancer cells. PLoS One 7, e39432.

Johnson, V.G., Schlom, J., Paterson, A.J., et al., 1986. Analysis of a human tumor associated glycoprotein (TAG-72) identified by monoclonal antibody B72.3. Cancer Res. 46, 850–857.

Kaiser, A., Gaidzik, N., Westerlind, U., et al., 2009. A synthetic vaccine consisting of a tumor-associated sialyl-T(n)-MUC1 tandem-repeat glycopeptide and tetanus toxoid: induction of a strong and highly selective immune response. Angew. Chem. Int. Ed. Engl. 48, 7551–7555.

Kalache, A., Maguire, A., and Thompson, S.G., 1993. Age at last full-term pregnancy and risk of breast cancer. Lancet 341, 33–36.

Kamata, M., Denda-Nagai, K., Kubota, N., et al., 2002. Vaccination of mice with MUC1 cDNA suppresses the development of lung metastases. Clin. Exp. Metastasis 19 (8), 689–696.

Karanikas, V., Hwang, L.A., Pearson, J., et al., 1997. Antibody and T cell responses of patients with adenocarcinoma immunized with mannan-MUC1 fusion protein. J. Clin. Invest. 100, 2783–2792.

Karanikas, V., Lodding, J., Maino, V.C., et al., 2000. Flow cytometric measurement of intracellular cytokines detects immune responses in MUC1 immunotherapy. Clin. Cancer Res. 6 (3), 829–837.

Kim, S.K., Ragupathi, G., Musselli, C., et al., 1999. Comparison of the effect of different immunological adjuvants on the antibody and T-cell response to immunization with MUC1-KLH and GD3-KLH conjugate cancer vaccines. Vaccine 18 (7–8), 597–603.

Kim, S.K., Ragupathi, G., Cappello, S., et al., 2000. Effect of immunological adjuvant combinations on the antibody and T-cell response to vaccination with MUC1-KLH and GD3-KLH conjugates. Vaccine 19 (4–5), 530–537.

Kimura, T., McKolanis, J.R., Dzubinski, L.A., et al., 2013. MUC1 vaccine for individuals with advanced adenoma of the colon: a cancer immunoprevention feasibility study. Cancer Prev. Res. (Phila.) 6 (1), 18–26.

Kinlough, C.L., Poland, P.A., Bruns, J.B., et al., 2004. MUC1 membrane trafficking is modulated by multiple interactions. J. Biol. Chem. 279 (51), 53071–53077.

Koido, S., Tanaka, Y., Chen, D., et al., 2002. The kinetics of in vivo priming of CD4 and CD8 T cells by dendritic/tumor fusion cells in MUC1-transgenic mice. J. Immunol. 168 (5), 2111–2117.

Kondo, H., Hazama, S., Kawaoka, T., et al., 2008. Adoptive immunotherapy for pancreatic cancer using MUC1 peptide-pulsed dendritic cells and activated T lymphocytes. Anticancer Res. 28 (1B), 379–387.

Kufe, D.W., 2013. MUC1-C oncoprotein as a target in breast cancer: activation of signaling pathways and therapeutic approaches. Oncogene 32 (9), 1073–1081.

Kurtenkov, O., Klaamas, K., Mensdorff-Pouilly, S., et al., 2007. Humoral immune response to MUC1and to the Thomsen-Friedenreich (TF) glycotope in patients with gastric cancer: relation to survival. Acta Oncol. 46, 316–323.

Lambe, M., Hsieh, C.C., Tsaih, S.W., et al., 1998. Parity, age at first birth and the risk of carcinoma in situ of the breast. Int. J. Cancer 77 (3), 330–332.

Lan, M.S., Batra, S.K., Qi, W.N., et al., 1990. Cloning and sequencing of a human pancreatic tumor mucin cDNA. J. Biol. Chem. 265 (25), 15294–15299.

Lees, C.J., Apostolopoulos, V., Acres, B., et al., 2000. Immunotherapy with mannan-MUC1 and IL-12 in MUC1 transgenic mice. Vaccine 19 (2–3), 158–162.

Lepisto, A.J., Moser, A.J., Zeh, H., et al., 2008. A phase I/II study of a MUC1 peptide pulsed autologous dendritic cell vaccine as adjuvant therapy in patients with resected pancreatic and biliary tumors. Cancer Ther. 6, 955–964.

Li, Q., Normolle, D.P., Sayre, D.M., et al., 2000. Immunological effects of BCG as an adjuvant in autologous tumor vaccines. Clin. Immunol. 94 (1), 64–72.

Ligtenberg, M.J., Vos, H.L., Gennissen, A.M., et al., 1990. Episialin, a carcinoma-associated mucin, is generated by a polymorphic gene encoding splice variants with alternative amino termini. J. Biol. Chem. 265 (10), 5573–5578.

Ligtenberg, M.J., Kruijshaar, L., Buijs, F., et al., 1992. Cell-associated episialin is a complex containing two proteins derived from a common precursor. J. Biol. Chem. 267, 6171–6177.

Liu, L.N., Shivakumar, R., Allen, C., et al., 2008. Delivery of whole tumor lysate into dendritic cells for cancer vaccination. Methods Mol. Biol. 423, 139–153. http://dx.doi.org/10.1007/978-1-59745-194-9_9.

Lloyd, K.O., Burchell, J., Kudryashov, V., et al., 1996. Comparison of O-linked carbohydrate chains in MUC-1 mucin from normal breast epithelial cell lines and breast carcinoma cell lines: Demonstration of simpler and fewer glycan chains in tumor cells. J. Biol. Chem. 271, 33325–33334.

Lohmueller, J.J., Sato, S., Popova, L., et al., 2016. Antibodies elicited by the first non-viral prophylactic cancer vaccine show tumor-specificity and immunotherapeutic potential. Sci. Rep. 6, 31740. http://dx.doi.org/10.1038/srep31740.

Lutsiak, M.E., Semnani, R.T., De Pascalis, R., et al., 2005. Inhibition of CD4(+)25+ T regulatory cell function implicated in enhanced immune response by low-dose cyclophosphamide. Blood 105 (7), 2862–2868.

Magnani, J.L., Steplewski, Z., Koprowski, H., et al., 1983. Identification of the gastrointestinal and pancreatic cancer-associated antigen detected by monoclonal antibody 19-9 in the sera of patients as a mucin. Cancer Res. 43, 5489–5492.

Menard, S., Tagliabue, E., Canevari, S., et al., 1983. Generation of monoclonal antibodies reacting with normal and cancer cells of human breast. Cancer Res. 43 (3), 1295–1300.

Miles, D., and Papazisis, K., 2003. Rationale for the clinical development of STn-KLH (Theratope) and anti-MUC-1 vaccines in breast cancer. Clin. Breast Cancer 3 (Suppl. 4), S134–S138.

Miles, D., Roche, H., Martin, M., et al., 2011. Phase III multicenter clinical trial of the sialyl-TN (STn)-keyhole limpet hemocyanin (KLH) vaccine for metastatic breast cancer. Oncologist 16, 1092–1100.

Mukherjee, P., Madsen, C.S., Ginardi, A.R., et al., 2003. Mucin 1-specific immunotherapy in a mouse model of spontaneous breast cancer. J. Immunother. 26 (1), 47–62.

Musselli, C., Ragpathi, G., Gilewski, T., et al., 2002. Reevaluation of the cellular immune response in breast cancer patients vaccinated with MUC1. Int. J. Cancer 97, 660–667.

Oei, A.L., Moreno, M., Verheijen, R.H., et al., 2008. Induction of IgG antibodies to MUC1 and survival in patients with epithelial ovarian cancer. Int. J. Cancer 123, 1848–1853.

Palmer, M., Parker, J., Modi, S., et al., 2001. Phase I study of the BLP25 (MUC1 peptide) liposomal vaccine for active specific immunotherapy in stage IIIB/IV non-small-cell lung cancer. Clin. Lung Cancer 3 (1), 49–57.

Pegram, M.D., Borges, V.F., Ibrahim, N., et al., 2009. Phase I dose escalation pharmacokinetic assessment of intravenous humanized anti-MUC1 antibody AS1402 in patients with advanced breast cancer. Breast Cancer Res. 11, R73.

Petrarca, C., Casalino, B., von Mensdorff-Pouilly, S., et al., 1999. Isolation of MUC1-primed B lymphocytes from tumour-draining lymph nodes by immunomagnetic beads. Cancer Immunol. Immunother. 47, 272–277.

Pillai, K., Pourgholami, M.H., Chua, T.C., et al., 2015. MUC1 as a potential target in anticancer therapies. Am. J. Clin. Oncol. 38, 108–118.

Pinheiro, S.P., Hankinson, S.E., Tworoger, S.S., et al., 2010. Anti-MUC1 antibodies and ovarian cancer risk: prospective data from the Nurses' Health Studies. Cancer Epidemiol. Biomarkers Prev. 19, 1595–1601.

Plunkett, T., Graham, R., Correa, I., et al., 2004. Protection against MUC1 expressing mouse tumours by intra-muscular injection of MUC1 cDNA requires functional $CD8^+$ and $CD4^+$ T cells but does not require the MUC1 tandem repeat domain. Int. J. Cancer 109 (5), 691–697.

Quoix, E., Ramlau, R., Westeel, V., et al., 2011. Therapeutic vaccination with TG4010 and first-line chemotherapy in advanced non-small-cell lung cancer: a controlled phase 2B trial. Lancet Oncol. 12 (12), 1125–1133.

Quoix, E., Lena, H., Losonczy, G., et al., 2016. TG4010 immunotherapy and first-line chemotherapy for advanced non-small-cell lung cancer (TIME): results from the phase 2b part of a randomised, double-blind, placebo-controlled, phase 2b/3 trial. Lancet Oncol. 17 (2), 212–223.

Ragupathi, G., Koide, F., Sathyan, N., et al., 2003. A preclinical study comparing approaches for augmenting the immunogenicity of a heptavalent KLH-conjugate vaccine against epithelial cancers. Cancer Immunol. Immunother. 52 (10), 608–616.

Rainone, V., Martelli, C., Ottobrini, L., et al., 2016. Immunological characterization of whole tumour lysate-loaded dendritic cells for cancer immunotherapy. PLoS One 11 (3), e0151008.

Ramanathan, R.K., Lee, K.M., McKolanis, J., et al., 2005. Phase I study of a MUC1 vaccine composed of different doses of MUC1 peptide with SB-AS2 adjuvant in resected and locally advanced pancreatic cancer. Cancer Immunol. Immunother. 54, 254–264.

Ramlau, R., Quoix, E., Rolski, J., et al., 2008. A phase II study of Tg4010 (Mva-Muc1-IL2) in association with chemotherapy in patients with stage III/IV non-small cell lung cancer. J. Thorac. Oncol. 3 (7), 735–744.

Rochlitz, C., Figlin, R., Squiban, P., et al., 2003. Phase I immunotherapy with a modified vaccinia virus (MVA) expressing human MUC1 as antigen-specific immunotherapy in patients with MUC1-positive advanced cancer. J. Gene Med. 5 (8), 690–699.

Sabbatini, P.J., Ragupathi, G., Hood, C., et al., 2007. Pilot study of a heptavalent vaccine-keyhole limpet hemocyanin conjugate plus QS21 in patients with epithelial ovarian, fallopian tube, or peritoneal cancer. Clin. Cancer Res. 13 (14), 4170–4177.

Scroll, S.M., Balloul, J., Le Goc, G., et al., 2000. Recombinant vaccinia virus encoding human MUC1 and IL2 as immunotherapy in patients with breast cancer. J. Immunother. 23 (5), 570–580.

Scroll, S.M., Squiban, P., Bizouarne, N., et al., 2003. Metastatic breast tumor regression following treatment by a gene-modified vaccinia virus expressing MUC1 and IL-2. J. Biomed. Biotechnol. 3, 194–201.

Shindo, Y., Hazama, S., Maeda, Y., et al., 2014. Adoptive immunotherapy with MUC1 mRNA transfected dendritic cells and cytotoxic lymphocytes plus gemcitabine for unresectable pancreatic cancer. J. Transl. Med. 12, 175. http://dx.doi.org/10.1186/1479-5876-12-175.

Shou, J., Massarweh, S., Osborne, C.K., Wakeling, A.E., Ali, S., Weiss, H., 2004. Mechanisms of tamoxifen resistance: increased estrogen receptor-Her2/neu cross-talk in ER/HER2-positive breast cancer. J. Natl. Cancer Inst. 96 (12), 926–935.

Singer, C.F., Pfeiler, G., Hubalek, M., et al., 2017. Abstract P6-10-01: efficacy and safety of the therapeutic cancer vaccine tecemotide (L-BLP25) in early breast cancer: results from a prospective, randomized, neoadjuvant phase-II study (ABCSG-34). Cancer Res. 77 (4 Suppl.), p6–10-01.

Snyder, L.A., Goletz, T.J., Gunn, G.R., et al., 2006. A MUC1/IL-18 DNA vaccine induces anti-tumor immunity and increased survival in MUC1 transgenic mice. Vaccine 24 (16), 3340–3352.

Tecalco-Cruz, A.C., and Ramirez-Jarquin, J.O., 2016. Mechanisms that increase stability of estrogen receptor alpha in breast cancer. Clin. Breast Cancer 17 (1), 1–10.

Vassilaros, S., Tsibanis, A., Tsikkinis, A., et al., 2013. Up to 15-year clinical follow-up of a pilot phase III immunotherapy study in stage II breast cancer patients using oxidized mannan-MUC1. Immunotherapy 5 (11), 1177–1182. http://dx.doi.org/10.2217/imt.13.126.

Von Mensdorff-Pouilly, S., Vertsraeten, A.A., Kenemans, P., et al., 2000. Survival in early breast cancer patients is favorably influenced by a natural humoral immune response to polymorphic epithelial mucin. J. Clin. Oncol. 18, 574–583.

Wei, X., Xu, H., and Kufe, D., 2006. MUC1 oncoprotein stabilizes and activates estrogen receptor alpha. Mol. Cell. 21 (2), 295–305.

Wreschner, D.H., Hareuveni, M., Tsarfaty, I., et al., 1990. Human epithelial tumor antigen cDNA sequences. Differential splicing may generate multiple protein forms. Eur. J. Biochem. 189, 463–473.

Wright, S.E., Rewers-felkins, K.A., Quinlin, I.S., et al., 2009. Tumor burden influences cytotoxic T cell development in metastatic breast cancer patients-a phase I/II study. Immunol. Invest. 38, 820–838.

Xia, J., Tanaka, Y., Koido, S., et al., 2003. Prevention of spontaneous breast carcinoma by prophylactic vaccination with dendritic/tumor fusion cells. J. Immunol. 170 (4), 1980–1986.

Yang, C., Trent, S., Ionescu-Tiba, V., et al., 2006. Identification of cyclin D1- and estrogen-regulated genes contributing to breast carcinogenesis and progression. Cancer Res. 66 (24), 11649–11658.

Yang, E., Hu, X.F., and Xing, P.X., 2007. Advances of MUC1 as a target for breast cancer immunotherapy. Histol. Histopathol. 22, 905–922.

Zhang, K., Sikut, R., and Hansson, G.C., 1997. A MUC1 mucin secreted from a colon carcinoma cell line inhibits target cell lysis by natural killer cells. Cell Immunol. 176, 158–165.

Zhang, S., Zhang, H., Shi, H., et al., 2008. Induction of immune response and anti-tumor activities in mice with a DNA vaccine encoding human mucin 1 variable-number tandem repeats. Hum. Immunol. 69 (4–5), 250–258.

Zhu, D., and Tuo, W., 2016. QS-21: a potent vaccine adjuvant. Nat. Prod. Chem. Res. 3 (4), e113.

Index

Note: Page numbers followed by "*f*" and "*t*" refer to figures and tables, respectively.

B

C

D

K

L

M

R

U

V

FSC
www.fsc.org
MIX
Papier | Fördert gute Waldnutzung
FSC® C083411

Kolibri 360 GmbH
Ferdinand-Jühlke-Straße 7
99095 Erfurt, Deutschland
produktsicherheit@kolibri360.de